MW00489389

QUATERNIONS

AND ROTATION SEQUENCES

QUATERNIONS
AND ROTATION SEQUENCES

A Primer with Applications to Orbits,
Aerospace, and Virtual Reality

JACK B. KUIPERS

PRINCETON UNIVERSITY PRESS
PRINCETON AND OXFORD

Copyright © 1999 by Princeton University Press
Published by Princeton University Press, 41 William Street,
Princeton, New Jersey 08540
In the United Kingdom: Princeton University Press, 3 Market Place,
Woodstock, Oxfordshire OX20 1SY

All Rights Reserved

Fifth printing, and first paperback printing, 2002
Paperback ISBN 0-691-10298-8

The Library of Congress has cataloged the cloth edition of
this book as follows

Kuipers, Jack B., 1921–
Quaternions and rotation sequences : a primer with applications to
orbits, aerospace, and virtual reality / Jack B. Kuipers.
p. cm.
Includes bibliographical references and index.
ISBN 0-691-05872-5 (cloth : alk. paper)
1. Quaternions. I. Title.
QA196.K85 1998
512.9′434—dc21   98-35389

British Library Cataloging-in-Publication Data is available

The publisher would like to acknowledge the author
of this volume for providing the camera-ready
copy from which this book was printed

Printed on acid-free paper. ∞

www.pupress.princeton.edu

Printed in the United States of America

10 9 8 7 6 5

Dedication

Although she still insists that a good murder mystery is a far more civilized excuse for the printed page than this book, I dedicate it to my literate, patient, and long-suffering spouse,
Lois Belle
— and to

Ben

&

Emily

&

Joel

&

Alison

&

Lynne

our children who, with purpose and competence, make worlds for themselves — thereby making our world so much larger. **jbk**

Aire, and ye Elements the eldest birth
Of Natures Womb, that in *quaternion* run
Perpetual Circle, multiform, and mix
And nourish all things, let your ceasless change
Varie to our great Maker still new praise.

Paradise Lost, Book V
John Milton

Contents

List of Figures

About This Book

This book is intended for all those mathematicians, engineers, and physicists who have to know, or who want to know, more about the modern theory of quaternions. Primarily, as the title page suggests, it is an exposition of the quaternion and its primary application in a rotation operator. In a parallel fashion, however, the conventional or more familiar matrix rotation operator is also presented. This parallel presentation affords the reader the opportunity for making comparative judgements about which approach is preferable, for very specific applications.

This book readily divides into three major areas of concern:

• The first three or four chapters present introductory material which establishes terminology and notation to be used later on.
• The mathematical properties of quaternions are then presented, including quaternion algebra and geometry. This is followed by more advanced special topics in spherical trigonometry, along with an introduction to quaternion calculus and perturbation theory, required in certain situations involving dynamics and kinematics.
• Lastly, state-of-the-art applications are discussed. A six degree-of-freedom electromagnetic *position and orientation* transducer is presented. With this we end with a discussion of computer graphics, necessary for the development of applications in *Virtual Reality*.

The writing of this book was early-on supported by the United States Air Force, whose objective was to provide a primer on quaternions, suitable for self-study. Our primary concern was that the book be written at a level such that much of the subject matter would be accessible to those with a modest background in mathematics. With this in mind, the quaternion is defined and its algebra is introduced and developed. Several applications of the quaternion, the quaternion rotation operator, and quaternion rotation sequences are presented. A perview of The Table of Contents will provide the reader with a measure of the intent and scope of this book.

xix

Acknowledgements

First of all, I am indebted to Dean F. Kocian, Project Engineer in The Paul M. Fitts Human Engineering Division, Armstrong Laboratory, United States Air Force, for his encouragement and support of this project. I appreciate especially his patience and forebearance during a time of personal and medical emergency which I encountered during this effort. The necessary time extensions granted to me by the Air Force Systems Command, Aeronautical Systems Division, at Wright-Patterson Air Force Base made possible my completion of the project. For this I am very grateful.

In May 1986, I gave a one-week tutorial on the application of quaternions to the Armstrong Laboratory's Virtual Panoramic Display Program that was supporting Army's Light Helicopter Experimental (LHX) predecessor to the Comanche Helicopter Program. This tutorial was designed to acquaint those engaged in this program, mostly engineers and mathematicians, with the use of the quaternion rotation operator as an alternative to the conventional matrix rotation operator. This tutorial was presented at the Laboratories of Hughes Aircraft Company in Albuquerque, NM. I found that experience to be challenging and stimulating, and I am again grateful to Dean Kocian and his offices for that experience.

I also greatly appreciate my colleagues in the Department of Mathematics and the Department of Physics at Calvin College. They have given me of their time during the development of the subject matter in this book. Our discussions invariably were stimulating and often generated useful material. In particular: in Mathematics, I appreciate the discussions with Tom Jager, a most competent algebraist, who helped me get some of the sticky places unstuck; and in Physics, Dave Van Baak, who has no peers, and who without fail can generate meaningful discourse and offer perspective on matters that many of us have never thought of or even heard of.

But very special appreciation, however, is reserved for my good friend and colleague, Carl Sinke, who read the entire document and throughout offered many comments and suggested insightful editorial changes which improved the overall readability and pedagogy.

There are many others. Ben Friedman and Scott Rabuka, both with Waterloo Maple, I thank for their friendly support and expertise helping me through some graphics programming using Maple. Also, several of my colleagues were often great comfort: Earl Fife, who was a very helpful resource for keeping many utilities, especially LaTeX, running smoothly on my Macintosh; Daryl Brink, who knows Mathematica's syntactical traps much better than I; George G. Harper, a very helpful doctor of the colon and semi-colon (and, I should add, the comma); David Laverell who helped me search and research; Paul Zwier, who called my attention to relevant documents; John Beebe who helped me maintain perspective with some meaningful quotes; and countless other colleagues from these and other disciplines — even other institutions. Although some would often and mercilessly ask the obvious but wrong question, I thank them all for reasons most of them may never know.

Also, I appreciate the kind comments by the reviewers, whoever they are, and for their suggestion that the book would have even greater appeal if some Calculus matters encountered in dynamics and kinematics were addressed. I agreed with them, so this I did in Chapter 11. It was difficult knowing where to stop, since the subject deserves much more attention and greater depth. The momentum generated in doing Chapter 11, however, led me to also include Chapter 12 — Rotations in Phase Space — an unpublished (to my knowledge) yet rather comprehensive perspective on the nature of the solution space of Ordinary Differential Equations.

And last but not least, I am grateful to Trevor Lipscombe, the Physical Sciences Editor of Princeton University Press, for his helpful suggestions while patiently leading me down a path where I have not been before.

Jack B. Kuipers
Calvin College
Grand Rapids, MI

QUATERNIONS
AND ROTATION SEQUENCES

Chapter 1

Historical Matters

1.1   Introduction

Mathematics, as with most subjects in science and engineering, has a long and varied history. Although it is an oversimplification of a complex subject, it is safe to say that in general the practice of mathematical techniques preceded the development of firm theoretical foundations for the subject. These tended to come later.

Historians of mathematics are generally agreed that in this connection three highly significant developments occurred during the nineteenth century. These were the development of non-Euclidean geometry, of a non-commutative algebra, and of a precise theoretical foundation for calculus.

Early on it was thought that geometry was limited to what was known as Euclidean Geometry, and it was only through extensive researches into the parallel line postulate as formulated by Euclid that other geometries were developed and found to be consistent. For calculus, and analysis in general, the formulation of a precise theory of limits was a significant step.

For many years algebra was thought simply to be generalized arithmetic, in the sense that although letters were used to represent the objects under study, all of the ordinary rules of operation for arithmetic were valid in the algebraic manipulations as well. Consistent with the spirit of the times, in which the search for more precise theoretical foundations in mathematical thinking received more emphasis, mathematicians working in the area of algebra began to focus more sharply on such matters of algebraic structure as closure, commutativity, and associativity for algebraic operations.

3

It was in this context that William Rowan Hamilton introduced his algebra of quaternions which, to the total surprise of the mathematical community, violated the law of commutativity for multiplication. What seemed to mathematicians of the day to be impossible was to have an otherwise consistent algebra for which this fundamental property of the algebra of real numbers did not hold.

According to Howard Eves [1], it was work such as that done by Hamilton which

> opened the floodgates of modern abstract algebra. By weakening or deleting various postulates of common algebra, or by replacing one or more of the postulates by others, which are consistent with the remaining postulates, an enormous variety of systems can be studied. As some of these systems we have groupoids, quasigroups, loops, semi-groups, monoids, groups, rings, integral domains, lattices, division rings, Boolean rings, Boolean algebras, fields, vector spaces, Jordan algebras, and Lie algebras, the last two being examples of non-associative algebras. It is probably correct to say that mathematicians have, to date, studied well over 200 such algebraic structures.

Our intent in these pages is to explore the use of Hamilton's quaternions in studying certain transformations in ordinary space of three dimensions. It must be said that it was not long after the publication of Hamilton's results that Josiah Willard Gibbs and others began to work out the details of what we know today as the algebra of vector spaces, and Hamilton's work seemed quickly to be eclipsed. Recently, however, interest in the use of quaternions has revived, and we want to consider ways in which quaternion algebra may still be more effective than the use of ordinary vector algebra.

1.2   Mathematical Systems

By the set of *real numbers* we shall mean all numbers which may be represented in decimal form. These include the natural numbers, the integers, and the rational numbers.

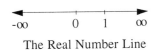

The Real Number Line

The set of *natural numbers* is the set of numbers whose arithmetic is studied in the elementary school, namely, the set

$$N = \{1, 2, 3, 4, \cdots\}$$

We mention that this set is *closed* under addition and multiplication, that is the sum of any two natural numbers is again a natural number, and the product of any two natural numbers is also a natural number. However, the set of natural numbers is *not closed* under subtraction or division, since there are natural numbers, say 3 and 5, whose difference

$$3 - 5 = -2$$

which is *not* a natural number. Clearly, the same is true for the quotient of natural numbers.

The set of *integers* is the set

$$Z = \{ \cdots - 3, -2, -1, 0, 1, 2, 3, \cdots \}$$

which includes the natural numbers as a subset. This set has the advantage that it is closed under addition, multiplication, and also subtration. Notice that the difference between any two integers is always another integer. However, the integers are not closed under division.

To obtain this fourth property we need the set of *rational numbers*, namely, the set of all possible quotients of the integers, except that division by zero is excluded. Thus the rational numbers, represented by the set,

$$Q = \{ p/q \mid \text{p,q integers with } q \neq 0 \}$$

This set includes both the natural numbers and the integers as subsets and is closed under all four of the ordinary operations of arithmetic. It exhibits the mathematical structure which earlier mathematicians began to recognize as what, in abstract algebra today, is called a *field*. The set R of real numbers, with the two ordinary operations of addition and multiplication, is another example of a field.

This mathematical system is one that is familiar to all of us, and exhibits the following *field properties*:

1. *Closure* under the operations, that is if a and b are real numbers so are $a + b$ and $a \cdot b$

2. Both of the operations are *associative*, that is, if a, b, and c are real numbers,

$$
\begin{aligned}
(a + b) + c &= a + (b + c) \\
\text{and} \qquad (a \cdot b) \cdot c &= a \cdot (b \cdot c)
\end{aligned}
$$

The Real Number Line

3. Both of the operations are *commutative*, that is,

$$a + b = b + a$$
$$\text{and} \quad a \cdot b = b \cdot a$$

4. There is an *identity for addition*, say, 0, with the property that

$$a + 0 = 0 + a = a$$

for every real number a

5. There are *inverses* for addition, that is, for every real number a there is a real number, say, $-a$, such that

$$a + (-a) = (-a) + a = 0$$

6. There is an *identity for multiplication*, say 1, such that

$$a \cdot 1 = 1 \cdot a = a$$

for every real number a. Further, $1 \neq 0$.

7. There are *inverses* for multiplication, that is, for every real number a, not equal to 0, there is a real number, say a^{-1}, such that

$$a^{-1} \cdot a = a \cdot a^{-1} = 1$$

8. Multiplication is *distributive* over addition, that is for any real numbers a, b, and c we have

$$a \cdot (b + c) = a \cdot b + a \cdot c$$

These properties for ordinary arithmetic of real numbers were thought to hold for all algebraic structures; hence it was so surprising that Hamilton's quaternions were a consistent algebraic system, yet violated property 3, the commutativity of multiplication.

1.3   Complex Numbers

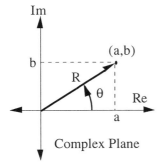

Complex Plane

Another example of a mathematical system which has the field properties is that of the *complex numbers*. It was in the 16th century that the Italian algebraist Bombelli, after Cardan and Tartaglia, had what he called a "wild thought." This wild thought seems to have been the precursor of the notion of complex numbers, including the idea of the complex conjugate. However it was not until the work of Gauss in the early 19th century that the algebra of complex numbers was given a firm mathematical basis. Since the square of a real number is always non-negative, there is, of course, no solution in the set of real numbers for the equation

$$x^2 + b^2 = 0$$

or perhaps more generally for quadratic equations of the form

$$x^2 - 2ax + a^2 + b^2 = 0$$

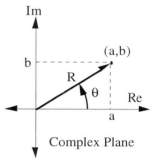

Complex Plane

for any non-zero real numbers a and b. But today the numbers which satisfy these two equations are well-known even to most high school students as

$$x = \pm\, \mathbf{i}b$$
$$\text{and} \quad x = a \pm \mathbf{i}b$$

respectively. Here $\mathbf{i}$ is the now familiar square root of -1, having the property that $\mathbf{i}^2 = -1$. In the complex number, $a + \mathbf{i}b$, we often identify a as the *real part* and the b as the *imaginary part*.

We must be clear, however, what we mean when we write $a + \mathbf{i}b$, since the symbol $\mathbf{i}$ does not represent a real number. More specifically, what does it mean to write the product $\mathbf{i}b$ when we know that $\mathbf{i}$ is *not* a real number. A mathematically acceptable way to answer this question is to identify our representation of the complex number

$$a + \mathbf{i}b$$

with the pair (a, b) of real numbers a and b. In this format, however, we must first define *equality* for complex numbers by stating

$$(a, b) = (c, d)$$

if and only if

$$a = c \quad \text{and} \quad b = d$$

That is to say, two complex numbers are equal if and only if their *real* parts are equal and their *imaginary* parts are equal.

Next, in this format, we define the sum and the product of two complex numbers by the equations

$$(a, b) + (c, d) = (a + c, b + d) \qquad (1.1)$$
$$\text{and} \quad (a, b)(c, d) = (ac - bd, ad + bc) \qquad (1.2)$$

We now relate this ordered pair format to the usual $a + \mathbf{i}b$ notation by making the following associations

$$(a, 0) \quad \leftrightarrow \quad a$$

and $\quad (0, 1) \quad \leftrightarrow \quad \mathbf{i}$

Using Equations 1.1 and 1.2 we may now compute

$$\mathbf{i}^2 \;=\; (0,1)(0,1) \;=\; (-1, 0) \;=\; -1 \qquad (1.3)$$

so that $\quad (0, 1) \;=\; \mathbf{i} \;=\; \sqrt{-1}$

We may then write $\quad (a, b) \;=\; (a, 0) + (0, 1)(b, 0) \qquad (1.4)$

$$= \quad a + \mathbf{i}b$$

Rigor

This rigorously establishes and justifies the notation

$$z = (a, b) = a + \mathbf{i}b$$

From this point on, for the most part, the arguments and proofs presented will be more intuitive and heuristic — an approach which is more appropriate for our purposes.

It is not difficult to show that the difference and quotient of complex numbers are given by the equations

$$(a, b) - (c, d) \;=\; (a - c, b - d)$$

$$\frac{(a, b)}{(c, d)} \;=\; \left(\frac{ac + bd}{c^2 + d^2}, \frac{bc - ad}{c^2 + d^2} \right)$$

This foregoing development may be used to define the arithmetic operations on complex numbers, and to verify that with these definitions the set of complex numbers does indeed satisfy all of the field properties. We wish, however, to use an approach which may be more familiar to the reader.

The algebra of complex numbers is usually defined by assuming that the algebra of real numbers holds for complex numbers as well, except that $\mathbf{i}^2 = -1$. This means that for addition and subtraction we have

$$(a + \mathbf{i}b) + (c + \mathbf{i}d) \;=\; (a + c) + \mathbf{i}(b + d) \qquad (1.5)$$

and $\quad (a + \mathbf{i}b) - (c + \mathbf{i}d) \;=\; (a - c) + \mathbf{i}(b - d) \qquad (1.6)$

In like fashion, multiplication is defined as

$$(a + \mathbf{i}b) \times (c + \mathbf{i}d) \;=\; ac + \mathbf{i}bc + \mathbf{i}ad + \mathbf{i}^2 bd$$

$$= \; (ac - bd) + \mathbf{i}(bc + ad) \qquad (1.7)$$

It should be mentioned that in the context of complex numbers, vector spaces, or even quaternions, a real number is often called a *scalar*. If in the above definition of the product of two complex numbers we set $b = 0$, then we get a product of the scalar a and the complex number $c + \mathbf{i}d$ as

$$a(c + \mathbf{i}d) = ac + \mathbf{i}ad$$

To compute the quotient of two complex numbers, we first note that the product

$$(c + \mathbf{i}d)(c - \mathbf{i}d) = c^2 + d^2$$

is always a *real* number. We then write

$$
\begin{aligned}
\frac{a + \mathbf{i}b}{c + \mathbf{i}d} &= \frac{(a + \mathbf{i}b)(c - \mathbf{i}d)}{(c + \mathbf{i}d)(c - \mathbf{i}d)} \\
&= \frac{ac + bd}{c^2 + d^2} + \mathbf{i}\frac{bc - ad}{c^2 + d^2}
\end{aligned}
\tag{1.8}
$$

In particular, we may now compute the *reciprocal* of $c + \mathbf{i}d$ as

$$\frac{1}{c + \mathbf{i}d} = (c + \mathbf{i}d)^{-1} = \frac{c}{c^2 + d^2} - \mathbf{i}\frac{d}{c^2 + d^2} \tag{1.9}$$

1.4   Polar Representation

In order to make a connection with Hamilton's quaternions, we next mention the geometric interpretation of complex numbers. This interpretation results from identifying the complex number (a, b), that is $a + \mathbf{i}b$, with the corresponding point (a, b) in a coordinatized plane, as in the figure in the margin.

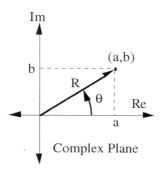

Complex Plane

For each complex number $z = (a, b)$ we can define a *magnitude* by

$$|z| = \sqrt{a^2 + b^2}$$

and an angle (often called the *argument*) by

$$arg(z) = \theta$$

where θ is an angle defined by

$$\tan \theta = b/a$$

Thus, for example, the magnitude of the complex number

$$z = (1, 1)$$

is

$$|z| = \sqrt{2}$$

while the corresponding angle is

$$\theta = \arctan(1) = 45 \text{ degrees}$$

or in radian measure, $\pi/4$.

Often a geometric interpretation of a complex number $z = (a, b)$ is made by identifying the number with the two dimensional vector (a, b), which we may think of as a vector from the origin to the point (a, b) in the plane. In that case the magnitude of the complex number is simply the length of the vector and the angle of the complex number is the angle between the vector and the positive X-axis. Geometrically, then, how are the addition and multiplication of complex numbers to be interpreted?

It is easy to see that addition of complex numbers corresponds exactly to the ordinary parallelogram law for adding vectors. In fact, if we think of a vector in two dimensions as the ordered pair (a, b) the rule for adding vectors is exactly the same as that for adding complex numbers, as shown by the example in the figure in the margin.

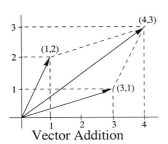

Vector Addition

In the case of multiplication for complex numbers, it turns out that the magnitude of the product is equal to the product of the magnitudes of the factors, and the angle of the product is the sum of the angles of the two factors. This fact is apparent if we represent the complex number in trigonometric form. If $z = (a, b)$, $|z| = r$, and $\arg(z) = \theta$, we learn from highschool trigonometry of a right triangle that

$$a = r \cos \theta \quad \text{and} \quad b = r \sin \theta$$

Thus we may write

$$
\begin{aligned}
z \;=\; (a, b) \;&=\; a + \mathbf{i}b \\
&=\; r \cos \theta + \mathbf{i}r \sin \theta \\
&=\; r(\cos \theta + \mathbf{i} \sin \theta)
\end{aligned}
$$

Since by definition $\quad e^{\mathbf{i}\theta} \;=\; \cos \theta + \mathbf{i} \sin \theta$

we may write $\quad z \;=\; re^{\mathbf{i}\theta} \hfill (1.10)$

where r is the magnitude or amplitude and θ is the angle or argument of z. Thus if we have

$$z_1 \;=\; r_1(\cos \alpha + \mathbf{i} \sin \alpha) \;=\; r_1 e^{\mathbf{i}\alpha} \hfill (1.11)$$

$$z_2 \;=\; r_2(\cos \beta + \mathbf{i} \sin \beta) \;=\; r_2 e^{\mathbf{i}\beta} \hfill (1.12)$$

then

$$
\begin{aligned}
z_1 z_2 \;&=\; r_1 r_2 (\cos \alpha + \mathbf{i} \sin \alpha)(\cos \beta + \mathbf{i} \sin \beta) \hfill (1.13) \\
&=\; r_1 r_2 (\cos \alpha \cos \beta + \mathbf{i}^2 \sin \alpha \sin \beta \\
&\quad + \mathbf{i} \sin \alpha \cos \beta + \mathbf{i} \cos \alpha \sin \beta) \\
&=\; r_1 r_2 [(\cos \alpha \cos \beta - \sin \alpha \sin \beta)
\end{aligned}
$$

$$+\mathbf{i}(\sin\alpha\cos\beta + \cos\alpha\sin\beta)]$$
$$= r_1 r_2[\cos(\alpha+\beta) + \mathbf{i}\sin(\alpha+\beta)]$$
$$= r_1 r_2 e^{\mathbf{i}(\alpha+\beta)} \qquad (1.14)$$

Hence it appears that magnitudes are multiplied while the angles are added. Were we to limit ourselves to complex numbers of magnitudes 1, then multiplication by a complex number amounts to a rotation in the plane. For example, suppose we multiply the vector $(1,1)$ by the vector $\mathbf{i} = (0,1)$. Note that

$$|\mathbf{i}| = |(0,1)| = \sqrt{0^2 + 1^2} = 1$$

and $\qquad \arg(i) = \pi/2$ radians

The result is the vector $(-1,1)$, which is the original vector $(1,1)$ rotated through an angle $\pi/2$, as in the figure in the margin.

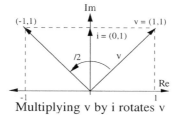

Multiplying v by i rotates v

Perhaps at this point it is worth noting that summing two complex numbers, say, z_1 and z_2, in polar form quite obviously is not the preferred way to merely get the desired sum. However, working through this summing process anyway does provide a proof of the well-known cosine law for triangles. We have

$$z_1 + z_2 = (r_1 e^{\mathbf{i}\alpha}) + (r_2 e^{\mathbf{i}\beta})$$
$$= (r_1\cos\alpha + r_2\cos\beta) + \mathbf{i}(r_1\sin\alpha + r_2\sin\beta)$$

and we can write

$$R = \sqrt{(r_1\cos\alpha + r_2\cos\beta)^2 + (r_1\sin\alpha + r_2\sin\beta)^2}$$
$$= \sqrt{r_1^2 + r_2^2 + 2r_1 r_2\cos(\alpha-\beta)}$$

or $\qquad R^2 = r_1^2 + r_2^2 + 2r_1 r_2\cos(\alpha-\beta) \qquad (1.15)$

where R is the magnitude of the vector sum of z_1 and z_2.

1.5 Hyper-complex Numbers

Although we have just interpreted the product of complex numbers geometrically as the rotation of vectors in a plane we might just as well have thought of it as a transformation of points in the plane. In our example above, multiplication by $\mathbf{i} = (0,1)$ rotates the initial point $(1,1)$ into its image point $(-1,1)$, where the rotation is through the angle $\pi/2$ and has the origin as its center. It was this connection between the algebra of complex numbers, or "double algebra" as it was then called, and rotations in the plane that intrigued Hamilton for years.

His problem was to discover an analogous relationship in three dimensional space, using triplets of real numbers. This he was unable to accomplish, for what later turned out to be very good reasons. The story goes that one morning, while walking with his wife along the Royal Canal, he had a flash of inspiration. Of an instant he saw that triples were not enough, but rather that four-tuples were required. He saw that he needed not just the complex number component $\mathbf{i}$, but rather three such components $\mathbf{i}$, $\mathbf{j}$, and $\mathbf{k}$, satisfying the relationship:

$$\mathbf{i}^2 = \mathbf{j}^2 = \mathbf{k}^2 = \mathbf{ijk} = -1$$

So struck was he by this discovery that supposedly he stopped and carved this equation into the stone of a nearby bridge. And so it was that the quaternion was born.

There are good reasons, buried somewhat deeply in the theory of abstract algebra, for Hamilton having such difficulty in coming up with this new idea. Some years later the mathematician Frobenius proved that what Hamilton was trying to do using triplets was not possible. It was actually the mathematics of the situation which finally forced him to turn to quaternions, sometimes called *complex numbers of rank 4*. We shall see later that these are particularly well suited for use as rotation operators in three dimensional space.

The set of all quaternions, along with the two operations of addition and multiplication (which we shall define shortly), form a mathematical system called a ring, more particularly a non-commutative division ring. This longer title merely emphasizes the fact that the product of quaternions is not commutative, and that multiplicative inverses do exist for every non-zero element in the set.

In summary, the set of quaternions under the operations of addition and multiplication satisfies all of the field properties which we discussed earlier, except for the commutative law for multiplication.

Chapter 2

Algebraic Preliminaries

2.1 Introduction

In Chapter 1 we have already introduced the set of complex numbers as an example of a set of numbers for which the field properties hold. Early on (in the 16th and 17th centuries) there was considerable suspicion, accusation of mystique and even ridicule directed at those who believed in the existence of these so-called "imaginary" numbers. The curious term *imaginary number* actually dates back to those times. And although today we know there is nothing "imaginary" about these numbers — they do indeed exist — the name stuck.

It was not until early in the 19th century, however, that mathematicians (particularly Gauss) had formally devised the complex number system in order to solve at least quadratic equations. Specifically, equations of the form

$$x^2 + b^2 = 0$$

$$x^2 - 2ax + a^2 + b^2 = 0$$

for any non-zero real numbers, a and b, were no longer declared as *not* having a solution!

Today, the numbers which satisfy these two equations are well known even to most high school students as being

$$x = \pm \mathbf{i}b$$
$$\text{and} \qquad x = a \pm \mathbf{i}b$$

respectively, where $\mathbf{i}^2 = -1$ or $\mathbf{i} = \sqrt{-1} = (-1)^{1/2}$.

In summary, numbers of the form

$$z \ = \ a + \mathbf{i}b \ = \ a + b\mathbf{i}$$

where a and b are real numbers ($\mathbf{i}^2 \ = \ -1$), has come to be known as a *complex number* or sometimes an *imaginary number*. The letter z, by common convention, is used to denote a complex number, and $z \ = \ a + \mathbf{i}b$ is read as "the complex number a plus $\mathbf{i}b$." In this so-called *Cartesian* representation for a complex number, the real number a is called the *real part*, while the product $\mathbf{i}b$ is called the *imaginary part*. The *polar* representation of this complex number is illustrated here in the margin. The notion of *uniqueness* of the polar representation of a complex number is presented in a margin note across the page.

Imaginary Numbers

The imaginary part of a complex number, namely, ib may also be written as bi.

The *real numbers* are a subset of the complex numbers obtained by setting $b = 0$; likewise, the set of all purely *imaginary numbers* is a subset of the complex numbers obtained by setting $a = 0$. In this context, the real numbers are said to be of rank 1, the complex numbers are of rank 2, and the quaternions are said to be hyper-complex numbers of rank 4.

**Complex Number
Polar Form**

$$a + \mathbf{i}b \ = \ Re^{\mathbf{i}\theta}$$
$$\text{where} \quad R \ = \ \sqrt{a^2 + b^2}$$
$$\text{and} \quad \theta \ = \ \arctan \frac{b}{a}$$

In Chapter 1 we mentioned that the set of rational numbers, the set of real numbers, and the set of complex numbers are all examples of a field, and we listed there the field properties that these numbers satisfy. The set of real numbers, that is, those numbers which may be written in decimal form and are the subject of study in elementary school arithmetic, is perhaps the most common and familiar to all of us — mathematicians and non-mathematicians alike. While we may not realize it, all of those operations which may legitimately be performed in our day-to-day housekeeping computations are in fact based on the field properties for those numbers.

We now show, in some detail, that the less-familiar set of complex numbers also satisfies these same field properties. In doing so we shall use, in general, the $a + \mathbf{i}b$ notation rather than the ordered pair (a, b) notation for a complex number. Our purpose in doing so is, first, to gain some appreciation for the algebraic structure which these numbers exhibit and, second, to provide important background for developing the properties and the algebra of quaternions.

2.2 Complex Number Operations

We now verify that the complex numbers, under addition and multiplication, satisfy all of the field properties. We begin by noting that (by definition) two complex numbers are *equal* if and only if both real parts and imaginary parts are the same. That is to say

$$a + \mathbf{i}b = c + \mathbf{i}d$$

if and only if

$$a = c \quad \text{and} \quad b = d$$

Note that here we have used the Cartesian representation for complex numbers. For equality of complex numbers in the polar form, somewhat more restrictive conditions are required, as noted in the margin.

Polar Form Equality

$$
\begin{aligned}
\text{For} \quad r_1, r_2 \; &\geq \; 0 \\
\text{and} \quad \alpha, \beta \; &\in \; (-\pi, \pi] \\
\text{Then} \quad r_1 e^{\mathbf{i}\alpha} \; &= \; r_2 e^{\mathbf{i}\beta} \\
\text{iff} \quad r_1 \; &= \; r_2 \\
\text{and} \quad \alpha \; &= \; \beta
\end{aligned}
$$

Proof:

$$
\begin{aligned}
r_1 e^{\mathbf{i}\alpha} e^{-\mathbf{i}\alpha} \; &= \; r_2 e^{\mathbf{i}\beta} e^{-\mathbf{i}\alpha} \\
r_1 e^{\mathbf{i}(\alpha-\alpha)} \; &= \; r_2 e^{\mathbf{i}(\beta-\alpha)} \\
r_1 e^{\mathbf{i}0} \; &= \; r_2 e^{\mathbf{i}(\beta-\alpha)} \\
r_1 \; &= \; r_2 e^{\mathbf{i}(\beta-\alpha)} \\
\Rightarrow \quad \beta - \alpha \; &= \; 0 \\
\text{or} \quad \alpha \; &= \; \beta \\
\text{and} \quad r_1 \; &= \; r_2
\end{aligned}
$$

QED

2.2.1 Addition and Multiplication

For every $a, b, c, d \; \epsilon \; R$ (the set of real numbers), we defined in Equation 1.2 addition for complex numbers by

$$(a + \mathbf{i}b) + (c + \mathbf{i}d) \; = \; (a + c) + \mathbf{i}(b + d)$$

Multiplication was defined in Equation 1.6 as

$$(a + \mathbf{i}b) \times (c + \mathbf{i}d) \; = \; (ac - bd) + \mathbf{i}(bc + ad)$$

We mention here that in the context of complex numbers, vector spaces, or even quaternions, a real number is often called a *scalar*. If in the above definition of the product of two complex numbers we set $b = 0$, then we get a product of the scalar a and the complex number $c + \mathbf{i}d$ as

$$a(c + \mathbf{i}d) = ac + \mathbf{i}ad$$

Since a, b, c, and d are all real numbers, and since the set of real numbers has the field properties, it follows that $a + c$, $b + d$, $ac - bd$, and $bc + ad$ are also real numbers. It then follows from our definitions for addition and multiplication that the sum and product of two complex numbers are also complex numbers; that is, we have the closure property for addition and multiplication of complex numbers.

In almost the same way it is clear that addition and multiplication for complex numbers are both associative and commutative.

For example, the details for commutativity of addition go this way. We have

$$(a + \mathbf{i}b) + (c + \mathbf{i}d) = (a + c) + \mathbf{i}(b + d)$$
$$\text{while} \quad (c + \mathbf{i}d) + (a + \mathbf{i}b) = (c + a) + \mathbf{i}(d + b)$$

Because addition for real numbers is commutative we know that

$$a + c = c + a \qquad \text{and} \qquad b + d = d + b$$

It follows that

$$(c + a) + \mathbf{i}(d + b) = (c + \mathbf{i}d) + (a + \mathbf{i}b)$$

and so addition for complex numbers is commutative.

Ordered Pairs Commute Under Addition

In this somewhat more rigorous proof we define

$$(a, b) = \text{complex number}$$
$$\text{Let} \quad z_1 = (a, b) \leftrightarrow a + \mathbf{i}b$$
$$\text{and} \quad z_2 = (c, d) \leftrightarrow c + \mathbf{i}d$$

then we may write

$$\begin{aligned} z_1 + z_2 &= (a, b) + (c, d) \\ &= (a + c, b + d) \end{aligned}$$

and because reals commute

$$\begin{aligned} &= (c + a, d + b) \\ &= (c, d) + (a, b) \\ &= z_2 + z_1 \end{aligned}$$

QED

Additive Identity

The identity for addition clearly is

$$0 = 0 + \mathbf{i}0$$

because for any complex number $a + \mathbf{i}b$ we have, according to the rule for addition

$$\begin{aligned} (a + \mathbf{i}b) + (0 + \mathbf{i}0) &= (a + 0) + \mathbf{i}(b + 0) \\ &= a + \mathbf{i}b \end{aligned}$$

as required.

Multiplicative Identity

Similarly, the identity for multiplication is $1 = 1 + \mathbf{i}0$, because for any complex number $a + \mathbf{i}b$ we have

$$\begin{aligned} (a + \mathbf{i}b)(1 + \mathbf{i}0) &= (a \cdot 1 - b \cdot 0 + \mathbf{i}(a \cdot 0 + b \cdot 1) \\ &= (a - 0) + \mathbf{i}(0 + b) \\ &= a + \mathbf{i}b \end{aligned}$$

as should be the case.

Additive Inverse

The additive inverse for the complex number $a + \mathbf{i}b$ is clearly $-a + \mathbf{i}(-b)$, since

$$\begin{aligned} (a + \mathbf{i}b) + (-a + \mathbf{i}(-b)) &= (a + (-a)) + \mathbf{i}(b + (-b)) \\ &= 0 + \mathbf{i}0 = 0 \end{aligned}$$

Multiplicative Inverse

In Equation 1.9 we have already shown that every non-zero complex number has a multiplicative inverse. More specifically, using Equation 1.7 we may show that the product of the complex number $a + \mathbf{i}b$ and the number

$$\frac{a}{a^2 + b^2} - \mathbf{i}\frac{b}{a^2 + b^2}$$

is equal to 1, as required.

Multiplication is Distributive over Addition

Finally, it is easy to show that multiplication is distributive over addition, that is:

$$
\begin{aligned}
z_1(z_2 + z_3) &= z_1z_2 + z_1z_3 \\
\text{For, if we let} \quad z_1 &= a_1 + \mathbf{i}b_1 \\
\text{and} \quad z_2 &= a_2 + \mathbf{i}b_2 \\
\text{and} \quad z_3 &= a_3 + \mathbf{i}b_3
\end{aligned}
$$

then

$$
\begin{aligned}
z_1(z_2 + z_3) &= (a_1 + \mathbf{i}b_1)[(a_2 + \mathbf{i}b_2) + (a_3 + \mathbf{i}b_3)] \\
&= a_1[(a_2 + \mathbf{i}b_2) + (a_3 + \mathbf{i}b_3)] \\
&\quad + \mathbf{i}b_1[(a_2 + \mathbf{i}b_2) + (a_3 + \mathbf{i}b_3)] \\
&= a_1(a_2 + \mathbf{i}b_2) + \mathbf{i}b_1(a_2 + \mathbf{i}b_2) \\
&\quad + a_1(a_3 + \mathbf{i}b_3) + \mathbf{i}b_1(a_3 + \mathbf{i}b_3) \\
&= (a_1 + \mathbf{i}b_1)(a_2 + \mathbf{i}b_2) \\
&\quad + (a_1 + \mathbf{i}b_1)(a_3 + \mathbf{i}b_3) \\
&= z_1z_2 + z_1z_3
\end{aligned}
$$

With this last step we have verified that, with our definition of addition and multiplication, all of the field properties hold for the set of complex numbers. However, as is the case for *any* set of numbers which satisfy the field properties, we can also define the operations of substraction and division. We do this in the next section.

2.2.2 Subtraction and Division

In a field, subtraction is usually defined in terms of addition of the additive inverse. This means that if we have numbers z_1 and z_2 we define the difference in this way:

$$z_1 - z_2 = z_1 + (-z_2) \tag{2.1}$$

where $-z_2$ is the additive inverse of z_2. For complex numbers this works out this way:

$$
\begin{aligned}
(a + \mathbf{i}b) - (c + \mathbf{i}d) &= (a + \mathbf{i}b) + (-c + \mathbf{i}(-d)) \\
&= (a + (-c)) + \mathbf{i}(b + (-d)) \\
&= (a - c) + \mathbf{i}(b - d)
\end{aligned}
$$

Thus subtraction for complex numbers proceeds much as does addition. For example,

$$
\begin{aligned}
(5 + 6\mathbf{i}) - (3 - 2\mathbf{i}) &= (5 - 3) + (6 - (-2))\mathbf{i} \\
&= 2 + 8\mathbf{i}
\end{aligned}
$$

Division in a field is usually defined in terms of multiplication by the multiplicative inverse. This means that if we have numbers z_1 and z_2 we define the quotient of these numbers in this way:

$$
\frac{z_1}{z_2} = z_1 z_2^{-1}
$$

where z_2^{-1} is the multiplicative inverse of z_2. For complex numbers, if we use the multiplicative inverse as given in Equation 1.9, division works out this way:

$$
\begin{aligned}
\frac{a + \mathbf{i}b}{c + \mathbf{i}d} &= (a + \mathbf{i}b)(c + \mathbf{i}d)^{-1} \\
&= (a + \mathbf{i}b)\left(\frac{c}{c^2 + d^2} - \mathbf{i}\frac{d}{c^2 + d^2}\right) \\
&= \frac{ac + bd}{c^2 + d^2} + \mathbf{i}\frac{bc - ad}{c^2 + d^2}
\end{aligned}
\tag{2.2}
$$

A numerical example will be useful here. Suppose

$$
\begin{aligned}
z_1 &= 2 + 3\mathbf{i} \\
\text{and} \quad z_2 &= 1 + 2\mathbf{i}
\end{aligned}
$$

Then we have

$$
\begin{aligned}
\frac{2 + 3\mathbf{i}}{1 + 2\mathbf{i}} &= (2 + 3\mathbf{i})(1 + 2\mathbf{i})^{-1} \\
&= (2 + 3\mathbf{i})\left(\frac{1}{1^2 + 2^2} - \mathbf{i}\frac{2}{1^2 + 2^2}\right) \\
&= (2 + 3\mathbf{i})\left(\frac{1}{5} - \mathbf{i}\frac{2}{5}\right) \\
&= \frac{8}{5} - \mathbf{i}\frac{1}{5}
\end{aligned}
$$

Up to this point we have defined the four ordinary operations for complex numbers and have confirmed that over the complex numbers we have a field. In the next section we introduce the idea of the *conjugate* of a complex number.

2.3   The Complex Conjugate

Associated with each complex number

$$z = a + \mathbf{i}b \qquad (2.3)$$

is a number called its *complex conjugate* which is designated

$$\overline{z} = a - \mathbf{i}b \qquad (2.4)$$

Using Equations (2.3) and (2.4) it is easy to show by direct substitution that the following properties hold:

$$
\begin{aligned}
z + \overline{z} &= (a + \mathbf{i}b) + (a - \mathbf{i}b) \\
&= (a + a) + \mathbf{i}(b + (-b)) \\
&= 2a
\end{aligned}
$$

and,

$$
\begin{aligned}
z\,\overline{z} &= (a + \mathbf{i}b)(a - \mathbf{i}b) \\
&= (aa - b(-b)) + \mathbf{i}(a(-b) + ba) \\
&= a^2 + b^2
\end{aligned}
$$

These next two properties

$$
\begin{aligned}
\overline{z_1 + z_2} &= \overline{z_1} + \overline{z_2} \\
\overline{z_1 z_2} &= \overline{z_1}\,\overline{z_2}
\end{aligned}
$$

may be shown to follow in similar fashion.

Earlier, we called the number $\sqrt{a^2 + b^2}$ the *absolute value or magnitude* of the complex number $z = a + ib$ and denoted it by

$$|z| = \sqrt{a^2 + b^2}$$

Now from the above results we have

$$z\,\overline{z} = a^2 + b^2 = |z|^2$$

or

$$|z| = \sqrt{z\,\overline{z}}$$

which is a very convenient formula for computing the absolute value of any complex number z. From this an important identity for the absolute value of a product of complex numbers follows.

Thm: For any two complex numbers, z_1 and z_2

$$|z_1 z_2| = |z_1||z_2|$$

Pf:

$$
\begin{aligned}
|z_1 z_2|^2 &= (z_1 z_2)\overline{(z_1 z_2)} \\
&= z_1 z_2 \overline{z_1}\,\overline{z_2} \\
&= z_1 \overline{z_1} z_2 \overline{z_2}
\end{aligned}
$$

$$= |z_1|^2 |z_2|^2$$

From this it follows that

$$|z_1 z_2| = |z_1||z_2| \qquad \textbf{QED}$$

Inverse Revisited

It is convenient to use the conjugate of the complex number in calculating its inverse. It is also convenient to use the conjugate in the division of complex numbers, as we now show. Earlier we showed that every non-zero complex number does have an inverse, and we showed how division may be accomplished in terms of this inverse. Suppose once again that we wish to compute the quotient

$$z = \frac{z_1}{z_2}$$

If we multiply both sides of this equation by z_2, finding this quotient amounts to finding a complex number z such that

$$z z_2 = z_1$$

And now if we multiply both sides of this equation by the complex conjugate $\overline{z_2}$ we get

$$z z_2 \overline{z_2} = z_1 \overline{z_2}$$

that is,

$$z(|z_2|)^2 = z_1 \overline{z_2}$$

Hence the quotient we want is given by

$$z = z_1 \frac{\overline{z_2}}{|z_2|^2}$$

If we compare this result with accomplishing division by z_2 by multiplying by $(z_2)^{-1}$ we notice that we have found the following formula for the *inverse* of a non-zero complex number:

$$z^{-1} = \frac{\overline{z}}{(|z|)^2} \tag{2.5}$$

To this point we have reviewed only a few of the basic properties of complex numbers. We remark, however, that it was this successful extension of the real number system which led mathematicians to ask

> *Can one construct a mathematical system which uses "higher" complex numbers represented, say, by triplets, 4-tuples, etc., of real numbers in the same way we now commonly represent a complex number as an ordered "pair" of real numbers, (a, b)?*

We now know, of course, that the answer is **yes** — at least for those 4-tuples we call *quaternions*. It is generally acknowledged that the primary application of the quaternion is as a rotation operator on objects in R^3. So before we discuss the properties and attributes of the quaternion and its related rotation operators we must first finish the algebraic preliminaries that relate to coordinatization of R^2 and R^3 and rotations in these spaces.

2.4 Coordinates

Consider an object (or set of points) in a plane. The relative location of each point is defined with respect to a coordinate frame fixed in the plane.

What this means is that in the plane an arbitrary but fixed point which we call the *origin* is specified. Two straight lines are drawn which intersect (perpendicularly, for our purposes) at this origin: one axis we call the *x-axis* (usually horizontal) the other the *y-axis* (usually vertical). These axes are actually real number lines, whose positive numbers lie to the right of and above the origin, respectively, for the usual orientation.

Note that, these two axes are ordered in the plane such that a 90 degree *counter-clockwise* rotation about the origin takes the positive x-axis into the positive y-axis.

The origin represents the zero point on each coordinate axis and is designated by the ordered pair of real numbers $(0, 0)$. On both the positive x-axis and the positive y-axis points are chosen which are a *unit* distance from the origin, labelled $(1, 0)$ and $(0, 1)$, respectively. Every distinct point, P, in the plane can now be uniquely specified by an ordered pair of real numbers, (x_1, y_1) as shown in Figure 2.1. This gives us a coordinatized plane, denoted R^2.

The ideas underlying this coordinatization of the plane can be extended, of course, to the coordinatization of 3-dimensional space, denoted R^3. This requires a third axis which, in context, we call the z-axis; this third coordinate axis is assumed to be perpendicular to the xy plane and directed positively out of the paper toward the reader. See the comment in the margin.

Our ultimate objective is to understand the use of quaternions for rotations in R^3. In order to do this we will need a coordinati-

n-tuples

It is now known that a mathematical system (with division) over a set of n-tuples of real numbers does not exist for $n = 3$ or $n > 4$.

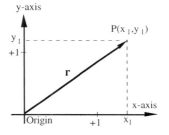

Figure 2.1
Coordinate Frame

Right Hand Rule

For rotations in the plane it is helpful to imagine a *z-axis* which is directed out of the plane toward the viewer and perpendicular to the x-axis and y-axis at the origin. Then, using the "right-hand" rule, with the thumb in the direction of the positive z-axis the fingers wrap in the direction for a positive rotation about this z-axis. A positive rotation rotates the positive x-axis toward the positive y-axis.

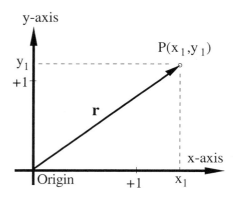

Figure 2.1: The Coordinatized Plane

zation of R^3. But before we define coordinates in R^3 we first state what we mean by rotations in the plane.

2.5   Rotations in the Plane

We shall distinguish between two perspectives on rotations in the plane, and shall determine the effect which each has on coordinates of points in the plane. The first is rotation of the coordinate frame with respect to fixed points (vectors) in the plane; the second is rotation of points (vectors) with respect to a fixed coordinate frame.

2.5.1   Frame Rotation - Points Fixed

In considering rotations in the plane we first determine what effect a rotation of a coordinate frame has on the coordinates of a point P fixed in the plane. In Figure 2.2 the fixed point P has coordinates (x_1, y_1) in the coordinate frame $\mathbf{X}, \mathbf{Y}$. We obtain a new coordinate frame $\mathbf{x}, \mathbf{y}$ by rotating the $\mathbf{X}, \mathbf{Y}$ frame about the origin, through a positive angle θ, as shown. We let the coordinates in the new frame be (x_2, y_2) and now determine a relationship between the new coordinates, (x_2, y_2), and the original coordinates, (x_1, y_1).

First we identify the point P with a vector $\mathbf{r}$ directed from the origin to the point P. Let r be the length of the vector $\mathbf{r}$ and α the angle between the vector and the positive X axis. Then, using ordinary right triangle trigonometry, we see that in triangle OPR we have

$$x_1 \quad = \quad r \cos \alpha$$

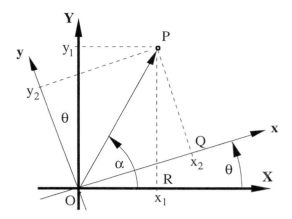

Figure 2.2: Rotation of Coordinates

$$\text{and} \qquad y_1 \;=\; r\sin\alpha$$

Next, from Figure 2.2 we notice that the angle between the vector **r** and the rotated **x**-axis is $\alpha - \theta$. In triangle OPQ we note

$$x_2 \;=\; r\cos(\alpha - \theta)$$
$$\text{and} \qquad y_2 \;=\; r\sin(\alpha - \theta)$$

Expanding these expressions gives

$$
\begin{aligned}
x_2 \;&=\; r\cos(\alpha - \theta)\\
&=\; r\cos\alpha\cos\theta + r\sin\alpha\sin\theta\\
\text{and} \qquad y_2 \;&=\; r\sin(\alpha - \theta)\\
&=\; r\sin\alpha\cos\theta - r\cos\alpha\sin\theta
\end{aligned}
$$

Thus the desired relationship is

$$x_2 \;=\; x_1\cos\theta + y_1\sin\theta \qquad\qquad (2.6)$$
$$\text{and} \qquad y_2 \;=\; y_1\cos\theta - x_1\sin\theta \qquad\qquad (2.7)$$

2.5.2   Point Rotation - Frame Fixed

In the preceeding derivation we thought of the point P (or vector **r**) as being fixed while the coordinate frame rotates about the origin through an angle θ. However, we may also think of the coordinate frame as being fixed, while the point P or vector **r** rotates about the origin through an angle θ from $\mathbf{r}_1$ to $\mathbf{r}_2$, as shown in Figure 2.3. Since the length of a vector is invariant under rotations, $|\mathbf{r}_1| = |\mathbf{r}_2| = r$. In this case, for triangle OP_1Q, we have

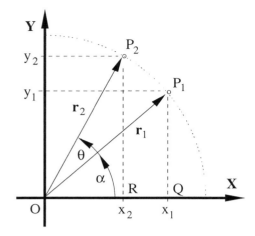

Figure 2.3: Rotation of Vector

$$x_1 = r\cos\alpha$$
$$\text{and}\qquad y_1 = r\sin\alpha$$

while in triangle OP_2R we have

$$\begin{aligned} x_2 &= r\cos(\alpha+\theta) \\ &= r\cos\alpha\cos\theta - r\sin\alpha\sin\theta \\ &= x_1\cos\theta - y_1\sin\theta \end{aligned} \qquad (2.8)$$

$$\begin{aligned} \text{and}\qquad y_2 &= r\sin(\alpha+\theta) \\ &= r\sin\alpha\cos\theta + r\cos\alpha\sin\theta \\ &= y_1\cos\theta + x_1\sin\theta \end{aligned} \qquad (2.9)$$

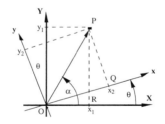

Figure 2.2
Rotated Frame

The (x_2,y_2) coordinate axes rotated thru an angle θ with respect to the (x_1,y_1) coordinate axes.

The reader should notice how Equations 2.8 and 2.9 differ from Equations 2.6 and 2.7, in that the *sine* terms in these pairs of equations *differ in sign*. The following example will illustrate and emphasize the significance of this difference.

Consider the point P with coordinates $(1,1)$, as in Figure 2.4 If we rotate *the coordinate frame* positively (that is, counter-clockwise) through an angle $\theta = \pi/4$, the resulting coordinates of the point P in the new frame, using Equation 2.6 and 2.7, are

$$\begin{aligned} x_2 &= 1\cos(\pi/4) + 1\sin(\pi/4) \\ &= \frac{\sqrt{2}}{2} + \frac{\sqrt{2}}{2} = \sqrt{2} \end{aligned}$$

$$\begin{aligned} \text{and}\qquad y_2 &= 1\cos(\pi/4) - 1\sin(\pi/4) \\ &= \frac{\sqrt{2}}{2} - \frac{\sqrt{2}}{2} = 0 \end{aligned}$$

Thus, in the rotated coordinate frame $\mathbf{x}, \mathbf{y}$ the point P lies on the $\mathbf{x}$-axis with coordinates $(\sqrt{2}, 0)$. If, however, we rotate the point P through an angle $\theta = \pi/4$, the resulting coordinates of P, using Equations 2.8 and 2.9, are $(0, \sqrt{2})$. The rotated point P now lies on

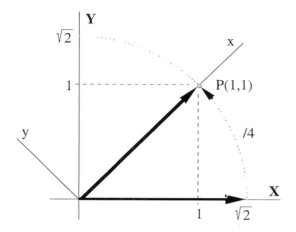

Figure 2.4: Frame Rotation

the $\mathbf{Y}$-axis! The perspective in the first case is that of an observer standing on the fixed point; in the second case the observer is seated in the fixed coordinate frame.

2.5.3 Equivalent Rotations

We have just discussed two rotations in the plane, one in which the points are fixed and the other in which the coordinate frame is fixed. In general these rotations produce quite different results. A moment's reflection, however, will convince one that a rotation of a coordinate frame through an angle θ results in exactly the same *vector-frame* relationship as a rotation of the vector through an angle $-\theta$. If θ is positive in Equations 2.8 and 2.9 the rotation is counter-clockwise, while the opposite is true when θ is negative. We illustrate the point by returning to our preceding example.

If the coordinate frame is rotated through a positive angle $\theta = \pi/4$, the fixed vector $(1,1)$ is transformed into the vector $(\sqrt{2}, 0)$ which lies along the new x-axis. If, however, the vector $(1,1)$ is rotated clockwise through the angle $-\pi/4$, relative to the fixed coordinate frame $\mathbf{X}, \mathbf{Y}$, the resulting vector is again $(\sqrt{2}, 0)$ along the $\mathbf{X}$-axis. This situation is illustrated more generally in Figure 2.5. It is easy to show that this is always so. Earlier we showed that if we rotate the coordinate frame through an angle θ, the new coordinates are given by

$$x_2 = x_1 \cos\theta + y_1 \sin\theta \qquad (2.10)$$

$$y_2 = y_1 \cos\theta - x_1 \sin\theta \qquad (2.11)$$

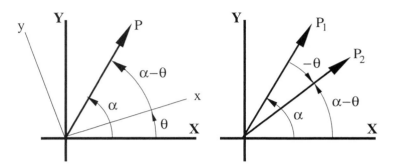

Figure 2.5: Rotation Perspectives

Suppose now the coordinate frame is fixed and we rotate the point (or vector) through an angle, $-\theta$. According to Equations 2.8 and 2.9 the new coordinates are given by

$$x_2 = x_1 \cos(-\theta) - y_1 \sin(-\theta)$$

$$y_2 = y_1 \cos(-\theta) + x_1 \sin(-\theta)$$

Since $\cos(-\theta) = \cos\theta$ and $\sin(-\theta) = -\sin\theta$ these become

$$x_2 = x_1 \cos\theta + y_1 \sin\theta$$

$$y_2 = y_1 \cos\theta - x_1 \sin\theta$$

which correspond exactly to Equations 2.10 and 2.11. Hence, these two rotations result in exactly the same *vector-frame relationship*.

2.5.4   Matrix Notation

Each of the two rotations discussed above is represented by a pair of equations. We remark here that each of these pairs may be written more concisely by using matrix notation. Those readers familiar with the product of matrices will recognize that Equations 2.6 and 2.7 may be written in the matrix form

$$\begin{bmatrix} x_2 \\ y_2 \end{bmatrix} = \begin{bmatrix} \cos\theta & \sin\theta \\ -\sin\theta & \cos\theta \end{bmatrix} \begin{bmatrix} x_1 \\ y_1 \end{bmatrix} \qquad (2.12)$$

If we define vectors $\mathbf{r}_1$ and $\mathbf{r}_2$ by

$$\mathbf{r}_1 = col[x_1, y_1] \quad \text{and} \quad \mathbf{r}_2 = col[x_2, y_2]$$

and a matrix A by

$$A = \begin{bmatrix} \cos\theta & \sin\theta \\ -\sin\theta & \cos\theta \end{bmatrix} \tag{2.13}$$

then the above matrix equation takes the simple form

$$\mathbf{r}_2 = A\mathbf{r}_1 \tag{2.14}$$

If the vector $\mathbf{r}$ has coordinates (x_1, y_1) with respect to an initial coordinate frame, this equation gives the pair (x_2, y_2) as the coordinates of $\mathbf{r}$ with respect to a new coordinate frame obtained from the initial frame by a rotation through an angle θ. In this context the matrix A is called the *rotation matrix* or sometimes the *rotation operator*. Such matrices will play an important role in our analysis of rotations, and we shall soon be making extensive use of both matrix notation and the algebra of matrices. Thus we turn next to an elementary review of that notation and of the algebra of matrices.

2.6 Review of Matrix Algebra

By an $m \times n$ matrix we shall mean any rectangular array of elements arranged in m rows and n columns, for suitable positive integers m and n. Such a matrix is said to have *order* $m \times n$. These elements may be scalars or functions or whatever is meaningful in the context of its use.

If $m = 1$ we have a *row matrix*. If $n = 1$ we have a *column matrix*. If $m = n$ we have a *square matrix*. We define matrices to be *equal* if they are of the same order and their corresponding elements are exactly the same.

In general, we shall represent a matrix A as

$$A = [a(i, j)]$$

In this notation "i" indicates the row, and "j" indicates the column in which the matrix element $a(i, j)$ is found. Therefore, in summary, when we say the matrix A is an $m \times n$ matrix — sometimes written $A(m, n)$ — we mean the matrix A has m rows and n columns, where $m \geq 1$ and $n \geq 1$. For appropriate values for m and n the matrix may be a row matrix, a column matrix, or a square matrix.

2.6.1   The Transpose

Rotation Transpose

It will be the case that for any *rotation matrix*, its **transpose** is the matrix which results if the signs of all the *angles* are changed. Alternatively, "flipping" any rotation matrix about its principal diagonal produces its transpose.

For any matrix $A = [a(i,j)]$, the *transpose* is defined

$$A^t = [a(i,j)]^t = [a(j,i)]$$

Notice that finding the transpose of a matrix amounts to interchanging its rows and columns. Thus, for the matrix A in Equation 2.13 the transpose is

$$A^t = \begin{bmatrix} \cos\theta & -\sin\theta \\ \sin\theta & \cos\theta \end{bmatrix}$$

For the reader acquainted with matrix multiplication it is easy to see that the matrix associated with the rotation described by Equations 2.8 and 2.9 (in which the coordinate frame is fixed while the vector is rotated through an angle θ) is given by

$$B = \begin{bmatrix} \cos\theta & -\sin\theta \\ \sin\theta & \cos\theta \end{bmatrix} \tag{2.15}$$

We note that this matrix is exactly the transpose of the matrix which represents the rotation described by Equations 2.6 and 2.7, in which the vector is fixed and the coordinate frame rotates through the angle θ; that is, we have $B = A^t$. It turns out that this will always be the case.

2.6.2   Addition and Subtraction

If two matrices, A and B, have the same order, that is, the same number of rows and columns, then their *sum* $A + B$ is computed simply by adding corresponding elements. In our matrix notation we write

$$A + B = [a(i,j) + b(i,j)]$$

An elementary example clearly illustrates the definition. If

$$A = \begin{bmatrix} 1 & 2 \\ 4 & -1 \end{bmatrix} \qquad \text{and} \qquad B = \begin{bmatrix} 3 & 1 \\ 2 & 3 \end{bmatrix}$$

then

$$A + B = \begin{bmatrix} 4 & 3 \\ 6 & 2 \end{bmatrix}$$

A matrix all of whose elements are 0 is called a *zero* matrix. Note that for any matrix A and a zero matrix, denoted O, of the same order we have

$$A + O = A$$

Further, the *negative* of a matrix A is defined by

$$-A = [-a(i,j)]$$

that is, $-A$ is a matrix whose elements are the negatives of the elements of the matrix A. Notice that for any matrix A we have

$$A + (-A) = O$$

Subtraction for matrices of the same order is usually defined as addition of the negative, that is

$$A - B = A + (-B)$$

2.6.3 Multiplication by a Scalar

If λ is a scalar and A is a matrix, the *product of the scalar and the matrix* is given by

$$\lambda A = [\lambda a(i,j)]$$

that is, we simply multiply each element of A by the scalar λ. For example, if

$$A = \begin{bmatrix} 1 & 3 \\ 1 & 4 \end{bmatrix}$$

then

$$2A = \begin{bmatrix} 2 & 6 \\ 2 & 8 \end{bmatrix}$$

Notice that we may now, for example, write $A + A = 2A$ and $-A = (-1)A$.

2.6.4 Product of Matrices

The product of two matrices is a bit more complicated to define. We begin by reviewing the dot product of two vectors. If we have two vectors in R^2, say,

$$\mathbf{a} = (a_1, a_2)$$
$$\text{and} \quad \mathbf{b} = (b_1, b_2)$$
$$\text{the dot product is} \quad \mathbf{a} \cdot \mathbf{b} = a_1 b_1 + a_2 b_2$$

Thus if

$$\mathbf{a} = (1,3)$$
$$\text{and} \quad \mathbf{b} = (-4, 2)$$
$$\text{we have} \quad \mathbf{a} \cdot \mathbf{b} = (1,3) \cdot (-4,2) = 1(-4) + 3(2)$$
$$= -4 + 6 = 2$$

Note that the result is a scalar.

If we have vectors in R^3 the definition is analogous. For example, if

$$
\begin{aligned}
\mathbf{a} &= (1, 3, 2) \\
\mathbf{b} &= (2, -1, 7)
\end{aligned}
$$

we have $\mathbf{a} \cdot \mathbf{b} = 1(2) + 3(-1) + 2(7) = 13$

The definition extends to vectors considered as n-tuples in R^n, for if

$$\mathbf{a} = (a_1, a_2, \cdots a_n)$$

and

$$\mathbf{b} = (b_1, b_2, \cdots b_n)$$

then $\mathbf{a} \cdot \mathbf{b} = a_1 b_1 + a_2 b_2 + \cdots + a_n b_n$

Using ordinary summation notation, we may nicely write this dot product as

$$\mathbf{a} \cdot \mathbf{b} = \sum_{k=1}^{n} a_k b_k$$

Now consider an $m \times p$ matrix $A(m, p)$ and a $p \times n$ matrix $B(p, n)$. Note that the number of columns, p, in A is the same as the number of rows in the matrix B. This must be the case if we are to be able to compute the matrix product AB. The result will be an $m \times n$ matrix C.

Matrix Product

Here we finally define the product of two matrices

The product is defined by this rule:

> *The element in the ith row and the jth column of the product AB is the dot product of the ith row vector of the matrix A and the jth column vector of the matrix B.*

Thus if the ith row of A is $(a_{i1}, a_{i2}, \cdots a_{in})$ and the jth column of B is $(b_{1j}, b_{2j}, \cdots b_{nj})$ then the element common to the ith row and the jth column of $C = AB$ is

$$c(i, j) = \sum_{k=1}^{p} a_{ik} b_{kj}$$

Notice that here we have used double subscript notation for the elements of a matrix. We give one numerical example. Suppose

$$
A = \begin{bmatrix} 1 & 3 \\ -2 & 7 \end{bmatrix} \qquad \text{and} \qquad B = \begin{bmatrix} 0 & 2 \\ 4 & 3 \end{bmatrix}
$$

The element in the 1st row and 1st column of the product, AB, is the dot product of the 1st row of the matrix A and the 1st column of the matrix B, that is,

$$(1,3) \cdot (0,4) \;=\; 1 \cdot 0 + 3 \cdot 4 \;=\; 12$$

Proceeding in exactly the same way with the remaining elements in the product, we obtain

$$AB \;=\; \begin{bmatrix} 1 & 3 \\ -2 & 7 \end{bmatrix} \begin{bmatrix} 0 & 2 \\ 4 & 3 \end{bmatrix} \;=\; \begin{bmatrix} 12 & 11 \\ 28 & 17 \end{bmatrix}$$

However, we also compute the matrix product commuted, that is

$$BA \;=\; \begin{bmatrix} 0 & 2 \\ 4 & 3 \end{bmatrix} \begin{bmatrix} 1 & 3 \\ -2 & 7 \end{bmatrix} \;=\; \begin{bmatrix} -4 & 14 \\ -2 & 33 \end{bmatrix}$$

and note that the results are not the same. From this we may conclude that

> Matrices *do not* commute under multiplication — and therefore the mathematical system which consists of the set of all 2×2 matrices is not a *field*.

Any square matrix with all 1's on the diagonal and zeroes elsewhere is an **Identity** matrix, usually denoted, I. It is called an Identity matrix because

$$AI = IA = A$$

2.6.5 Rotation Matrices

We return now to the rotation in the plane in which the coordinate frame rotates while the points (or vectors) remain fixed. The transformed coordinates were given by

$$x_2 \;=\; x_1 \cos \theta + y_1 \sin \theta$$
$$y_2 \;=\; y_1 \cos \theta - x_1 \sin \theta$$

If we write these equations in the slightly altered form

$$x_2 \;=\; (\cos \theta) x_1 + (\sin \theta) y_1$$
$$y_2 \;=\; (-\sin \theta) x_1 + (\cos \theta) y_1$$

we may recognize that both equations are contained in the single matrix equation

$$\begin{bmatrix} x_2 \\ y_2 \end{bmatrix} \;=\; \begin{bmatrix} \cos \theta & \sin \theta \\ -\sin \theta & \cos \theta \end{bmatrix} \begin{bmatrix} x_1 \\ y_1 \end{bmatrix}$$

Commutivity

Notwithstanding what has just been said about commutivity under multiplication, for any 2×2 rotation matrix A we can always write

$$A^t A = A A^t = I$$

If in terms of vectors we write $\mathbf{r}_1 = \text{col}[x_1, y_1]$ and $\mathbf{r}_2 = \text{col}[x_2, y_2]$ and define the matrix A by

$$A = \begin{bmatrix} \cos\theta & \sin\theta \\ -\sin\theta & \cos\theta \end{bmatrix}$$

then the above matrix equation has the simple form

$$\mathbf{r}_2 = A\mathbf{r}_1$$

In this equation, A is called the *rotation matrix* or equivalently a *rotation operator* which takes $\mathbf{r}_1$ into $\mathbf{r}_2$.

In this same fashion we may verify that for the second type of rotation in the plane, in which the coordinate frame is fixed and the point (or vector) rotates, the appropriate rotation matrix is

$$B = \begin{bmatrix} \cos\theta & -\sin\theta \\ \sin\theta & \cos\theta \end{bmatrix} \tag{2.16}$$

Note that this matrix B is the transpose of the matrix A, that is

$$B = A^t$$

Consider once again the rotation matrix A associated with the rotation of the coordinate frame through an angle θ while the point (or vector) remains fixed. We will have occasion to consider how such a rotation can be "undone," that is, how the effect of the rotation can be negated. If we think about it for a moment, it seems clear that there are two ways to do this. First, we may simply rotate the coordinate frame through an angle $-\theta$. Surely this will return things to their original position. Second, though this is not quite so clear, we may rotate the point (or vector) through an angle θ while the coordinate frame remains fixed. Now an interesting thing happens. The matrix which represents the first of these possibilities may be obtained by replacing θ by $-\theta$ in the matrix A. We get

$$B = \begin{bmatrix} \cos(-\theta) & \sin(-\theta) \\ -\sin(-\theta) & \cos(-\theta) \end{bmatrix} \tag{2.17}$$

Since we know, $\cos(-\theta) = \cos\theta$ and $\sin(-\theta) = -\sin\theta$, the matrix B may be written

$$B = \begin{bmatrix} \cos\theta & -\sin\theta \\ \sin\theta & \cos\theta \end{bmatrix} \tag{2.18}$$

But this is exactly the matrix B of Equation 2.16, namely, the transpose A^t. From this we learn that the rotation matrix needed to "undo," that is, to *invert* the rotation represented by the matrix A is exactly the matrix A^t. Or we may say that the inverse of a rotation matrix is its transpose. The inverse of a matrix A in general we consider in following sections.

We indicated that there was a second way to invert the rotation we are considering, and that was to follow it by a rotation of the point (or vector) through an angle θ while the coordinate frame remains fixed. As we noted above, the matrix associated with this rotation is exactly the matrix B of Equation 2.14, which in fact is A^t. Thus we obtain exactly the same result as before, so the two possibilities are in fact equivalent.

We note finally that these results hold whether the angle θ is positive or negative, that is, the result does not depend on the direction of the rotation. We turn now to the procedure for finding the inverse of a matrix in general.

Rotation Inverse

The *inverse* of a rotation matrix A is its transpose, A^t, that is

$$A^t A \ = \ AA^t \ = \ I$$

where I is an Identity matrix

2.7   The Determinant

In our discussion so far we have considered the sum, difference, transpose, and product of two matrices. We must yet consider the quotient of two matrices, that is, how division by a matrix is accomplished. As with complex numbers, we shall define division by a matrix in terms of multiplication by the inverse of the matrix.

By the *inverse* of a square matrix A we mean a matrix B, of the same order as A, such that $AB = BA = I$, where I is an identity matrix. As an example of the use of the inverse of a matrix, consider the following system of two equations in two unknowns

$$
\begin{aligned}
x + 3y &= 7 \\
2x + 7y &= 16
\end{aligned}
$$

Almost by inspection we see the solution to the system to be $x = 1$ and $y = 2$. However, if we use the product of matrices we may write this system in the matrix form

$$AX \ = \ B$$

where

$$
A \ = \ \begin{bmatrix} 1 & 3 \\ 2 & 7 \end{bmatrix}
\qquad
B \ = \ \begin{bmatrix} 7 \\ 16 \end{bmatrix}
\qquad
X \ = \ \begin{bmatrix} x \\ y \end{bmatrix}
$$

If we have the inverse of A available, say A^{-1}, we may multiply the equation $AX = B$ on the left by this inverse to obtain

$$A^{-1}AX \;=\; A^{-1}B$$

that is

$$X \;=\; A^{-1}B$$

which gives us the solution to the system. It turns out in our case that the solution is

$$X \;=\; \begin{bmatrix} 7 & -3 \\ -2 & 1 \end{bmatrix} \begin{bmatrix} 7 \\ 16 \end{bmatrix} \;=\; \begin{bmatrix} 1 \\ 2 \end{bmatrix}$$

We consider now ways in which the inverse of a matrix may be found, beginning with the idea of the determinant of a matrix.

Determinant

The determinant of every **rotation matrix** is equal to **one**. As an example, note that for Equation 2.16 the det(B) is equal to 1.

Associated with every *square* matrix, A, is a scalar called its *determinant*, denoted $\det(A)$ or $|A|$.

The determinant of a 1×1 matrix, say $B = [b]$, is equal to the single element b.

The determinant of the 2×2 matrix

$$A \;=\; \begin{bmatrix} a & b \\ c & d \end{bmatrix}$$

is the scalar defined by

$$\det(A) \;=\; |A| \;=\; ad - bc$$

In order to compute the value of the determinant for an $n \times n$ matrix where $n > 2$ we first introduce and define, by example, some preliminary matters relating to $n \times n$ matrices.

2.7.1   Minors

Minor

A **Minor** is a determinant.

Cofactor

The Minor and its related Cofactor are both determinants and therefore they are both numbers.
    Cofactor = $\pm$ Minor
    Cofactor = $(-1)^{i+j}$ Minor

Consider the square matrix

$$A \;=\; \begin{bmatrix} a_{11} & a_{12} & a_{13} \\ a_{21} & a_{22} & a_{23} \\ a_{31} & a_{32} & a_{33} \end{bmatrix} = \begin{bmatrix} 1 & 0 & 1 \\ 2 & 1 & -1 \\ 0 & 1 & 2 \end{bmatrix} \qquad (2.19)$$

Associated with each element a_{ij} of the matrix A is a *minor* denoted A_{ij}. Minor A_{ij} is a number which is equal to the value of the determinant of the submatrix obtained by deleting row i and column

j of the matrix A. For example, in the matrix of Equation 2.19, the minor of the element a_{32} is

$$A_{32} = \begin{vmatrix} a_{11} & a_{13} \\ a_{21} & a_{23} \end{vmatrix} = \begin{vmatrix} 1 & 1 \\ 2 & -1 \end{vmatrix} = -3$$

2.7.2 Cofactors

The *cofactor* associated with the element a_{ij} is denoted and defined by

$$A_{ij}^c = (-1)^{i+j} A_{ij}$$

that is, it is a *signed* minor. For the above example, then, the cofactor is,

$$A_{32}^c = (-1)^{3+2} A_{32} = (-1)(-3) = 3$$

2.7.3 Determinant of an $n \times n$ Matrix

We now can calculate the determinant of any $n \times n$ matrix in terms of any selected row (or column) and the associated cofactors of this selected row (or column).

For example, let $B = n \times n$ matrix. Then expanding over any selected row, say the ith row, we have

$$det(B) = \sum_{k=1}^{n} b_{ik} B_{ik}^c$$

Or again, if we expand over, say some j^{th} column, we have

$$det(B) = \sum_{k=1}^{n} b_{kj} B_{kj}^c$$

Choosing a row (column) with some elements equal to zero obviously simplifies the required computations in the indicated sums. There are methods for generating the desired zeroes in order to simplify the computation of a determinant, but we shall not review these here.

Using Equation 2.19 and expanding about the third column we can write

$$det(A) = a_{13} \begin{vmatrix} a_{21} & a_{22} \\ a_{31} & a_{32} \end{vmatrix} - a_{23} \begin{vmatrix} a_{11} & a_{12} \\ a_{31} & a_{32} \end{vmatrix} + a_{33} \begin{vmatrix} a_{11} & a_{12} \\ a_{21} & a_{22} \end{vmatrix}$$

Minor ↔ Cofactor

The scalar value of a Minor (a determinant), of course, may be positive or negative. In any event, the related **Cofactor** has a sign which is opposite that of the Minor, if $i+j$ is odd. The **Cofactor** is equal to the Minor if $i + j$ is even.

Rank ≥ 4

We emphasize, if the Cofactors in an expansion still have rank ≥ 4 then a further expansion may be required for these cofactors.

or if we expand about the second row

$$det(A) = -a_{21} \begin{vmatrix} a_{12} & a_{13} \\ a_{32} & a_{33} \end{vmatrix} + a_{22} \begin{vmatrix} a_{11} & a_{13} \\ a_{31} & a_{33} \end{vmatrix} - a_{23} \begin{vmatrix} a_{11} & a_{12} \\ a_{31} & a_{32} \end{vmatrix}$$

Matrix Elements

$$A = \begin{bmatrix} a_{11} & a_{12} & a_{13} \\ a_{21} & a_{22} & a_{23} \\ a_{31} & a_{32} & a_{33} \end{bmatrix}$$

In either case we get the expected result we were taught in secondary school,

$$det(A) = a_{11}a_{22}a_{33} + a_{12}a_{23}a_{31} + a_{13}a_{21}a_{32}$$
$$-a_{13}a_{22}a_{31} - a_{11}a_{23}a_{32} - a_{12}a_{21}a_{33} \qquad (2.20)$$

In fact, the determinant can be found by expanding about any row or column.

If we replace these elements by their numerical values in Equation 2.19 we get $det(A) = 5$.

If $det(A) = 0$ we say A is *singular*. Otherwise, A is said to be *non-singular*. It is only a non-singular matrix that is invertible, that is, it has an inverse.

Determinant Rules

It may be that the Cofactor matrices themselves, that is the $A^c(ij)$'s, will also require expansion. This rapidly results in computational congestion. The properties of determinants provide some useful means for simplifying the work involved.

In summary, the value of the determinant of any $n \times n$ square matrix, B, can be found by the following procedure:

1. Choose any row (column).

2. Compute cofactors for elements in chosen row (column).

3. $\det(B) = \sum_{i=1}^{n} a_{ij} \times B_{ij}^c$ for chosen column j, or

4. $\det(B) = \sum_{j=1}^{n} a_{ij} \times B_{ij}^c$ for chosen row i

2.8 The Cofactor Matrix

The *cofactor matrix*, denoted $A^c(i,j)$ or simply as A^c, for the matrix $A = [a(i,j)]$ is

$$A^c(i,j) = [A_{ij}^c]$$

where each of the elements of A_{ij}^c are the corresponding cofactors of the matrix A. That is, the cofactor matrix A^c is constructed from the matrix A by replacing each of its elements by its cofactor.

As an example, consider the 3×3 matrix A given by

$$A = \begin{bmatrix} 1 & 0 & 1 \\ 2 & 1 & -1 \\ 0 & 1 & 2 \end{bmatrix}$$

Earlier we calculated the cofactor of the element $a_{32} = 1$ to be $A^c_{32} = 3$, so in the cofactor matrix the element 1 is replaced by its cofactor 3. The reader should check that if this is done for each of the elements in the matrix A the result is the cofactor matrix

$$A^c = \begin{bmatrix} 3 & -4 & 2 \\ 1 & 2 & -1 \\ -1 & 3 & 1 \end{bmatrix}$$

2.9 Adjoint Matrix

The *adjoint matrix*, denoted A^a, of the matrix A is simply the transpose of the cofactor matrix of A. That is

$$A^a = (A^c)^t$$

Thus the adjoint of our matrix A above is the matrix

$$A^a = \begin{bmatrix} 3 & 1 & -1 \\ -4 & 2 & 3 \\ 2 & -1 & 1 \end{bmatrix}$$

We next review two ways in which the inverse of any square non-singular matrix may be computed.

2.10 The Inverse Matrix - Method 1

The first method is based on the ideas we have just presented. If $\det(A)$ is not zero, that is A is non-singular, we may compute the inverse, A^{-1}, simply by dividing the adjoint matrix A^a by the determinant of A. That is

$$A^{-1} = \frac{A^a}{\det(A)} \qquad \det(A) \neq 0$$

In our example the result is

$$A^{-1} = \frac{1}{5} \begin{bmatrix} 3 & 1 & -1 \\ -4 & 2 & 3 \\ 2 & -1 & 1 \end{bmatrix}$$

The reader should now verify that the product $A^{-1}A$ is indeed a 3×3 identity matrix.

NOTATION SUMMARY

- Matrix: $A = [a(i,j)]$

- Minor: A_{ij}

- Cofactor: $\quad A^c_{ij} \quad = (-1)^{i+j} A_{ij}$

- Cofactor Matrix:
  $$A^c(ij) = [(-1)^{i+j} A_{ij}]$$

- Adjoint: $A^a = (A^c)^t$

2.11   The Inverse Matrix - Method 2

The second method for finding the inverse of a matrix is quite different, and is based on the **Cayley-Hamilton Theorem** which states that

Every non-singular matrix satisfies its own characteristic equation.

Every $n \times n$ matrix A has a *characteristic equation* defined by

$$\det(A - \lambda I) = p(\lambda) = 0$$

Note

In Method 2, note, there is no need to solve for the characteristic values or characteristic vectors (which can be tedious) — find only the characteristic equation.

In general, the characteristic equation of an $n \times n$ matrix has the form

$$p(\lambda) = \lambda^n + a_{n-1}\lambda^{n-1} + \cdots + a_1\lambda + a_0 = 0$$

Therefore, invoking this theorem, we have

$$p(A) = A^n + a_{n-1}A^{n-1} + \cdots + a_1A + a_0I = 0$$

Multiplying both sides of this polynomial by A^{-1} and solving for the inverse yields,

$$A^{-1} = -\frac{1}{a_0}[A^{n-1} + a_{n-1}A^{n-2} + \cdots + a_2A + a_1I]$$

This method for finding the inverse is quite simple for 2×2, and even 3×3, matrices.

As an example, we will find the inverse for the 2×2 matrix,

$$M = \begin{bmatrix} 3 & 4 \\ 1 & 2 \end{bmatrix}$$

The characteristic equation for the matrix M is defined by

$$p(\lambda) = \det \begin{vmatrix} 3 - \lambda & 4 \\ 1 & 2 - \lambda \end{vmatrix} = 0$$

$$= \lambda^2 - 5\lambda + 2 = 0$$

$$= M^2 - 5M + 2I = 0$$

Multiplying this equation by M^{-1} gives

$$M - 5I + 2M^{-1} = 0$$

From this we can write

$$M^{-1} = \frac{1}{2}(5I - M)$$

$$= \frac{1}{2}\left\{\begin{bmatrix} 5 & 0 \\ 0 & 5 \end{bmatrix} - \begin{bmatrix} 3 & 4 \\ 1 & 2 \end{bmatrix}\right\}$$

$$= \frac{1}{2}\begin{bmatrix} 2 & -4 \\ -1 & 3 \end{bmatrix}$$

As a second example, we find the inverse for the 3×3 matrix

$$A = \begin{bmatrix} 1 & 0 & 1 \\ 2 & 1 & -1 \\ 0 & 1 & 2 \end{bmatrix}$$

The characteristic equation for the matrix A is a polynomial equation found by solving the determinant

$$p(\lambda) = det(A - \lambda I) = 0$$

$$= \begin{vmatrix} (1 - \lambda) & 0 & 1 \\ 2 & (1 - \lambda) & -1 \\ 0 & 1 & (2 - \lambda) \end{vmatrix}$$

$$= 5 - 6\lambda + 4\lambda^2 - \lambda^3 = 0$$

and because matrix A satisfies its own characteristic equation

$$5I - 6A + 4A^2 - A^3 = 0$$

Multiplying both sides by A^{-1} and rearranging terms gives

$$A^{-1} = \frac{1}{5}(6I - 4A + A^2)$$

$$= \frac{1}{5}\begin{bmatrix} 3 & 1 & -1 \\ -4 & 2 & 3 \\ 2 & -1 & 1 \end{bmatrix}$$

The intermediate computational details, that is solving for A^2 and summing multiples of matrices, are straight-forward. This result, of course, is the same as that obtained by Method 1.

2.12 Rotation Operators Revisited

In Section 2.6.5 we derived rotation operators for rotations in R^2 of the coordinate frame and of the points (or vectors). In the case of a rotation of the coordinate frame through an angle θ the rotation operator is

$$A \; = \; \begin{bmatrix} \cos\theta & \sin\theta \\ -\sin\theta & \cos\theta \end{bmatrix}$$

while in the case of a rotation of the points (or vectors) through an angle θ the rotation operator is given by

$$B \; = \; \begin{bmatrix} \cos\theta & -\sin\theta \\ \sin\theta & \cos\theta \end{bmatrix}$$

We noted that the operator in the one case is exactly the transpose of the operator in the other. In fact, each is the inverse of the other. It should also be mentioned that the matrix representation of a rotation is unique. The only matrices that work are those we have found.

We want to consider two properties which these matrices have, as well as to determine conditions under which a given 2×2 matrix is a rotation operator. The first property we note is that each of these operators has determinant $+1$, as the reader may easily verify. It turns out that this is always the case with a rotation operator. The second property is that each of these rotation matrices is *orthogonal*. We say a matrix A is *orthogonal* if the product of the matrix and its transpose is an identity matrix. That is, an $n \times n$ matrix A is *orthogonal* when

$$A^t A \; = \; AA^t = I$$

Equivalently, we may say a square matrix A is orthogonal if it is invertible and its inverse is exactly its transpose. With the use of the familiar trigonometric identity

$$\cos^2\theta + \sin^2\theta \; = \; 1$$

the reader may easily verify that the rotation operators A and B above are indeed orthogonal.

We mention, without proof, that the determinant of the product of two matrices is the product of the determinants of the individual matrices. In particular if matrices A and B both have determinant $+1$, so does their product AB. Further, if both A and B are orthogonal matrices, the product AB is also orthogonal. This is fairly easy to show, once we recall that in general

$$(AB)^t \; = \; B^t A^t$$

Rotation Matrix?

Are there conditions on a matrix M that will guarantee that it is a rotation operator? The answer is **yes**. The conditions are:

1. $\det M = |M| = +1$

2. $M^t M = MM^t = I$

Note

All rotation operators are orthogonal and have determinant $+1$.

Then we may write

$$
\begin{aligned}
(AB)(AB)^t &= ABB^tA^t \\
&= AIA^t \\
&= AA^t \\
&= I
\end{aligned}
$$

which is what we needed to show.

Finally we show that if a 2×2 matrix is orthogonal and has determinant $+1$, then it must be a rotation matrix. The details of the argument are somewhat tedious, but it may be well to consider them, since we are dealing with concepts important to our work. Let the matrix A be given by

$$
A = \begin{bmatrix} a_{11} & a_{12} \\ a_{21} & a_{22} \end{bmatrix}
$$

SO(2)

The group **SO**(2) is comprised of all those special orthogonal $2X2$ matrices whose determinant is $+1$.

We assume that A is orthogonal and has determinant $+1$. These two conditions yield the following equations

$$
\begin{aligned}
a_{11}a_{12} + a_{21}a_{22} &= 0 \\
a_{11}a_{22} - a_{21}a_{12} &= 1 \\
a_{11}^2 + a_{21}^2 &= 1 \\
a_{12}^2 + a_{22}^2 &= 1
\end{aligned}
$$

Now if we add the third and fourth of these equations, then, subtract twice the second equation we obtain

$$
(a_{11}^2 - 2a_{11}a_{22} + a_{22}^2) + (a_{12}^2 + 2a_{12}a_{21} + a_{21}^2) = 0
$$

that is, we have

$$
(a_{11} - a_{22})^2 + (a_{12} + a_{21})^2 = 0
$$

From this equation it follows that

$$
a_{11} = a_{22}
$$
$$
\text{and} \qquad a_{12} = -a_{21}
$$

Hence the matrix A must be of the form

$$
A = \begin{bmatrix} a & b \\ -b & a \end{bmatrix}
$$

where

$$
a^2 + b^2 = 1
$$

This implies that the point (a,b) lies somewhere on a circle of radius 1, so there is an angle θ such that $a = \cos\theta$ and $b = \sin\theta$. Therefore the matrix must be of the form

$$A \;=\; \left[\begin{array}{cc} \cos\theta & \sin\theta \\ -\sin\theta & \cos\theta \end{array} \right]$$

Note

Remember that the determinant of a rotation matrix is +1

But this is exactly the rotation matrix for a coordinate frame rotation through an angle θ.

In this analysis, the angle θ may be replaced by the angle $-\theta$, which would produce the rotation matrix associated with the rotation of *points (or vectors)* through an angle θ. Hence we have characterized rotation operators in R^2 as exactly those 2×2 matrices which are orthogonal and have determinant +1. Since the product of orthogonal matrices is orthogonal and the determinant of a product is the product of the determinants, it follows that the product of two rotation operators is always another rotation operator. We shall explore these matters further in the next chapter, where we extend these ideas to three-dimensional space, designated R^3.

A Polar Domain

It will be convenient to choose θ so that

$$-\pi < \theta \le \pi$$

Exercises for Chapter 2

Some exercises which follow are intended merely to help readers recall matters which may have been long since forgotten. They should provide the reader not only an opportunity to brush away some dust collected on ideas from disuse but also may encourage an effort to understand and apply the material presented. The exercises in the following chapters also, hopefully, will continue this reclamation process.

Given $v_1 = a + ib$ and $v_2 = c + id$

1. Find the product $w_1 = v_1 * v_2$.

2. Find the quotient $w_2 = v_1/v_2$.

3. Find the argument and magnitude of w_1 and w_2.

4. Express the following in exponential form:

$$\begin{aligned} (a) \quad & z_1 = -2 - i2 & (b) \quad & z_2 = 4 - i4 \\ (c) \quad & z_3 = -3 + i3 & (d) \quad & z_4 = 3 + i4 \end{aligned}$$

5. Find z_1^{-1} and v_1^{-1}

Given $A = \begin{bmatrix} 2 & -1 \\ 3 & 1 \end{bmatrix}$ and $B = \begin{bmatrix} \cos\theta & -\sin\theta \\ \sin\theta & \cos\theta \end{bmatrix}$

6. Compute AB and BA

7. Use Method 1 to find A^{-1} and B^{-1}

8. Find A^{-1} and B^{-1} using Cayley-Hamilton Theorem.

Chapter 3

Rotations in 3-space

3.1   Introduction

In Chapter 2 we considered rotations in two-dimensional space, i.e. R^2, and how such rotations are represented by matrices of the group $SO(2)$. In particular, we made a careful distinction between a rotation of the coordinate frame with respect to fixed points (or vectors) as opposed to a rotation of the points (or vectors) with respect to a fixed coordinate frame. In fact, if in both cases the angle of rotation is $+\theta$, then it was noted that the rotation operator for the one case was simply the transpose of the rotation operator for the other case. We noted further that the one is the inverse of the other. In this chapter we extend these ideas to the three-dimensional case, that is, to R^3.

Our ultimate objective is to show how quaternions may be used as rotation operators in R^3. In this chapter we consider first a simple sequence of rotations in R^2, that is, in the plane. We then develop matrices as rotation operators in R^3, and consider sequences of rotations in R^3, *all in terms of matrix algebra*. We apply these results to the aerospace application of *tracking* a remote object, and conclude with a *geometric* analysis of the single axis equivalent of the tracking transformation.

3.2   Rotation Sequences in the Plane

We now consider what happens when one rotation is followed by another, that is, what is the result of a sequence of two rotations. Suppose a rotation of the initial coordinate frame (points fixed) through an angle α is followed by a rotation of the resulting coordinate frame through an angle β. Clearly the result is a rotation of

SO(2)

The group **SO(2)** is comprised of all those special orthogonal $2X2$ matrices whose determinant is $+1$.

the initial coordinate frame (points fixed) through an angle $\alpha + \beta$, as illustrated in Figure 3.1.   The axes of the reference frame are

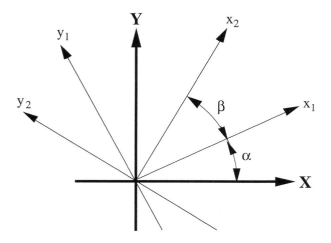

Figure 3.1: Rotation Sequence in R^2

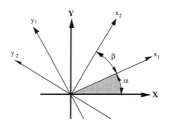

labelled **X** and **Y**. Consider any vector **v** defined in this reference frame. We define a new frame, labelled, x_1 and y_1, which is related to the reference frame by a rotation through an angle α. The vector **v** defined in this new frame we denote $\mathbf{v}_1$ and we write

$$\mathbf{v}_1 = R_\alpha \mathbf{v}$$

where the matrix rotation operator is

$$R_\alpha = \left[\begin{array}{cc} \cos \alpha & \sin \alpha \\ -\sin \alpha & \cos \alpha \end{array} \right]$$

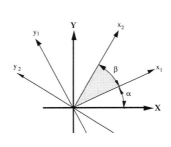

Next, we define another new frame, which is rotated through an angle β with respect to the x_1, y_1 frame; the axes of this new frame are labelled x_2 and y_2, respectively. We then can write

$$\mathbf{v}_2 = R_\beta \mathbf{v}_1$$

where the matrix rotation operator is

$$R_\beta = \left[\begin{array}{cc} \cos \beta & \sin \beta \\ -\sin \beta & \cos \beta \end{array} \right]$$

Then, using rules of matrix algebra, we may write

$$\begin{aligned} \mathbf{v}_2 &= R_\beta \mathbf{v}_1 \\ &= R_\beta (R_\alpha \mathbf{v}) \\ &= (R_\beta R_\alpha) \mathbf{v} \end{aligned}$$

This equation shows that the rotation operator for the sequence of rotations is exactly the product of the two individual rotation operators R_β and R_α; that is, we may obtain the vector $\mathbf{v}_2$ directly from the vector $\mathbf{v}$ by multiplying $\mathbf{v}$ by the product $R_\beta R_\alpha$. Using the rules for calculating this product, as well as two familiar trigonometric identities, we may write

$$R_\beta R_\alpha = \begin{bmatrix} \cos\beta & \sin\beta \\ -\sin\beta & \cos\beta \end{bmatrix} \begin{bmatrix} \cos\alpha & \sin\alpha \\ -\sin\alpha & \cos\alpha \end{bmatrix}$$

$$= \begin{bmatrix} a_{11} & a_{12} \\ a_{21} & a_{22} \end{bmatrix}$$

where
$$\begin{aligned}
a_{11} &= \cos\beta\cos\alpha - \sin\beta\sin\alpha = \cos(\alpha+\beta) \\
a_{12} &= \cos\beta\sin\alpha + \sin\beta\cos\alpha = \sin(\alpha+\beta) \\
a_{21} &= -\sin\beta\cos\alpha - \cos\beta\sin\alpha = -\sin(\alpha+\beta) \\
a_{22} &= -\sin\beta\sin\alpha + \cos\beta\cos\alpha = \cos(\alpha+\beta)
\end{aligned}$$

Thus we have

$$R_\beta R_\alpha = \begin{bmatrix} \cos(\alpha+\beta) & \sin(\alpha+\beta) \\ -\sin(\alpha+\beta) & \cos(\alpha+\beta) \end{bmatrix}$$

We notice immediately that this final matrix is just the rotation operator representing a rotation of the coordinate frame through an angle $\alpha + \beta$, while the points (or vectors) remain fixed. Thus we have shown, algebraically, that a rotation of the frame through an angle α, followed by another rotation of the frame through an angle β is equivalent to a single rotation through an angle $\alpha + \beta$, as we asserted earlier.

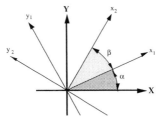

For rotations in the plane, the *product* of two such rotations,

$$R(\alpha)R(\beta) = R(\alpha+\beta)$$

is another rotation which represents the algebraic sum of the angular rotations.

3.3 Coordinates in R^3

Three-dimensional space, designated R^3, may be coordinatized in a way which is entirely analogous to the way in which we introduced coordinates in R^2. In R^3 an arbitrary but fixed point is specified which we call the *origin*. Three mutually perpendicular lines passing through this origin are specified, called the *X-axis*, the *Y-axis*, and the *Z-axis*, respectively. Each of these axes is again a real number line, with the zero-point at the origin. These axes are oriented so as to form a positive or right-handed coordinate frame. By a *right-handed* coordinate frame we mean: with the fingers of the right hand pointing positively along the x-axis, then as the fingers wrap toward the direction of the positive y-axis the upright thumb points positively in the direction of the z-axis. The three mutually perpendicular axes, however, may be pointed or oriented in any convenient manner with respect to the viewer, consistent with established conventions.

Given such a three-dimensional coordinate system, points in R^3 are now represented by triplets (x,y,z) of real numbers. The *origin*, in particular, has coordinates (0,0,0). In R^3 we may represent any given point, say $P = (x, y, z)$ as a vector $\mathbf{v}$ from the origin O to the point P, as in Figure 3.2.

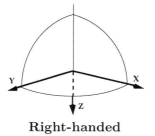

**Right-handed
Coordinate Frame**

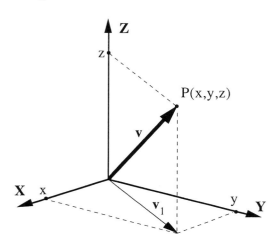

Figure 3.2: Vector in R^3

We begin, as we did in R^2, with a rotation of the $\mathbf{XYZ}$ coordinate frame while the point (or vector) remains fixed. Our problem is to determine the coordinates of the point relative to the rotated frame. More than that, we wish to determine a rotation matrix A, necessarily in this case a 3×3 matrix, such that if the coordinates relative to the rotated frame are given by $\mathbf{v}_2 = (x_2, y_2, z_2)$ we have, just as in R^2

$$\mathbf{v}_2 = A\mathbf{v}_1$$

Although in R^2 we simply rotated the coordinate frame about the origin through some angle θ, in R^3 simple rotation about the origin is not well defined. We need also to specify an *axis* about which the rotation is to occur. For example, in R^3 we may have a rotation of the coordinate frame through an angle $\pi/2$ about the $\mathbf{Z}$-axis, where the x and y coordinates change while the z coordinate does not. Such a rotation clearly is quite different from a rotation of the coordinate frame through an angle $\pi/2$ about, say, the $\mathbf{X}$-axis where y and z coordinates change but the x coordinate does not.

More specifically, suppose that a point P (or vector $\mathbf{v}$) has co-ordinates (x_1, y_1, z_1) relative to the $(\mathbf{XY\ Z})$ coordinate frame. We rotate the frame *about the* $\mathbf{Z}$*-axis* through an angle ψ. Let the co-ordinates of P relative to the rotated frame be (x_2, y_2, z_2). It seems clear that rotation about the $\mathbf{Z}$-axis will not change the z-coordinate of the point, and so we must have

$$z_2 = z_1$$

In order to determine x_2 and y_2, consider the vector $\mathbf{v}_1$ pictured in the margin. The vector $\mathbf{v}_1$ is called the *projection* of the vector $\mathbf{v}$ onto the $\mathbf{XY}$ plane. In R^2, $\mathbf{v}_1$ has coordinates (x_1, y_1). Further, the rotation through the angle ψ about the $\mathbf{Z}$-axis is clearly just a rotation in R^2 of the $\mathbf{XY}$ coordinate frame through an angle ψ, a rotation which is described by Equations 2.6 and 2.7. Hence we must have

$$
\begin{aligned}
x_2 &= x_1 \cos\psi + y_1 \sin\psi \\
y_2 &= y_1 \cos\psi - x_1 \sin\psi
\end{aligned}
$$

If we combine these results into a single set of equations we get

$$
\begin{aligned}
x_2 &= x_1 \cos\psi + y_1 \sin\psi + 0 \cdot z_1 \\
y_2 &= -x_1 \sin\psi + y_1 \cos\psi + 0 \cdot z_1 \\
z_2 &= 0 \cdot x_1 + 0 \cdot y_1 + 1 \cdot z_1
\end{aligned}
$$

This set of equations may be written in matrix form as

$$
\begin{bmatrix} x_2 \\ y_2 \\ z_2 \end{bmatrix} = \begin{bmatrix} \cos\psi & \sin\psi & 0 \\ -\sin\psi & \cos\psi & 0 \\ 0 & 0 & 1 \end{bmatrix} \begin{bmatrix} x_1 \\ y_1 \\ z_1 \end{bmatrix}
$$

We now recognize immediately that the 3×3 rotation matrix associated with the rotation in R^3 about the $\mathbf{Z}$-axis through an angle ψ is the matrix R_ψ given by

$$
R_\psi = \begin{bmatrix} \cos\psi & \sin\psi & 0 \\ -\sin\psi & \cos\psi & 0 \\ 0 & 0 & 1 \end{bmatrix} \tag{3.1}
$$

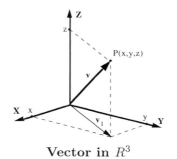

Vector in R^3

An example will be helpful at this point. Consider the vector $\mathbf{v} = (1, 1, 1)$ in an $\mathbf{XYZ}$ coordinate frame. We rotate the frame about the $\mathbf{Z}$-axis through an angle $\psi = \pi/4$, to get a new $\mathbf{xyz}$ coordinate frame. According to the above results, the coordinates of the vector $\mathbf{v}$ relative to this new frame are

$$
\begin{bmatrix} x_2 \\ y_2 \\ z_2 \end{bmatrix} = \begin{bmatrix} \sqrt{2}/2 & \sqrt{2}/2 & 0 \\ -\sqrt{2}/2 & \sqrt{2}/2 & 0 \\ 0 & 0 & 1 \end{bmatrix} \begin{bmatrix} 1 \\ 1 \\ 1 \end{bmatrix} = \begin{bmatrix} \sqrt{2} \\ 0 \\ 1 \end{bmatrix}
$$

The reader should now consider the geometric nature of this rotation of the coordinate frame to see that we have in fact obtained the correct result.

3.3.1   Successive Same-axis Rotations

Before leaving rotations of the coordinate frame through certain angles about the **Z**-axis, we note that if we have a sequence of two such rotations about the same axis, the first through an angle α, followed by a second through an angle β, the composite rotation matrix is obtained by multiplying the individual rotation matrices (just as in the case of rotations of the frame in R^2); that is, we have

$$R_\alpha = \begin{bmatrix} \cos\alpha & \sin\alpha & 0 \\ -\sin\alpha & \cos\alpha & 0 \\ 0 & 0 & 1 \end{bmatrix}$$

$$R_\beta = \begin{bmatrix} \cos\beta & \sin\beta & 0 \\ -\sin\beta & \cos\beta & 0 \\ 0 & 0 & 1 \end{bmatrix}$$

and

$$R_{\alpha+\beta} = \begin{bmatrix} \cos(\alpha+\beta) & \sin(\alpha+\beta) & 0 \\ -\sin(\alpha+\beta) & \cos(\alpha+\beta) & 0 \\ 0 & 0 & 1 \end{bmatrix}$$

Note, that if the angles in the above example are equal to zero then these rotation matrices become identity matrices. In this context therefore, it makes sense that for a non-zero rotation about the z-axis that the $r(3,3)$ entry should be a "1" because the direction of the z-axis remains unchanged and the values of the z-components of any vectors, whatever their values, also remain unchanged.

From the foregoing it follows that if the rotation in R^3 were about, say, the **Y**-axis then the "1" should appear in the $r(2,2)$ position of the rotation matrix, because only the **Z**-axis and the **X**-axis components are affected by a rotation about the **Y**-axis. Hence, for a positive coordinate frame rotation through an angle θ about the **Y**-axis the rotation matrix is

$$R_\theta = \begin{bmatrix} \cos\theta & 0 & -\sin\theta \\ 0 & 1 & 0 \\ \sin\theta & 0 & \cos\theta \end{bmatrix} \tag{3.2}$$

Similarly, for a positive coordinate frame rotation through an angle,

ϕ, about the **X**-axis the rotation matrix is

$$R_\phi = \begin{bmatrix} 1 & 0 & 0 \\ 0 & \cos\phi & \sin\phi \\ 0 & -\sin\phi & \cos\phi \end{bmatrix} \tag{3.3}$$

3.3.2 Signs in Rotation Matrices

There is a useful, if not important, point here which the reader might not want to ignore, and that is the proper placement of the sine and -sine terms in the matrices of Equations 3.1, 3.2, and 3.3. This placement is dictated by our use of a right-handed coordinate frame. A convenient device for determining the proper placement of these terms is the sequence

X Y Z X Y

Then, reading this sequence from left to right, a positive rotation about the **X**-axis through an angle ϕ rotates the **Y**-axis into the **Z**-axis, and thus the $\sin\phi$ term is associated with the y-coordinate while the -$\sin\phi$ term is associated with the z-coordinate, as written in the matrix R_ϕ above.

Next, again reading the sequence from left to right, a rotation about the **Y**-axis (in a positive direction in the right-handed coordinate frame, using the right-hand rule) through an angle θ rotates the **Z**-axis into the **X**-axis. Hence the $\sin\theta$ term is associated with the z-coordinate while the -$\sin\theta$ term is associated with the x-coordinate, as in the matrix R_θ above.

Finally, a positive rotation about the **Z**-axis through an angle ψ rotates the **X**-axis into the **Y**-axis, so the term $\sin\psi$ is associated with the x-coordinate while the -$\sin\psi$ term is associated with the y-coordinate, as in the matrix R_ψ of Equation 3.1.

Sign of Sine Terms

It is helpful to devise and memorize a **simple rule** for placement of the minus sine term in the two equations, for a positive rotation; the **sign** on the two sine terms alternate if the rotation is negative.

It is important to understand how the rotation operator for a rotation about a coordinate frame axis is written because we next develop the general rotation operator in R^3 as a sequence of rotations about successive coordinate frame axes. The reader should verify, or at least note, that each of the rotation operators R_ψ, R_θ, and R_ϕ, which represent rotations of the coordinate frame about a coordinate axis, is orthogonal and has determinant +1. We make use of this fact in the next section.

So far in this section we have considered only rotations of the coordinate frame while the points (or vectors) remain fixed. We may also have rotations of the points (or vectors) through a certain angle about a coordinate axis, while the coordinate frame remains fixed. We do not discuss the details here, but the resulting rotation operators turn out to be the transpose of the operators we have just considered. This is, of course, entirely analogous to the case in R^2.

3.4   Rotation Sequences in R^3

In the preceding section we showed that any sequence of two successive rotations in R^3 about the same coordinate axis, whether the frame rotates or whether the points rotate, amounts to a single rotation about that axis through an angle which is the sum for the angles of the rotations in the sequence. We now wish to show that a sequence of two rotations, say,

$$A = R_2 R_1$$

in which the rotations are not about the same coordinate axes, is also a rotation through some angle about some axis, usually not a coordinate axis. In fact, we shall determine the rotation operator for such a sequence, that is, the 3×3 matrix which represents this rotation. We shall also determine this *fixed axis* of rotation, as well as the *angle* of rotation about that axis.

Rotation Matrix

We should note here that the theory of linear algebra tells us the matrix representation for a rotation in R^3 is unique. That is, for a given rotation in R^3 there is *one and only one* matrix which represents that rotation relative to the initial coordinate frame.

The Euler Angles are, however, not unique. This we will show in what follows.

3.4.1   Some Rotation Geometry

From geometric considerations it should be clear that for a rotation in R^3 about some axis through the origin the axis itself is fixed under the rotation, that is, it does not change. Further, a plane, passing through the origin, which is perpendicular to this axis rotates into itself, that is, a vector in that plane is rotated into some other vector in that plane. And, of course, the length of vectors, and for that matter, the angle between two vectors is not changed by a rotation, which means that the scalar product of two vectors remains unchanged by the rotation. We describe this fact by saying that the scalar product is *invariant* under the rotation. We wish first to argue that the characterization of rotation operators in R^2 also holds in R^3, that is

SO(3)

The group **SO**(3) is comprised of all those special orthogonal 3X3 matrices whose determinant is +1.

A 3×3 matrix is a rotation operator in R^3 if and only if it is an orthogonal matrix and has determinant +1.

We need first to show that a rotation matrix must be orthogonal and have determinant +1. Suppose that the rotation is through some angle θ about the vector $\mathbf{v}$ as the axis. As we have just mentioned, the scalar product of two vectors is invariant under the rotation, which means that for any two vectors $\mathbf{v}_1$ and $\mathbf{v}_2$

$$(A\mathbf{v}_1)^t(A\mathbf{v}_2) = \mathbf{v}_1^t\mathbf{v}_2$$

But then we have $\quad \mathbf{v}_1^t A^t A\mathbf{v}_2 = \mathbf{v}_1^t\mathbf{v}_2$

or $\quad \mathbf{v}_1^t A^t A\mathbf{v}_2 - \mathbf{v}_1^t\mathbf{v}_2 = 0$

that is $\quad \mathbf{v}_1^t(A^t A - I)\mathbf{v}_2 = 0$

Since this last equation holds for any pair of vectors $\mathbf{v}_1$ and $\mathbf{v}_2$ it must be the case that

$$A^t A - I = 0$$

that is $\quad A^t A = I$

This shows that the matrix A is orthogonal.

Next, it is rather easily shown that the determinant of an orthogonal matrix is either +1 or -1 . We know that $det(A^t) = det A$ and that the determinant of a product is the product of the determinants, so if A is orthogonal we have

$$
\begin{aligned}
(det(A))^2 &= det(A^t)det(A) \\
&= det(A^t A) \\
&= det(I) \\
&= 1
\end{aligned}
$$

Hence $det(A)$ must be either +1 or -1. Now, it should be clear that a rotation preserves the right-handedness of a coordinate frame; hence we cannot have $det(A) = -1$. For example, if the matrix is

$$
B = \begin{bmatrix} 1 & 0 & 0 \\ 0 & 1 & 0 \\ 0 & 0 & -1 \end{bmatrix}
$$

$\longleftarrow$

A very simple check will show that the matrix B is orthogonal and that

$$det(B) = -1$$

then the standard right-handed coordinate frame is mapped into a left-handed frame. In fact, the transformation represented by the matrix B is a *reflection* in the **XY** plane, which leaves the **X** and **Y** axes unchanged but reverses the **Z** axis. Thus the frame becomes a left-handed coordinate frame. From all this it follows that we must have $det(A) = +1$. So far we have shown that if a matrix A represents a rotation, then A is orthogonal with $det(A) = +1$.

3.4.2   Rotation Eigenvalues & Eigenvectors

Eigenvectors

The literal meaning of the term *eigen*-vector is something like *same*-vector or *self*-vector. Or, expressed mathematically

$$A\mathbf{v} = \lambda\mathbf{v}$$

which says that a real-valued, nonsingular $3x3$ matrix A takes some vector $\mathbf{v}$ into a scalar multiple of itself. The scalar, λ in this case, is called the **eigenvalue** and the vector $\mathbf{v}$ is called the **eigenvector**. The equation

$$[A - \lambda I]\mathbf{v} = \mathbf{0}$$

has, of course, an uninteresting trivial solution, namely, $\mathbf{v} = \mathbf{0}$. The non-trivial solutions must satisfy the matrix equation $[A - \lambda I] = 0$, which is to say

$$det|A - \lambda I| = 0$$

Moreover, if the matrix A is a rotation, this equation will always produce only one real eigenvalue, namely, $\lambda = +1$, along with a pair of complex conjugates, in general.

Now suppose that we have a 3×3 matrix A which is orthogonal and has determinant $+1$. We wish to show that it represents a rotation through some angle about some fixed axis. Since the axis of any rotation is fixed under rotation, our first step is to show that there must be a fixed vector, $\mathbf{v}_0$ such that

$$A\mathbf{v}_0 = \mathbf{v}_0$$

This is equivalent to saying that the matrix A must have a *characteristic value* or *eigenvalue* of $+1$. The eigenvalues of the matrix A are exactly the scalars λ which satisfy the characteristic equation

$$det(A - \lambda I) = 0$$

Thus there will be an eigenvalue of $+1$ if and only if the determinant of the matrix $(A - I)$ is equal to 0. The following computation shows that this is so.

$$
\begin{aligned}
det(A - I) &= det A^t det(A - I) \\
&= det(A^t A - A^t) \\
&= det(I - A^t) \\
&= det((I - A)^t) \\
&= det(I - A) \\
&= (-1)det(A - I)
\end{aligned}
$$

It follows that

$$det(A - I) = 0$$

$\longrightarrow$

Notice that in this computation we have used the condition that the matrix A is orthogonal, and that

$$det(A) = det(A^t) = 1$$

Thus our transformation does have a fixed vector, say $\mathbf{v}_0$, which is the axis of the rotation. We next show that the plane, which contains the origin (and is perpendicular to this fixed vector $\mathbf{v}_0$) is fixed in the sense that any vector in that plane is mapped by A into another vector in that plane.

We first remark that since $\mathbf{v}_0$ is fixed under A, it is also fixed under A^t. For if

$$
\begin{aligned}
A\mathbf{v}_0 &= \mathbf{v}_0 \\
\text{we may write}\quad A^t\mathbf{v}_0 &= A^t A\mathbf{v}_0 \\
&= I\mathbf{v}_0 \\
&= \mathbf{v}_0
\end{aligned}
$$

Now consider a vector $\mathbf{v}$ in the plane perpendicular to the fixed vector $\mathbf{v}_0$, which means that $\mathbf{v}^t \mathbf{v}_0 = 0$. The vector $A\mathbf{v}$ must also be perpendicular to $\mathbf{v}_0$, for we have

$$
\begin{aligned}
\mathbf{v}^t \mathbf{v}_0 \;=\; (A\mathbf{v})^t \mathbf{v}_0 \;&=\; (\mathbf{v}^t A^t)\mathbf{v}_0 \\
&=\; \mathbf{v}^t A^t A \mathbf{v}_0 \\
&=\; \mathbf{v}^t I \mathbf{v}_0 \\
&=\; \mathbf{v}^t \mathbf{v}_0 \;=\; 0
\end{aligned}
$$

as we wished to show.

Orthogonal

If A is orthogonal we have

$$
\begin{aligned}
(A\mathbf{v}_1)^t (A\mathbf{v}_2) \;&=\; \mathbf{v}_1^t A^t A \mathbf{v}_2 \\
&=\; \mathbf{v}_1^t I \mathbf{v}_2 \\
&=\; \mathbf{v}_1^t \mathbf{v}_2
\end{aligned}
$$

Finally, if the matrix A is orthogonal, then the scalar product is invariant under this transformation, and it follows that lengths of vectors and the angle between vectors are unchanged under A as we show in the margin. This confirms that the transformation represented by A must be a rotation.

To this point we have established that every rotation may be defined by a unique fixed axis and angle of rotation about this axis. We now solve for this fixed axis of rotation.

3.5 The Fixed Axis of Rotation

We use the standard "Eigenvalue-Eigenvector Method" to determine the rotation axis, $\mathbf{v}$, for a given rotation operator, A. However, since we have already established that $+1$ is the only real eigenvalue for any rotation operator, it remains that we must solve

$$
[A - I]\mathbf{v} \;=\; \mathbf{0}
$$

which upon expansion gives the homogeneous equations

$$
\begin{aligned}
(a_{11} - 1)v_1 + a_{12}v_2 + a_{13}v_3 \;&=\; 0 \\
a_{21}v_1 + (a_{22} - 1)v_2 + a_{23}v_3 \;&=\; 0 \\
a_{31}v_1 + a_{32}v_2 + (a_{33} - 1)v_3 \;&=\; 0
\end{aligned}
$$

The standard technique is to set, say, $v_3 = 1$ and then use any two of the equations to solve for v_1 and v_2. We use

$$
\begin{aligned}
a_{21}v_1 + (a_{22} - 1)v_2 \;&=\; -a_{23} \\
a_{31}v_1 + a_{32}v_2 \;&=\; -(a_{33} - 1)
\end{aligned}
$$

to get one possible expression for the composite rotation axis

$$\mathbf{v} = [\, v_1, \ v_2, \ v_3 \,]$$

where
$$v_1 = a_{12}a_{23} - (a_{22} - 1)a_{13}$$
$$v_2 = a_{21}a_{13} - (a_{11} - 1)a_{23}$$
$$v_3 = (a_{11} - 1)(a_{22} - 1) - a_{12}a_{21}$$

Now that we are able to determine the fixed axis of rotation for any given rotation matrix, we must determine the magnitude and direction of the rotation about this fixed axis.

3.5.1 Rotation Angle about the Fixed Axis

Trace Property

The *trace* of a product of two square matrices is invariant under commutation. That is,

$$Tr(AB) = Tr(BA)$$

Proof:

$$\text{Let} \quad C = AB = [c_{ij}]$$

$$c_{ij} = \sum_{k=1}^{n} a_{ik}b_{kj}$$

$$\text{then} \quad c_{ii} = \sum_{k=1}^{n} a_{ik}b_{ki}$$

$$\text{and} \quad Tr(C) = \sum_{i=1}^{n} c_{ii}$$

$$= tr(AB)$$

$$= \sum_{i=1}^{n}\sum_{k=1}^{n} a_{ik}b_{ki}$$

but scalar products commute, so

$$= \sum_{i=1}^{n}\sum_{k=1}^{n} b_{ki}a_{ik}$$

interchanging order of summation

$$= \sum_{k=1}^{n}\sum_{i=1}^{n} b_{ki}a_{ik}$$

$$= tr(BA) \quad \textbf{QED}$$

There remains the problem of determining the angle of rotation. To calculate the angle of rotation we need the concept of the *trace* of a matrix. The *trace* of a square matrix A, denoted $Tr(A)$, is simply the sum of the elements lying on the main diagonal of the matrix. Thus the trace of the matrix

$$R_\phi = \begin{bmatrix} 1 & 0 & 0 \\ 0 & \cos\phi & \sin\phi \\ 0 & -\sin\phi & \cos\phi \end{bmatrix}$$

is simply

$$Tr(R_\phi) = 1 + 2\cos\phi$$

We need also to know that for any square matrices A and B (of the same order) we always have

$$Tr(AB) = Tr(BA)$$

as may rather easily be shown when we recall the rule for multiplying matrices. See proof in margin.

Now consider the rotation through an angle ϕ about the axis $\mathbf{v}_0$, represented by the matrix A. We rewrite this rotation as a sequence of rotations about coordinate frame axes, as follows. First, we rotate the frame about the $\mathbf{Z}$-axis so that the $\mathbf{X}$-axis coincides with the projection of $\mathbf{v}_0$ onto the $\mathbf{XY}$ plane. We follow with a rotation about the new y-axis so that the new x-axis coincides with the vector $\mathbf{v}_0$. See Figures in the margin on the previous page. We may represent the product of these two rotations by the matrix Q, and we know that Q is orthogonal and has determinant $+1$. Then we rotate the resulting frame about the vector $\mathbf{v}_0$, that is, about the new x-axis, through the angle ϕ. Call this rotation R_ϕ. From our earlier work we know that

$$R_\phi \;=\; \begin{bmatrix} 1 & 0 & 0 \\ 0 & \cos\phi & \sin\phi \\ 0 & -\sin\phi & \cos\phi \end{bmatrix}$$

We note, incidentally, that

$$Tr(R_\phi) = 1 + 2\cos\phi$$

Finally, we follow that rotation with the inverse of the rotation Q, which, since Q is orthogonal, is just Q^t. Although it is not too easy to see geometrically, the sequence of these rotations is exactly equivalent to our original rotation represented by the matrix A, so we must have

$$A \;=\; Q^t R_\phi Q$$

where ϕ is the angle of rotation. But then we have

$$
\begin{aligned}
Tr(A) &= Tr(Q^t R_\phi Q) \\
&= Tr(Q(Q^t R_\phi)) \\
&= Tr((QQ^t)R_\phi) \\
&= Tr(I R_\phi) \\
&= Tr(R_\phi)
\end{aligned}
$$

Hence we obtain the equation

$$Tr(A) = 1 + 2\cos\phi$$

which, if we solve for ϕ, gives the following formula for the angle of the rotation

$$\phi = \arccos \frac{Tr(A) - 1}{2} \tag{3.4}$$

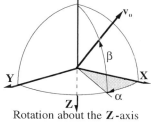

Rotation about the **Z**-axis

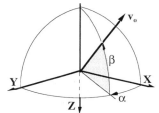

Rotation about the new y-axis

The Transformation Q which locates the fixed axis of rotation.

3.6   A Numerical Example

No doubt a simple numerical example will help the reader to understand the material in the preceding section. Consider an **XYZ** coordinate frame, with the vector $\mathbf{v}_0 = (1, 1, 1)$ in that frame. We consider a rotation of the frame about the vector $\mathbf{v}_0$ through an angle $\phi = 2\pi/3$. It is geometrically clear that such a rotation results in a new frame in which the new x-axis now coincides with the former **Y**-axis, the new y-axis coincides with the former **Z**-axis, and the new z-axis coincides with the initial **X**-axis. It is interesting to note that a sequence of three such rotations simply carries the intial coordinate frame into itself.

We now wish to find the matrix A which represents this rotation. Consider the initial **X**-axis, and in particular the point (1,0,0) on that axis. Since the points (or vectors) are fixed while the frame rotates, after the rotation this point will lie on the new z-axis, and hence the new coordinates of the point must be (0,0,1). In the same way we determine that the rotation we are considering must change coordinates of points on the coordinate axes of the initial coordinate frame in this way

$$
\begin{aligned}
(1,0,0) &\rightarrow (0,0,1) \\
(0,1,0) &\rightarrow (1,0,0) \\
(0,0,1) &\rightarrow (0,1,0)
\end{aligned}
$$

This tells us that we are looking for a matrix

$$
A \;=\; \begin{bmatrix} a_{11} & a_{12} & a_{13} \\ a_{21} & a_{22} & a_{23} \\ a_{31} & a_{32} & a_{33} \end{bmatrix}
$$

such that

$$
A \begin{bmatrix} 1 \\ 0 \\ 0 \end{bmatrix} = \begin{bmatrix} 0 \\ 0 \\ 1 \end{bmatrix}
$$

It is not difficult to see that this equation implies

$$
a_{11} = 0, \quad a_{21} = 0, \quad a_{31} = 1
$$

Proceeding in exactly the same way with the remaining two vectors above, we obtain the rotation matrix A as

$$
A \;=\; \begin{bmatrix} 0 & 1 & 0 \\ 0 & 0 & 1 \\ 1 & 0 & 0 \end{bmatrix}
$$

Given the matrix A, we can now find the coordinates of any point $\mathbf{v} = (x, y, z)$ relative to the rotated frame simply by calculating the product $A\mathbf{v}$.

If we think of the three points (1,0,0), (0,1,0), and (0,0,1) on the axes of the intial coordinate frame as forming the columns in an identity matrix, then the matrix A is simply this identity matrix with its column shifted one column to the right. In general, the matrix which represents a rotation has columns which give the new coordinates of these points in the rotated frame.

Now suppose we are given the matrix A above, and are asked what sort of transformation of R^3 it represents. We check that A is orthogonal and has determinant $+1$, which tells us that A does represent a rotation. What is the axis of the rotation? To find the axis we simply look for the fixed vector, that is a vector $\mathbf{v}_0 = (x_0, y_0, z_0)$ such that $\mathbf{v}_0$ does not change under the rotation. We must have

$$\begin{bmatrix} 0 & 1 & 0 \\ 0 & 0 & 1 \\ 1 & 0 & 0 \end{bmatrix} \begin{bmatrix} x_0 \\ y_0 \\ z_0 \end{bmatrix} = \begin{bmatrix} x_0 \\ y_0 \\ z_0 \end{bmatrix}$$

Here, it is very easy to check that the matrix A indeed is orthogonal and has determinant $+1$.

If we solve this system, we find that

$$x_0 = y_0 = z_0 = k$$

for any real number k. This means the vector (k, k, k), for any non-zero real number k, is fixed. In particular we might designate $(1, 1, 1)$ as the axis of the rotation.

Finally, the trace of the matrix A is

$$Tr(A) = 0 + 0 + 0 = 0$$

No surprise here!

so that the angle of the rotation is given by

$$\begin{aligned} \phi &= \arccos \frac{0-1}{2} \\ &= \arccos(-1/2) \\ &= 2\pi/3 \end{aligned}$$

which of course is just what we expected.

3.7 An Application - Tracking

In this section we discuss an application of rotations in R^3, a familiar rotation sequence called the *Tracking Transformation*. Consider a remote object, such as an aircraft, which is to be tracked from some point on the surface of the earth. The *Local Tangent Plane* is simply a plane tangent to the surface of the earth at this point. We define an initial coordinate frame with the **X** and **Y** axes lying in this tangent plane, pointing in directions *North* and *East* respectively. The **Z**-axis is *geocentric*, that is, it points toward the center of the earth. We then have a right-handed coordinate frame.

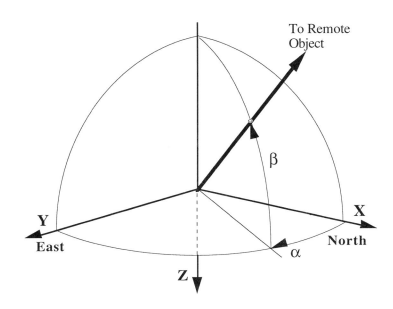

Figure 3.3: Tracking Transformation

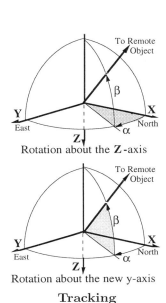

**Tracking
Rotation Sequence**

We define an angle α, called *Heading*, which is the angle in the tangent plane between North and the projected direction to the remote object. We also define an angle β, called *Elevation*, which is the angle between the tangent plane and the direction to the remote object being tracked, as in Figure 3.3. The *Tracking Transformation* is first a rotation about the **Z**-axis through the angle α, followed by a rotation about the new y-axis through the angle β. Notice that in the resulting coordinate frame the new x-axis is pointing directly toward the object being tracked.

Rotation Notation

Here we introduce a new notation

$$R^z_{-\theta}$$

By this symbol, in this case, we mean a rotation through an angle equal to minus theta, taken about the z-axis of the *current* frame.

If R is the 3×3 matrix representing this rotation, our preceding results tell us that

$$R = R^y_\beta R^z_\alpha \qquad \text{(See note in margin)}$$

$$= \begin{bmatrix} \cos\beta & 0 & -\sin\beta \\ 0 & 1 & 0 \\ \sin\beta & 0 & \cos\beta \end{bmatrix} \begin{bmatrix} \cos\alpha & \sin\alpha & 0 \\ -\sin\alpha & \cos\alpha & 0 \\ 0 & 0 & 1 \end{bmatrix}$$

$$= \begin{bmatrix} \cos\alpha\cos\beta & \sin\alpha\cos\beta & -\sin\beta \\ -\sin\alpha & \cos\alpha & 0 \\ \cos\alpha\sin\beta & \sin\alpha\sin\beta & \cos\beta \end{bmatrix} \qquad (3.5)$$

This composite tracking transformation may be represented as an equivalent transformation which consists of single rotation through some angle about some axis. The axis for this single rotation is found by finding the fixed vector for the rotation operator, that is, a vector $\mathbf{v} = (x_1, y_1, z_1)$, say, such that

$$R\mathbf{v} = \mathbf{v}$$

Thus we have to solve the equation

$$\begin{bmatrix} \cos\alpha\cos\beta & \sin\alpha\cos\beta & -\sin\beta \\ -\sin\alpha & \cos\alpha & 0 \\ \cos\alpha\sin\beta & \sin\alpha\sin\beta & \cos\beta \end{bmatrix} \begin{bmatrix} x_1 \\ y_1 \\ z_1 \end{bmatrix} = \begin{bmatrix} x_1 \\ y_1 \\ z_1 \end{bmatrix}$$

Our rules for matrix multiplication tell us that we must then have

$$\begin{aligned} x_1\cos\alpha\cos\beta + y_1\sin\alpha\cos\beta - z_1\sin\beta &= x_1 \\ -x_1\sin\alpha + y_1\cos\alpha + z_1\cdot 0 &= y_1 \\ x_1\cos\alpha\sin\beta + y_1\sin\alpha\sin\beta + z_1\cos\beta &= z_1 \end{aligned}$$

These equations are easily rewritten in the form

$$\begin{aligned} x_1(\cos\alpha\cos\beta - 1) + y_1\sin\alpha\cos\beta - z_1\sin\beta &= 0 \\ -x_1\sin\alpha + y_1(\cos\alpha - 1) + z_1\cdot 0 &= 0 \\ x_1\cos\alpha\sin\beta + y_1\sin\alpha\sin\beta + z_1(\cos\beta - 1) &= 0 \end{aligned}$$

Now this is a system of *homgeneous* equations, which always has the trivial solution $\mathbf{v} = (0,0,0)$. In order to get a non-trivial solution we may, if we like, set

$$x_1 = k$$

for any non-zero real number k. Then from the second equation above we get

$$y_1 = \frac{k\sin\alpha}{\cos\alpha - 1}$$

Finally from the third equation we calculate

$$\begin{aligned} z_1 &= \frac{-k\cos\alpha\sin\beta - y_1\sin\alpha\sin\beta}{\cos\beta - 1} \\ &= \frac{-k\cos^2\alpha\sin\beta + k\cos\alpha\sin\beta - k\sin^2\alpha\sin\beta}{(\cos\beta - 1)(\cos\alpha - 1)} \end{aligned}$$

$$= \frac{-k\sin\beta(\cos^2\alpha + \sin^2\alpha) + k\cos\alpha\sin\beta}{(\cos\beta - 1)(\cos\alpha - 1)}$$

$$= \frac{k\sin\beta(\cos\alpha - 1)}{(\cos\beta - 1)(\cos\alpha - 1)}$$

$$= \frac{k\sin\beta}{\cos\beta - 1}$$

Thus our tracking transformation has axis of rotation given by

$$\mathbf{v} = \left(k, \; \frac{k\sin\alpha}{\cos\alpha - 1}, \; \frac{k\sin\beta}{\cos\beta - 1} \right)$$

Notice that in this computation we determine only the *direction* of the axis of rotation. Should we wish to obtain a specific vector as the axis of rotation we may, for instance, choose k = -1, to obtain

$$\mathbf{v} = \left(-1, \; \frac{\sin\alpha}{1 - \cos\alpha}, \; \frac{\sin\beta}{1 - \cos\beta} \right) \qquad (3.6)$$

Singularities

Equation 3.6 for the vector **v** is clearly not valid for $\alpha = 0$, in which case Equation 3.7 is appropriate.

We note that by using the trigonometric identity

$$1 - \cos\alpha = 2\sin^2\frac{\alpha}{2}$$

we may write the following expression for the axis of the rotation

$$\mathbf{v} = \left(-\sin\frac{\alpha}{2}\sin\frac{\beta}{2}, \; \cos\frac{\alpha}{2}\sin\frac{\beta}{2}, \; \sin\frac{\alpha}{2}\cos\frac{\beta}{2} \right) \qquad (3.7)$$

We shall look at this relationship later.

In order to find the angle of rotation we compute the trace of the rotation operator R as

Trace

Remember that the trace is merely the sum of the elements on the main diagonal of a square matrix.

$$tr\,R = \cos\alpha\cos\beta + \cos\alpha + \cos\beta$$

We then obtain the angle of rotation, say ϕ, from the equation

$$tr\,R = \cos\alpha\cos\beta + \cos\alpha + \cos\beta = 1 + 2\cos\phi$$

Solving this equation for ϕ gives

$$\phi = \arccos\left(\frac{\cos\alpha\cos\beta + \cos\alpha + \cos\beta - 1}{2} \right) \qquad (3.8)$$

We may make the example even more specific if we consider a numerical version of it. Suppose a remote object has heading

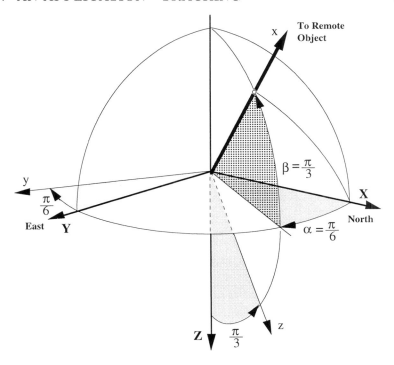

Figure 3.4: Rotations to Tracking Frame

$\alpha = \pi/6$ and elevation $\beta = \pi/3$, as in Figure 3.4. Then we have

$$\sin \alpha = \cos \beta = 1/2 \qquad \text{and} \qquad \sin \beta = \cos \alpha = \sqrt{3}/2$$

Thus, according to Equation 3.6, the axis of the rotation is

$$
\begin{aligned}
\mathbf{v} &= \left(-1, \ \frac{1/2}{1 - \sqrt{3}/2}, \ \frac{\sqrt{3}/2}{1 - 1/2} \right) \\
&= \left(-1, \ \frac{1}{2 - \sqrt{3}}, \ \sqrt{3} \right) \\
&= (-1.000, \ 3.7321, \ 1.732)
\end{aligned}
$$

given by

$$
\begin{aligned}
\phi &= \arccos \left(\frac{\cos \alpha \cos \beta + \cos \alpha + \cos \beta - 1}{2} \right) \\
&= \arccos \left(\frac{\frac{\sqrt{3}}{2}\frac{1}{2} + \frac{\sqrt{3}}{2} + \frac{1}{2} - 1}{2} \right) \\
&= \arccos \left(\frac{3\sqrt{3} - 2}{8} \right)
\end{aligned}
$$

$$= \arccos{(.3995)}$$
$$= 1.159 \ radians$$
$$= 66.41 \ degrees$$

The two coordinate frames, namely, the reference frame **XYZ** and the rotated frame xyz are represented in Figure 3.5, along with the rotation axis **V**. A rotation about this axis (through an angle 66.41 degrees, in our example) takes the reference frame into the rotated frame.

A Single Rotation

We emphasize: The vector **V** is the direction of the axis about which a *single* rotation takes the **XYZ** reference frame into the xyz rotated tracking frame.

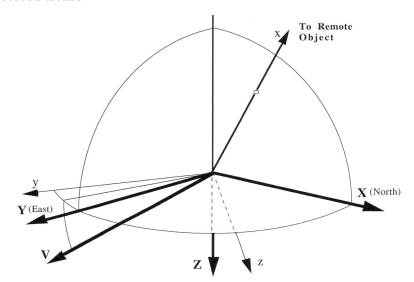

Figure 3.5: Frame Rotation Example

There is one more numerical result which we may verify in the preceding example. With α and β chosen as above it is easy to verify that the object being tracked lies in the direction of the vector

$$(\sqrt{3}/2, \ 1/2, \ -\sqrt{3})$$

This vector has length 2, again easily verified. Under the tracking transformation, in the final coordinate frame this vector lies along the final x-axis, so in this last coordinate frame it must be the vector

$$(2, \ 0, \ 0)$$

With the given values for α and β, the rotation matrix (see Equation 3.5) is

$$A = \begin{bmatrix} \sqrt{3}/4 & 1/4 & -\sqrt{3}/2 \\ -1/2 & \sqrt{3}/2 & 0 \\ 3/4 & \sqrt{3}/4 & 1/2 \end{bmatrix}$$

The reader may now verify that

$$
\begin{bmatrix} \sqrt{3}/4 & 1/4 & -\sqrt{3}/2 \\ -1/2 & \sqrt{3}/2 & 0 \\ 3/4 & \sqrt{3}/4 & 1/2 \end{bmatrix} \begin{bmatrix} \sqrt{3}/2 \\ 1/2 \\ -\sqrt{3} \end{bmatrix} = \begin{bmatrix} 2 \\ 0 \\ 0 \end{bmatrix}
$$

and so the expected result has been obtained.

3.7.1 A Simpler Rotation-Axis Algorithm

As suggested in the foregoing, the axis of rotation, $\mathbf{v}$, is the eigenvector corresponding to the eigenvalue $+1$. That is, the equation

$$[A - \lambda I]\mathbf{v} = \mathbf{0}$$

has a non-trivial solution *iff*

$$det|A - \lambda I| = 0$$

This eigenvalue equation, which is a cubic in λ for any rotation matrix A, always has one real solution, namely $\lambda = +1$, and two other solutions which are complex conjugates. This means the eigenvector we seek is the solution of the matrix equation

$$[A - I]\mathbf{v} = \mathbf{0}$$

After a bit of tedious algebra a solution vector may be found. The components of the resulting fixed vector are, of course, functions of the elements of the given rotation matrix, A.

An alternative, considerably less tedious, method for determining the fixed axis of rotation, $\mathbf{v}$, is based upon the fact that any real-valued matrix, and therefore in particular a rotation matrix A, may be written as

$$
\begin{aligned}
A &= \frac{1}{2}[S + Q] \\
\text{where} \quad S &= [A + A^t] \quad &\text{symmetric part of } 2A \\
\text{and} \quad Q &= [A - A^t] \quad &\text{skew-symmetric part of } 2A
\end{aligned}
$$

Since, from our earlier results in rotations we have

$$\mathbf{v} = A\mathbf{v} = A^t\mathbf{v} \quad \text{therefore, we may write}$$

$$
\left[A - A^t\right]\mathbf{v} = Q\mathbf{v} = \begin{bmatrix} 0 & q_3 & -q_2 \\ -q_3 & 0 & q_1 \\ q_2 & -q_1 & 0 \end{bmatrix} \begin{bmatrix} v_1 \\ v_2 \\ v_3 \end{bmatrix} = \begin{bmatrix} 0 \\ 0 \\ 0 \end{bmatrix}
$$

a homogeneous equation where $\mathbf{v}$ is more easily obtained.

Expanding this matrix equation (along with a convenient adjustment of signs) we write

$$0 \cdot v_1 + q_3 v_2 - q_2 v_3 \;=\; 0 \tag{3.9}$$
$$q_3 v_1 + 0 \cdot v_2 - q_1 v_3 \;=\; 0 \tag{3.10}$$
$$q_2 v_1 - q_1 v_2 + 0 \cdot v_3 \;=\; 0 \tag{3.11}$$
$$\text{where} \qquad q_1 \;=\; a_{23} - a_{32}$$
$$q_2 \;=\; a_{31} - a_{13}$$
$$q_3 \;=\; a_{12} - a_{21}$$

Since these last three equations, 3.9, 3.10, and 3.11, are not independent, we choose, say, the first two. With this choice we let, $v_3 = 1$. Then we may write

$$q_3 v_2 - q_2 \;=\; 0$$
$$q_3 v_1 - q_1 \;=\; 0$$

from which we solve for the composite axis of rotation

$$\mathbf{v} \;=\; [\frac{q_1}{q_3},\; \frac{q_2}{q_3},\; 1] \;\equiv\; [q_1,\; q_2,\; q_3]$$

We summarize this result more concisely:

If a rotation is defined by the matrix

$$A \;=\; \begin{bmatrix} a_{11} & a_{12} & a_{13} \\ a_{21} & a_{22} & a_{23} \\ a_{31} & a_{32} & a_{33} \end{bmatrix}$$

then its fixed axis of rotation is given by

$$\begin{aligned} \mathbf{v} \;=\; & \mathbf{i}(a_{23} - a_{32}) \\ & + \mathbf{j}(a_{31} - a_{13}) \\ & + \mathbf{k}(a_{12} - a_{21}) \end{aligned} \tag{3.12}$$

This formulation for the fixed-rotation-axis for any arbitrary rotation matrix, $A[a_{ij}]$, is easy to derive and apply.

Many readers may find it difficult to visualize the results obtained in the last several sections. Because it is quite useful to be able to do so, in the next section we analyze the Tracking Transformation example again, this time from a geometric point of view.

3.8 A Geometric Analysis

In an effort to gain further insight into how a tracking rotation sequence is related to its equivalent composite rotation, consider the

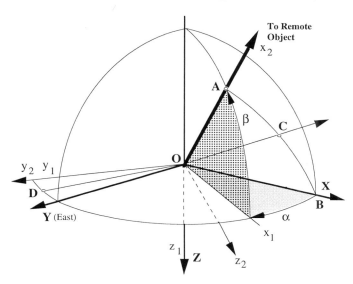

Figure 3.6: Euler Rotations R_α^z and R_β^y

intrinsic geometry of the tracking rotations shown in Figure 3.6. In this tracking application the two rotations, R_α^z and R_β^y, are taken about mutually orthogonal axes: first, about the **Z**-axis, and then about the *new* y-axis, y_1.

With respect to the *fixed* reference frame, **XYZ**, the indicated x_2-axis is directed towards, that is, *tracks* the remote object. The $x_2y_2z_2$ frame is related to the **XYZ** reference frame by the indicated rotations: a rotation through the *heading* angle α about the reference **Z**-axis, followed by a rotation through the *elevation* angle β about the new y-axis y_1, as shown in Figure 3.6.

As we already know, a single rotation about the axis of the composite rotation takes all three of the coordinate axes of the two frames: **X** into x_2, **Y** into y_2, and **Z** into z_2. We will now find the direction of this single rotation-axis geometrically from the angles and the geometry which relate the two frames, that is, the Tracking frame and the Reference frame.

We begin by considering rotations which will take **X** into x_2; two such rotations are shown in Figure 3.7 and in Figure 3.8. The first rotation, through $\angle BOA$ (see Figure 3.7), is about the axis OE.

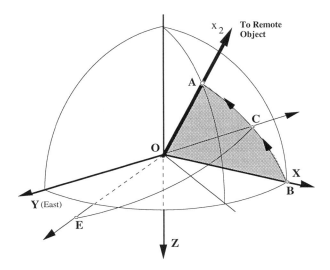

Figure 3.7: Rotation Axis **OE** takes **X** into x_2

Axis OE

Note that the direction of the axis **OE** is found by calculating the *cross-product* $\mathbf{X} \times x_2$.

This axis is normal to the plane containing **X** and x_2. Note, this rotation takes point **B** into point **A** along the great circle arc BCA. This axis OE, however, is not the only axis of rotation which will take **X** into x_2. The second rotation which takes **X** into x_2 is about the axis OC through an angle π as shown in Figure 3.8. The axis OC lies in the plane containing **X** and x_2, and it bisects $\angle AOB$. In this case, the rotation takes point **B** into point **A** along the circular path BLA.

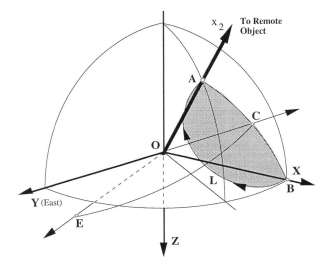

Figure 3.8: Rotation Axis **OC** also takes **X** into x_2

In summary, we have found two rotations, with axes OE and OC, respectively, both of which will rotate $\mathbf{X}$ into x_2. However, neither of these rotations, in general, will also take the axes $\mathbf{Y}$ into y_2 and $\mathbf{Z}$ into z_2 (see Figure 3.6). The two distinct rotation axes, OC and OE, each take $\mathbf{X}$ into x_2. These two axes (vectors) define the plane shown in Figure 3.9. This plane through the origin represents the locus of all possible directions for axes about which rotations which will take $\mathbf{X}$ into x_2. One of these directions must also take $\mathbf{Y}$ into y_2 and $\mathbf{Z}$ into z_2, as we will show.

The reader should convince herself that for every axis through the origin and lying in this plane, there is some rotation which will take $\mathbf{X}$ into x_2. For instance, let F be any point distinct from the origin in the plane OCE. It should be clear that the point F is equidistant from the points A and B. Therefore, OA (that is, x_2) and OB (that is, $\mathbf{X}$) are generators of a circular cone whose axis is OF; hence, some rotation with axis OF takes $\mathbf{X}$ into x_2.

What we must do, therefore, is to find that particular axis in the plane OCE which takes not only $\mathbf{X}$ into x_2 but also $\mathbf{Y}$ into y_2 and, consequently, $\mathbf{Z}$ into z_2. To do this we must find the intersection of two planes. The first plane, OCE, which we have already described above, contains all axes which rotate $\mathbf{X}$ into x_2. A second plane, which we will find in a similar fashion, contains all possible axes which take $\mathbf{Y}$ into y_2. The intersection of these two planes defines the axis about which a single rotation will take the reference coordinate frame $\mathbf{XYZ}$ into the tracking coordinate frame $x_2 y_2 z_2$.

We now find this desired axis of rotation by first finding the equations of these two planes. The direction of the line of intersection is the direction of the rotation axis for the composite rotation, that is, for the tracking transformation.

Important Geometry

The planes EOC and AOB are perpendicular to each other and the segment OC bisects angle AOB. Moreover, since points A and B lie on the sphere, OA is equal to OB.

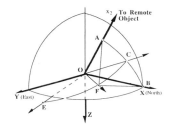

3.8.1 All Axes which take X into x₂

We first find the equation of the plane containing all possible directions for rotation axes which take $\mathbf{X}$ into x_2. This is the plane OCE shown in Figure 3.9. As usual, we find this plane by finding its normal vector, which we denote $\mathbf{n}_x$.

We find $\mathbf{n}_x$ by taking the cross-product of the vector $\vec{OE}$ with the vector $\vec{OC}$. Since the vector $\vec{OE}$ is the normal vector which

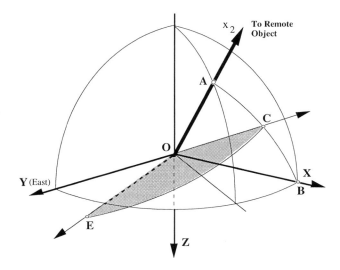

Figure 3.9: Locus of Axes which take **X** into x

defines the plane OAB, we have

$$\vec{OE} \;=\; \vec{OB} \times \vec{OA}$$

Do it!

We leave it to the reader to verify that the entries in the 3rd row of this determinant are the components of the vector $\vec{OA}$ shown in Figure 3.9

$$= \; \begin{vmatrix} \mathbf{i} & \mathbf{j} & \mathbf{k} \\ 1 & 0 & 0 \\ \cos\alpha\cos\beta & \sin\alpha\cos\beta & -\sin\beta \end{vmatrix}$$

$$= \; \mathbf{j}\sin\beta + \mathbf{k}\sin\alpha\cos\beta$$

Note

Here we use "$\cong$" to indicate that the vectors have the same direction — not necessarily the same length.

Since $\vec{OC}$ bisects $\angle AOB$ the *direction* of the vector $\vec{OC}$ is the *direction* of the vector sum $\vec{OB} + \vec{OA}$. This we may write as

$$\vec{OC} \;\cong\; \vec{OB} + \vec{OA}$$
$$\cong\; \mathbf{i}(1 + \cos\alpha\cos\beta) + \mathbf{j}\sin\alpha\cos\beta - \mathbf{k}\sin\beta$$

We now compute the normal to the plane OCE as

$$\mathbf{n}_x \;=\; \vec{OC} \times \vec{OE}$$

$$= \; \begin{vmatrix} \mathbf{i} & \mathbf{j} & \mathbf{k} \\ (1 + \cos\alpha\cos\beta) & \sin\alpha\cos\beta & -\sin\beta \\ 0 & \sin\beta & \sin\alpha\cos\beta \end{vmatrix}$$

$$= \; \mathbf{i}(1 - \cos\alpha\cos\beta) - \mathbf{j}\sin\alpha\cos\beta + \mathbf{k}\sin\beta$$

In summary, this vector, $\mathbf{n}_x$, is the normal vector to a plane which contains the origin. It is also this plane which contains all possible rotation axes for taking $\mathbf{X}$ into x_2.

3.8.2   All Axes which take Y into y_2

We now find the equation for the plane DOZ which is the locus of all possible directions for rotation axes which take $\mathbf{Y}$ into y_2 (see Figure 3.10). Notice that the $\mathbf{Z}$-axis is normal to the plane of $\mathbf{Y}$ and y_2, and that OD bisects the angle between $\mathbf{Y}$ and y_2. Clearly, a rotation about the $\mathbf{Z}$-axis takes $\mathbf{Y}$ into y_2. Moreover, a rotation about the axis OD through an angle π, also takes $\mathbf{Y}$ into y_2. Hence, just as before, the plane DOZ which contains both of these two axes must therefore contain *all* axes which take $\mathbf{Y}$ into y_2.

The normal to the plane DOZ, which we denote $\mathbf{n}_y$, is defined by the cross-product of vectors $\vec{OD}$ and $\mathbf{k}$ (along the $\mathbf{Z}$-axis). The reader may verify the direction of OD is

$$\vec{OD} \cong -\mathbf{i}\sin\alpha + \mathbf{j}(1+\cos\alpha)$$

Therefore
$$\mathbf{n}_y = \vec{OD} \times \mathbf{k}$$

$$= \begin{vmatrix} \mathbf{i} & \mathbf{j} & \mathbf{k} \\ -\sin\alpha & (1+\cos\alpha) & 0 \\ 0 & 0 & 1 \end{vmatrix}$$

$$= \mathbf{i}(1+\cos\alpha) + \mathbf{j}\sin\alpha$$

In summary, this vector, $\mathbf{n}_y$, is the normal vector to the plane which contains the origin and also all possible rotation axes for taking $\mathbf{Y}$ into y_2.

3.8.3   Rotation Axis for both X into x_2 and Y into y_2

The two normal vectors, $\mathbf{n}_x$ and $\mathbf{n}_y$ computed above, define two intersecting planes. These two planes, both of which contain the origin, represent the locus of all possible axes which take $\mathbf{X}$ into x_2 and $\mathbf{Y}$ into y_2, respectively. The line of intersection of these two planes defines an axis of rotation in *each* plane. A rotation about this common axis, therefore, not only takes $\mathbf{X}$ into x_2 but also takes $\mathbf{Y}$ into y_2, and therefore it must also take $\mathbf{Z}$ into z_2. This common axis is the intersection of the two planes as shown in Figure 3.11.

Orthogonality

Note: The orthogonality of coordinate axes is preserved under rotations.

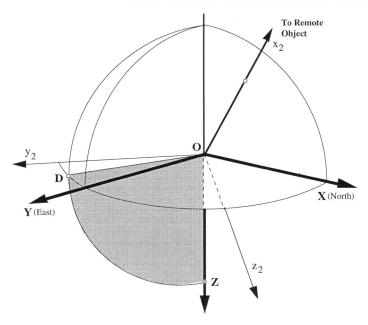

Figure 3.10: Locus of Axes which take $\mathbf{Y}$ into y

The cross-product, $\mathbf{n}_y \times \mathbf{n}_x$, gives the vector direction, $\mathbf{V}$, of this line of intersection. To compute this cross-product we write

$$\mathbf{V} = \begin{vmatrix} \mathbf{i} & \mathbf{j} & \mathbf{k} \\ (1 + \cos\alpha) & \sin\alpha & 0 \\ (1 - \cos\alpha\cos\beta) & -\sin\alpha\cos\beta & \sin\beta \end{vmatrix}$$

$$\mathbf{V} = \mathbf{i}\sin\alpha\sin\beta - \mathbf{j}(1 + \cos\alpha)\sin\beta - \mathbf{k}\sin\alpha(1 + \cos\beta)$$

or if expressed in terms of half-angles

$$\mathbf{V} = -\mathbf{i}\sin\frac{\alpha}{2}\sin\frac{\beta}{2} + \mathbf{j}\cos\frac{\alpha}{2}\sin\frac{\beta}{2} + \mathbf{k}\sin\frac{\alpha}{2}\cos\frac{\beta}{2}$$

But this is exactly the result we obtained in Equation 3.7. This completes our preliminary algebraic and geometric analysis of rotations in R^3.

Up to this point we have considered rotations in R^3, as well as sequences of two such rotations. We also need to consider sequences of several such rotations, which we do in the next chapter. Before we go to these more generalized rotation sequences, however, we introduce the useful notion of *Incremental Rotations* for a two-rotation sequence and then we briefly consider the reality of the potentially troublesome *singularities* encountered in rotation sequences.

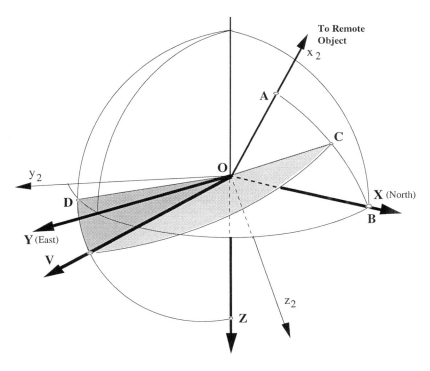

Figure 3.11: Composite Rotation Axis

3.9 Incremental Rotations in R^3

By *incremental rotations* we mean those rotations in R^3 through angles, say θ, which are so small that we may write

$$\sin\theta \simeq \theta \quad \text{and} \quad \cos\theta \simeq 1$$

Using these equivalencies we then can write the tracking incremental matrix representations, as

$$R_z^\alpha = \begin{bmatrix} \cos\alpha & \sin\alpha & 0 \\ -\sin\alpha & \cos\alpha & 0 \\ 0 & 0 & 1 \end{bmatrix} \simeq \begin{bmatrix} 1 & \alpha & 0 \\ -\alpha & 1 & 0 \\ 0 & 0 & 1 \end{bmatrix}$$

$$R_y^\beta = \begin{bmatrix} \cos\beta & 0 & -\sin\beta \\ 0 & 1 & 0 \\ \sin\beta & 0 & \cos\beta \end{bmatrix} \simeq \begin{bmatrix} 1 & 0 & -\beta \\ 0 & 1 & 0 \\ \beta & 0 & 1 \end{bmatrix}$$

An important and useful property of incremental rotations is

Incremental rotation matrices commute under multiplication. That is,

$$R_y^\beta R_z^\alpha \;=\; R_z^\alpha R_y^\beta \;=\; \begin{bmatrix} 1 & \alpha & -\beta \\ -\alpha & 1 & 0 \\ \beta & 0 & 1 \end{bmatrix}$$

Rotation Axis Algorithm

A rotation operator

$$A = \begin{bmatrix} a_{11} & a_{12} & a_{13} \\ a_{21} & a_{22} & a_{23} \\ a_{31} & a_{32} & a_{33} \end{bmatrix}$$

has a composite rotation axis given by

$$\begin{aligned} \mathbf{v} \;=\; & \mathbf{i}(a_{23} - a_{32}) \\ & + \mathbf{j}(a_{31} - a_{13}) \\ & + \mathbf{k}(a_{12} - a_{21}) \end{aligned}$$

And so long as the angles are small, we may write

$$R_y^{-\beta} R_z^\alpha R_z^\alpha R_y^\beta R_z^\alpha R_x^\gamma R_y^\beta R_y^\beta = \begin{bmatrix} 1 & 3\alpha & -2\beta \\ -3\alpha & 1 & \gamma \\ 2\beta & -\gamma & 1 \end{bmatrix}$$

We now take a brief look at the composite axis of rotation for the incremental rotation matrix

$$R_x^\gamma R_y^\beta R_z^\alpha \;=\; \begin{bmatrix} 1 & \alpha & -\beta \\ -\alpha & 1 & \gamma \\ \beta & -\gamma & 1 \end{bmatrix}$$

Using Equation 3.9 we may write

$$\mathbf{v} \;=\; \mathbf{i}2\gamma + \mathbf{j}2\beta + \mathbf{k}2\alpha$$

These matters will be considered in greater depth later in the context of the quaternion rotation operator.

3.10   Singularities in SO(3)

Those who design coordinate transformations know well that

SO(3)

The group **SO(3)** is comprised of all 3×3 orthogonal matrices whose determinant is $+1$.

Inherent in every minimal Euler angle rotation sequence in SO(3) — the group whose elements are the Special Orthogonal matrices in R^3 — is at least one singularity.

In order to gain some insight and understanding of what this statement means, we consider again the now familiar tracking sequence described in Section 3.6. That particular (perhaps the most common) tracking example involved a rotation through an angle α about the z-axis, followed by a rotation through an angle β about the new y-axis, such that the resulting x-axis points toward the remote object being tracked. The domain for these tracking angles is

Gimbal Lock

The angle β can never achieve to set $\beta = \frac{\pi}{2}$, especially in mechanical systems such as certain gyroscopes, results in a condition called *Gimbal Lock*.

$$-\pi < \alpha \leq \pi \qquad \text{and} \qquad |\beta| < \frac{\pi}{2}$$

As the angle β increases from $\frac{\pi}{2} - \epsilon$ toward, say, $\frac{\pi}{2} + \epsilon$, the angle α, when $\beta = \frac{\pi}{2}$, must instantaneously go from α to $\alpha + \pi$. This is necessary in order to insure the desired uniqueness of (α, β) over the unit sphere. The (α, β) domain, as specified above, must avoid $\beta = \frac{\pi}{2}$ since α is not defined at $\beta = \frac{\pi}{2}$. For this reason $(\alpha, \frac{\pi}{2})$ is called the *singular point* for this particular tracking sequence, and is deleted. Every *Euler angle-axis sequence* in SO(3) has at least one such singularity. Moreover, in the neighborhood of its *singular point* the systemic behaviour of tracking sequence applications is, in general, erratic and often troubled with serious transient errors.

Exercises for Chapter 3

In Section 3.6, one of the several alternatives to the tracking sequence $R_\beta^y R_\alpha^z$ might have been chosen, say, $R_\gamma^z R_\delta^x$. For this particular alternative sequence:

1. Find the expression for the composite single rotation axis.

2. Find the expression for the single angle of rotation.

3. Determine the composite single rotation axis using geometric methods similar to those employed in Section 3.7

Chapter 4

Rotation Sequences in R^3

4.1 Introduction

In the preceding chapter we showed that a 3×3 matrix represents a rotation in R^3 if and only if it is an orthogonal matrix and has determinant $+1$. We also showed that the product of two rotation operators is another rotation operator, and we developed methods for finding the axis and the angle for any rotation represented by a matrix. Further, we found the matrices representing rotations about each of the coordinate axes, and we used these in our tracking example. However, up to this point we have actually used only a sequence of two such rotation operators. In the work that follows it will be important to analyze sequences of several such rotation operators, and so in this chapter we introduce a convenient notation which allows us to do just that. We shall also introduce the important concept of *Euler angles*, and shall consider three important examples: an *Aerospace Sequence*, an *Orbit Ephemeris Sequence*, and a *Great Circle Navigation Sequence*.

4.2 Equivalent Rotations

Earlier, in Section 2.5.3, we introduced the idea of *equivalent rotations*, that is, different rotations which result in the same final *vector-frame* relationship. In particular, we mentioned that a rotation of the reference frame through an angle θ while the points (or vectors) remain fixed results in the same vector-frame relationship as does rotating the points (or vectors) through the angle $-\theta$ while the reference frame remains fixed. We noted also that the *inverse* of a rotation through an angle θ is a rotation about the same axis through the angle $-\theta$, and further that the rotations of the reference frame and rotations of the points (or vectors) are exactly the inverses of each other.

In this section we consider several sequences of rotation operators which result in the same final vector-frame relationship. Such rotation sequences are said to be *equivalent*. It is helpful here to introduce a convenient notation, pictorial in nature, for representing and even constructing these sequences. This notation will allow us to represent sequences of rotation operators in a sort of *flow chart* form.

4.2.1   New Rotation Symbol

Up to this point we have used the symbol R_α^z to represent, say, a rotation through an angle α about the **Z**-axis. Our new symbol for

Figure 4.1: Our New Rotation Symbol

this same rotation is given in Figure 4.1. Also see the note in the margin.

The notation for a sequence of rotation operators is then a string of these symbols, the order of the sequence being from *left to right*. Thus the sequence of two coordinate frame rotations which are represented by the matrix in Equation 3.5 is symbolically represented as shown in Figure 4.2. Proceeding from left to right, the first rotation is through an angle α about the **Z**-axis, followed by a second rotation through an angle β about the new y-axis.

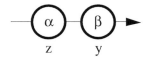

Figure 4.2: The Tracking Rotation Sequence

It is clear, as we have remarked before, that the inverse of a rotation through an angle α about some axis is simply a rotation through the angle $-\alpha$ about that same axis. Further, it is clear, particularly in the case of the tracking example, that the inverse of a sequence of rotation operators is simply the product of the inverses of the individual rotations in the sequence, written in reverse order. Thus the inverse of the tracking sequence is a rotation through

---

$c_4 \longrightarrow \!\!\!\bigcirc\!\!\!\!\alpha\!\!\! \blacktriangleright c_5$

z_4

Rotation Symbol

This symbol represents a rotation of, in this case, the 4th coordinate frame, denoted c_4 about, in this case, the z-axis (then, of course, $z_5 = z_4$) thru a positive angle α to coordinate frame c_5. The new frame is related to the old frame by the equations

$$
\begin{aligned}
x_5 &= x_4 \cos\alpha + y_4 \sin\alpha \\
y_5 &= y_4 \cos\alpha - x_4 \sin\alpha \\
z_5 &= z_4
\end{aligned}
$$

Clearly, the coordinate frame c_n consists of coordinate axes x_n, y_n, and z_n, etc. A positive rotation is always a *right-handed* rotation about the indicated axis.

the angle $-\beta$ about the y-axis, followed by a rotation through the angle $-\alpha$ about the **Z**-axis. This certainly is clear geometrically, if

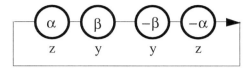

Figure 4.3: Tracking Sequence and its Inverse

not algebraically. Our new notation nicely represents this fact as is shown in Figure 4.3.

In this figure, the closed-loop merely indicates that the final frame is the same as the initial frame; that is, this sequence is an identity. Such a sequence is said to be *closed*. This attribute of a sequence we will find very useful in our analysis of coordinate relationships. The solid line which closes the loop emphasizes that the output frame of this sequence of rotations is exactly the same as its input frame. That is to say, the final vector-frame relationship produced by the closed sequence is exactly the same as the initial vector-frame relationship to which the sequence is applied.

4.2.2 A Word of Caution

For any sequence of rotations, we must understand that the axis of rotation, indicated below each rotation symbol, is always one of the axes of the input frame to that particular rotation symbol. In any

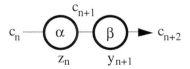

Figure 4.4: Rotation Sequence Notation

sequence of rotations if, as shown in Figure 4.4, we let c_n denote the nth coordinate frame whose axes are x_n, y_n, and z_n then, the frame, denoted c_{n+1} is the output frame after the α-rotation and it is the input frame to the β-rotation. Therefore, in this case,

$$z_{n+1} = z_n$$
$$\text{and} \quad y_{n+2} = y_{n+1}$$

In the light of this explanation, there should be no ambiguity as to the axes indicated in the notation used in Figures 4.2.

Further, it should be clear that, in general, subscripts are not necessary in the notation for rotation sequences — especially, closed sequences. To minimize congestion, from this point on subscripts will be used only when absolutely necessary.

4.2.3   Another Word of Caution

In the applications which follow we will often encounter angles which have a well established and accepted meaning. These angles have a sense, positive or negative, determined by some convention independent of the coordinate frame. For example, by convention we consider East Longitudes to be positive, so that West Longitudes are negative. Similarly, we take North Latitudes to be positive and South Latitudes to be negative.

On the other hand, the sense of the actual rotation angle itself will always be determined by the right-hand rule as it is applied to the coordinate frame at that point. The problem is that the sense of the rotation angle may or may not agree with that of the angle defined by convention. (For example, consider Longitude or Latitude angles.)

Our new notation is designed to alleviate this difficulty, in the following way. In this notation, the circle always displays the rotation angle. If, say, α is the angle defined by convention, our new notation will display α in the circle whenever these two senses (convention sense and right-hand rule sense) agree; if the senses do not agree, then we write $-\alpha$ in the circle. Therefore, it is very important, at each point in a rotation sequence, to be certain of the coordinate frame right-hand rule orientation, as well as the sense of the angle as conventionally defined. In the examples which follow, we shall be careful to do this. The Great Circle Navigation application considered later in Section 4.6 is an important case in point.

4.2.4 Equivalent Sequence Pairs

We now consider a closed sequence of n *Euler angle-axis rotations*, R_i, for $n \geq 4$. We write the product of these rotations as

$$\prod_{i=1}^{n} R_i = R_1 R_2 R_3 \cdots R_n = I \qquad (4.1)$$

We emphasize that closure means that the sequence must be an identity, as indicated. This entire identity sequence may be partitioned at any point in the sequence into two contiguous sub-sequences as follows

$$\underbrace{R_1 R_2 \cdots R_k}_{M} \overbrace{R_{k+1} R_{k+2} \cdots R_n}^{N} = I = \text{Identity} \qquad (4.2)$$

Several such sub-sequence pairs are, in general, possible. The first sub-sequence, comprising the first k rotations, we will call sequence M; the second sub-sequence, which comprises the remaining successive $n - k$ rotations, we will call sequence N. Then, of course, we have

$$MN = I = \text{Identity}$$
$$\text{That is,} \quad M = N^{-1} = N^t$$

More explicitly this means that M and N^t are equivalent Euler angle-axis rotation sequences. Thus, given a closed sequence, it is easy to find equivalent sequence pairs, as we do in the next section.

4.2.5 An Application

We will now use the ideas in the foregoing sections to determine a sequence of three rotations which we will show is exactly equivalent to the tracking sequence, in the sense we have just been discussing.

For this new sequence we arbitrarily let the first rotation be about the **X**-axis through an angle ψ. The second rotation we take about the new y-axis through an angle θ, and the third rotation we take about the newest x-axis through an angle ϕ. We note, incidentally, that we could have used any sequence of three rotation axes, as long as successive axes are distinct.

Our claim is that there are appropriate angles for each of the three indicated axes, represented in Figure 4.5, such that the composite rotation is equivalent to the tracking rotation in Equation 3.5.

Sub-sequence Length

In applications, sub-sequences will consist of, at least, two rotations.

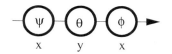

**Equivalent
Rotation Sequence
Figure 4.5**

Incidentally, the sequence of rotations shown in Figure 4.2 followed by the *inverse* of the equivalent sequence shown in Figure 4.5 must be equivalent to the identity. In other words, the resulting output

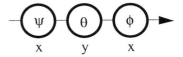

Figure 4.5: An Equivalent Rotation Sequence

frame must be the same as the input frame. This fact is indicated in Figure 4.6 by the line connecting the output to the input of the rotation sequence.

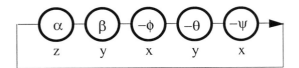

Figure 4.6: An Identity Rotation Sequence

We justify our claim by showing that the required angles ψ, θ, and ϕ indeed do exist. In fact, we find expressions from which the required angles may actually be calculated. In terms of rotation matrices, we must have

$$R = R_\phi^x R_\theta^y R_\psi^x = R_\beta^y R_\alpha^z$$

Using the matrices of the form given in Equations 3.2 and 3.3 we calculate the product of the rotation operators on the left side of the above equation. Then, equating the elements in this matrix with the corresponding elements of the matrix in Equation 3.5, we obtain the following set of equations

$$
\begin{aligned}
r_{11} &= \cos\theta & &= \cos\alpha\cos\beta \\
r_{12} &= \sin\psi\sin\theta & &= \sin\alpha\cos\beta \\
r_{13} &= -\cos\psi\sin\theta & &= -\sin\beta \\
r_{21} &= \sin\theta\sin\phi & &= -\sin\alpha \\
r_{22} &= \cos\psi\cos\phi - \sin\psi\cos\theta\sin\phi & &= \cos\alpha & \quad (4.3) \\
r_{23} &= \sin\psi\cos\phi + \cos\psi\cos\theta\sin\phi & &= 0 \\
r_{31} &= \sin\theta\cos\phi & &= \cos\alpha\sin\beta \\
r_{32} &= -\cos\psi\sin\phi - \sin\psi\cos\theta\cos\phi & &= \sin\alpha\sin\beta \\
r_{33} &= -\sin\psi\sin\phi + \cos\psi\cos\theta\cos\phi & &= \cos\beta
\end{aligned}
$$

From r_{11} we get

$$\cos \theta = \cos \alpha \cos \beta$$

Dividing r_{21} by r_{31} gives

$$\tan \phi = -\frac{\tan \alpha}{\sin \beta}$$

Dividing r_{12} by r_{13} gives

$$\tan \psi = \frac{\sin \alpha}{\tan \beta}$$

Thus we have obtained expressions for the angles ψ, θ, and ϕ as functions of angles α and β. Hence, given values for α and β we may compute values for the required three angles. The solutions, in general, will not be unique, as is the case for most trigonometric equations. Uniqueness can be obtained by carefully specifying the allowable domains for these angles.

Many other rotation sequences are also equivalent to the tracking rotation sequence. We have merely demonstrated the existence of such sequences of rotations through three *Euler angles*. In the next section we explore such sequences in more detail.

4.3 Euler Angles

Leonard Euler (1707-1783) was one of the giants in mathematics. Among the myriad of noteworthy contributions he made to the Physical Sciences in general, and mathematics in particular, many of which bear his name, is his work in Mechanics and Dynamics. In connection with his work in celestial mechanics Euler stated and proved a theorem which is closely related to our work here. The theorem states that

> *Any two independent orthonormal coordinate frames can be related by a sequence of rotations (not more than three) about coordinate axes, where no two successive rotations may be about the same axis.*

Euler's Theorem

This theorem, which we do not prove, guarantees the existence of sequences of three such rotations which properly relate two independent coordinate frames. We will find such sequences for a number of interesting examples.

When we say "two independent coordinate frames are *related*" we mean that a sequence of rotations about successive coordinate axes will rotate the first frame into the second.

The angle of rotation about a coordinate axis is called an *Euler Angle*. A sequence of such rotations is often called an *Euler Angle*

Sequence, or more precisely an *Euler Angle-axis Sequence*, since the sequential order of the axes about which each rotation is taken is an important matter.

The restriction that *successive* axes of rotation be *distinct* still permits at least *twelve* Euler angle-axis sequences. That is to say: Any two arbitrary but distinct coordinate frames may be completely related using any one of twelve different angle-axis sequences. These twelve *axis*-sequences are

xyz	yzx	zxy
xzy	yxz	zyx
xyx	yzy	zxz
xzx	yxy	zyz

We read these sequences from left to right; that is, the sequence xzy means a rotation about the **X**-axis, followed by a rotation about the new z-axis, followed by a rotation about the newer y-axis. Notice that in our example of the preceding section we used the sequence of axes xyx. We could as well have used any one of the other sequences.

These angles are called Euler angles because it was Leonard Euler who first used angle sequences to determine orbit relationships in celestial mechanics. In the remaining sections in this chapter we give several examples and applications of specific Euler sequences which are of current interest. The first example is the well-known Aerospace sequence, described in the next section. A second example we call the Orbit Sequence. Then, as an application of these two sequences we show how one might compute the *Orbit Ephemeris* of, say, a near-earth satellite. Finally, we use an Euler angle sequence in an application for navigation which determines the great-circle course between two points on the surface of the earth.

Heading & Attitude Indicator

The Aircraft Heading and Attitude Indicator, which early-on has been commonly referred to as the 'Eight-Ball', is perhaps the primary flight instrument in the cockpit of every aircraft.

It is helpful if in the presentation of the Aerospace sequence we identify intuitively with the Heading and Attitude Indicator in an aircraft.

4.4   The Aerospace Sequence

In the Euler Angle-axis Sequences tabulated in the preceeding section, zyx, is the sequence commonly used in Aircraft and Aerospace applications. For example, a primary flight instrument in virtually every cockpit, called the Heading and Attitude Indicator, continuously relates the orientation of the Aircraft frame to a Reference frame (Earth's local tangent plane and North) in accordance with Figure 4.7. In this Aerospace application the positive x-axis of the aircraft body frame points forward along the aircraft longitudinal axis; the positive y-axis is directed along the right wing; the positive z-axis is normal to the x and y axes, pointing downward. This defines a right-handed orthonormal *body coordinate frame* for the aircraft.

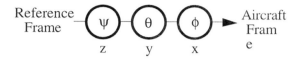

Figure 4.7: Aircraft Euler Angle Sequence

The *reference coordinate frame* is defined in terms of the Earth's *local tangent plane* and North. That is, the positive **Z**-axis is directed geocentrically (downward). The positive **X**-axis points north in the tangent plane and the positive **Y**-axis points east. This "local" Earth-referenced triad defines a right-handed orthonormal reference frame.

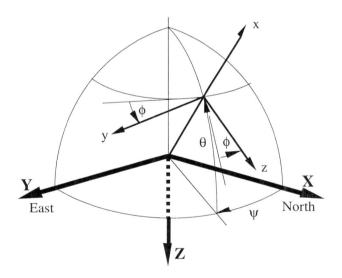

Figure 4.8: Aerospace Euler Sequence

Heading & Attitude Conventions

As indicated in Figure 4.7, the sense of the angles, ψ, θ, and ϕ, by convention agrees with the right-hand rule. That is, a positive heading angle, ψ, is directed to the right; a positive elevation angle, θ, indicates climb; a positive bank angle, ϕ, corresponds to a right turn. These angles define the *orientation* or *heading and attitude state*, say, of an aircraft relative to a 'fixed' reference frame.

These two frames are related by the Heading and Attitude sequence of rotations specified in Figure 4.7: From the *Reference coordinate frame*, first a rotation through the angle ψ about the **Z**-axis defines the aircraft *Heading*. This is followed by a rotation about the new y-axis through an angle θ which defines the aircraft *Elevation*. Finally, the aircraft *Bank angle*, ϕ, is a rotation about the

Yaw Pitch Roll

The terms *Yaw*, *Pitch*, and *Roll*, are often incorrectly used to refer to the heading and attitude of an aircraft. Properly used these terms refer to incremental *deviations* or *perturbations* of or about the nominal aircraft heading and attitude *state*.

newest x-axis. These three Euler angle rotations relate the *Body coordinate frame* of the aircraft to the local *Reference coordinate frame* of the Earth.

The relative *orientation* of these two coordinate frames is illustrated geometrically in Figure 4.8. The Aerospace rotation sequence of Figure 4.7, which is illustrated in Figure 4.8, is expressed as the matrix product

$$R = R_\phi^x R_\theta^y R_\psi^z \tag{4.4}$$

$$= R_\phi^x \begin{bmatrix} \cos\theta & 0 & -\sin\theta \\ 0 & 1 & 0 \\ \sin\theta & 0 & \cos\theta \end{bmatrix} \begin{bmatrix} \cos\psi & \sin\psi & 0 \\ -\sin\psi & \cos\psi & 0 \\ 0 & 0 & 1 \end{bmatrix}$$

$$= \begin{bmatrix} 1 & 0 & 0 \\ 0 & \cos\phi & \sin\phi \\ 0 & -\sin\phi & \cos\phi \end{bmatrix} \begin{bmatrix} \cos\theta\cos\psi & \cos\theta\sin\psi & -\sin\theta \\ -\sin\psi & \cos\psi & 0 \\ \sin\theta\cos\psi & \sin\theta\sin\psi & \cos\theta \end{bmatrix}$$

$$= \begin{bmatrix} \cos\psi\cos\theta & \sin\psi\cos\theta & -\sin\theta \\ \left(\begin{array}{c}\cos\psi\sin\theta\sin\phi \\ -\sin\psi\cos\phi\end{array}\right) & \left(\begin{array}{c}\sin\psi\sin\theta\sin\phi \\ +\cos\psi\cos\phi\end{array}\right) & \cos\theta\sin\phi \\ \left(\begin{array}{c}\cos\psi\sin\theta\cos\phi \\ +\sin\psi\sin\phi\end{array}\right) & \left(\begin{array}{c}\sin\psi\sin\theta\cos\phi \\ -\cos\psi\sin\phi\end{array}\right) & \cos\theta\cos\phi \end{bmatrix}$$

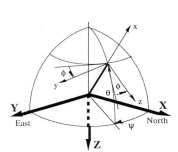

**Figure 4.8
Aerospace
Heading & Attitude**

This product of rotation matrices is itself a rotation matrix which represents the Aerospace Sequence in terms of Heading & Attitude. This matrix may be viewed as a single rotation about some axis (in general, not a coordinate axis) which takes the reference coordinate frame into the body coordinate frame. This is another important consequence of Euler's theorem. Using our methods developed earlier we could, if we so desired, find the axis and the angle of this single rotation, which would be equivalent to this Aerospace Euler sequence of rotations.

4.5   An Orbit Ephemeris Determined

In Section 4.3 we listed the twelve possible Euler angle-axis rotation sequences. Each rotation in these sequences occurs about the indicated coordinate frame axes. In this section we consider the application of two such sequences to the determination of an *orbit ephemeris* associated with a near-earth orbiting satellite.

An *orbit ephemeris* of such an orbiting body is simply a tabulation of the earth longitude and latitude of the remote body as a function of time. It specifies, for a given point in time, the location, on the surface of the earth, of the geocentric radial direction to the satellite. In Figure 4.9 the radial direction is specified by the point R, and the longitude and latitude of the point P, which is on the surface of the earth.

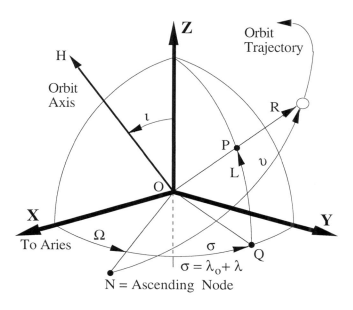

Figure 4.9: Orbit Euler Angle Sequence

In order to compute values for the parameters which define the orbit ephemeris, we define both an *inertially fixed reference frame* and an *orbit frame*. We then define two Euler angle-axis sequences of rotations, each of which relates the inertially fixed frame and the orbit frame. It is clear that these two sequences must be equivalent. This allows us to equate corresponding elements in the matrices which represent these two sequences, and in turn allows us to derive algorithms of interest. We begin with an Euler angle-axis sequence for orbits.

4.5.1 Euler Angle-Axis Sequence for Orbits

It is quite likely that the third or fourth sequence in column three of the tabulation in Section 4.3, namely, zxz or zyz, was the first sequence employed by Euler in connection with his work in orbit mechanics. We consider the orbit of a "near-earth" satellite, and begin by defining an *inertially fixed* reference frame. In this reference frame the **X** and **Y** axes are contained in the equatorial plane of the earth, that is, a plane which contains the earth equator. The

Z-axis is normal to this **XY** plane such that the **XYZ** frame is right-handed. Finally, the **X**-axis is "fixed" in the direction of the constellation Aries. See Figure 4.9. This **XYZ** frame we adopt for our purposes as our *inertially fixed reference* frame.

Vernal Equinox

Actually, the *relatively fixed* direction of the reference **X**-axis, which points to the constellation *Aries*, is established by the intersection of the Earth's Equatorial plane and the Earth's Orbital plane (also known as the Plane of the Ecliptic).

The orbit trajectory of the near-earth satellite is illustrated in Figure 4.9. The plane NOR is the *orbital plane*. This orbital plane and the equatorial plane intersect in a line ON, called the *line of nodes*, on which the point N is the *ascending node*. We define the *orientation* of the orbital plane by means of two Euler angle rotations from the inertially fixed reference frame, as shown in Figure 4.9. The first rotation is about the reference frame **Z**-axis through an angle Ω (Omega) such that the new positive x-axis contains the orbit ascending node, N. The second rotation is about this

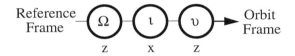

Figure 4.10: Euler Angle Sequence for Orbits

new x-axis through an angle ι (iota) such that the new y-axis lies in the orbital plane. The angle ι is usually called the *inclination angle* of the orbital plane, and the new z-axis is normal to this orbital plane.

We now define the *orbit frame* by using a third rotation about the new z-axis through an angle ν (nu), so that the resulting new x-axis points toward the orbiting body. This sequence of three Euler angle rotations is called the *Euler angle Sequence for Orbits*. See Figure 4.10. In summary, the orbit frame has a geocentric origin, the x-axis contains the orbiting body, and the z-axis is normal to the plane of the orbit (directed such that it is in the direction of the orbit angular rate vector). The y-axis is directed such that the orbit frame is a right-handed coordinate frame.

Proceeding just as before, we may now calculate the rotation matrix which takes the inertially fixed reference frame into the orbit frame. We have

$$
\begin{aligned}
S &= S_\nu^z S_\iota^x S_\Omega^z \\
&= S_\nu^z \begin{bmatrix} 1 & 0 & 0 \\ 0 & \cos\iota & \sin\iota \\ 0 & -\sin\iota & \cos\iota \end{bmatrix} \begin{bmatrix} \cos\Omega & \sin\Omega & 0 \\ -\sin\Omega & \cos\Omega & 0 \\ 0 & 0 & 1 \end{bmatrix}
\end{aligned}
\tag{4.5}
$$

$$
= \begin{bmatrix} \cos\nu & \sin\nu & 0 \\ -\sin\nu & \cos\nu & 0 \\ 0 & 0 & 1 \end{bmatrix} \begin{bmatrix} \cos\Omega & \sin\Omega & 0 \\ -\sin\Omega\cos\iota & \cos\Omega\cos\iota & \sin\iota \\ \sin\Omega\sin\iota & -\cos\Omega\sin\iota & \cos\iota \end{bmatrix}
$$

$$
= \begin{bmatrix} \begin{pmatrix} \cos\Omega\cos\nu- \\ \sin\Omega\cos\iota\sin\nu \end{pmatrix} & \begin{pmatrix} \sin\Omega\cos\nu+ \\ \cos\Omega\cos\iota\sin\nu \end{pmatrix} & \sin\iota\sin\nu \\ \begin{pmatrix} -\cos\Omega\sin\nu- \\ \sin\Omega\cos\iota\cos\nu \end{pmatrix} & \begin{pmatrix} -\sin\Omega\sin\nu+ \\ \cos\Omega\cos\iota\cos\nu \end{pmatrix} & \sin\iota\cos\nu \\ \sin\Omega\sin\iota & -\cos\Omega\sin\iota & \cos\iota \end{bmatrix}
$$

Once again, we recognize that this composite rotation matrix may also be viewed as a single rotation which will take the reference frame into the orbit frame. And, should we so desire, we could find the axis and the angle of this rotation.

4.5.2 The Orbit Ephemeris Sequence

The sequence shown in Figure 4.10, namely, the Euler angle Sequence for Orbits, takes the inertially fixed reference frame into the orbit frame, with the newest x-axis through point P, which is on the geocentric line to the orbiting body (see Figure 4.9).

In order to determine the orbit ephemeris, that is the Latitude and Longitude of point P on the surface of the earth, we invoke a zyx sequence, namely, an Aerospace Euler angle sequence, such as the one shown in Figure 4.7. More specifically, we define this sequence so that it is exactly equivalent to the sequence for orbits shown in Figure 4.10. This sequence, shown in Figure 4.11, we call the *Orbit Ephemeris Sequence*. The first rotation in this sequence is

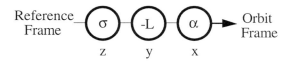

Figure 4.11: Orbit Ephemeris Sequence

a rotation about the **Z**-axis through an angle σ so that the new x-axis coincides with the line OQ in Figure 4.9. The second rotation is about the new y-axis through an angle $-L$ (see margin), so that the new x-axis coincides with the line OR. (This was the case with the orbit frame in the preceding section.) The third rotation is about the newest x-axis, through an appropriate angle α, so as to rotate the y-axis into the orbital plane NOR. The result is then exactly the orbit frame of the preceding section. Thus the orbit ephemeris sequence of rotations, shown in Figure 4.11 must be equivalent to

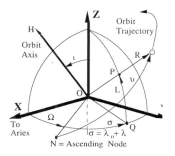

Figure 4.9
Orbit Geometry

The relevant spherical triangles are reproduced here and on subsequent pages as a reference to the rotations.

Sign of the Angle

Note that the sense of the angle σ agrees with that of the frame. However, the angle L in this sequence is the latitude of point P. Its sense differs from that specified by the right-hand rule. Thus the rotation angle is $-L$, as shown in the Figure 4.11.

the Euler angle sequence for orbits, of Figure 4.10.

In summary, we have the following notation:

$$
\begin{aligned}
\Omega &= \text{angle to the orbit ascending node} \\
\iota &= \text{the orbit } \textit{angle of inclination} \\
\nu &= \text{the argument of the latitude to orbiting body} \\
\alpha &= \text{ephemeris path direction angle} \\
L &= \text{Earth-latitude of orbiting body} \\
\lambda &= \text{Earth-longitude of orbiting body} \\
\lambda_0 &= \text{Greenwich wrt } \mathbf{X}\text{-axis} \\
\sigma &= \lambda + \lambda_0
\end{aligned}
$$

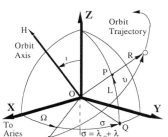

Figure 4.9
Orbit Geometry

The relevant spherical triangles are reproduced here and on subsequent pages as a reference to the rotations.

The angle λ_0 locates *zero* longitude (Greenwich), on the surface of the rotating Earth, with respect to the $\mathbf{X}$-axis (Aries) of the inertially fixed frame. Clearly, the parameters, ν, α, L, and σ, are time-varying functions of the orbital angular rate, ω_o, and the Earth's angular rate, ω_e. In what follows, however, we simply compute the ephemeris in terms of the orbit parameters.

The orbit ephemeris sequence is shown in Figure 4.11 and the matrix product which represents this rotation sequence is

$$
\begin{aligned}
R &= R_\alpha^x R_{-L}^y R_\sigma^z \qquad\qquad\qquad\qquad\qquad\qquad (4.6) \\[4pt]
&= R_\alpha^x
\begin{bmatrix}
\cos L & 0 & \sin L \\
0 & 1 & 0 \\
-\sin L & 0 & \cos L
\end{bmatrix}
\begin{bmatrix}
\cos \sigma & \sin \sigma & 0 \\
-\sin \sigma & \cos \sigma & 0 \\
0 & 0 & 1
\end{bmatrix} \\[4pt]
&=
\begin{bmatrix}
1 & 0 & 0 \\
0 & \cos \alpha & \sin \alpha \\
0 & -\sin \alpha & \cos \alpha
\end{bmatrix}
\begin{bmatrix}
\cos L \cos \sigma & \cos L \sin \sigma & \sin L \\
-\sin \sigma & \cos \sigma & 0 \\
-\sin L \cos \sigma & -\sin L \sin \sigma & \cos L
\end{bmatrix} \\[4pt]
&=
\begin{bmatrix}
\cos \sigma \cos L & \sin \sigma \cos L & \sin L \\
\begin{pmatrix} -\cos \sigma \sin L \sin \alpha \\ -\sin \sigma \cos \alpha \end{pmatrix} & \begin{pmatrix} -\sin \sigma \sin L \sin \alpha \\ +\cos \sigma \cos \alpha \end{pmatrix} & \cos L \sin \alpha \\
\begin{pmatrix} -\cos \sigma \sin L \cos \alpha \\ +\sin \sigma \sin \alpha \end{pmatrix} & \begin{pmatrix} -\sin \sigma \sin L \cos \alpha \\ -\cos \sigma \sin \alpha \end{pmatrix} & \cos L \cos \alpha
\end{bmatrix}
\end{aligned}
$$

Since these two angle-axis sequences, namely the Orbit sequence and the Ephemeris sequence, are equivalent, we may equate corresponding elements in their matrix representations, as given in Equations 4.5 and 4.6. If we equate the elements of the first row and third column, we obtain

$$\sin L = \sin \iota \sin \nu$$

If in each matrix we divide the element in the second row and third column by the element in the third row and third column, and then equate the results, we have

$$\frac{\cos L \sin \alpha}{\cos L \cos \alpha} = \tan \alpha = \frac{\sin \iota \cos \nu}{\cos \iota} = \tan \iota \cos \nu$$

Similarly, if in each matrix we divide the second element in the first row by the first element in that row and equate the resulting fractions we obtain

$$\frac{\sin \sigma \cos L}{\cos \sigma \cos L} = \tan \sigma = \frac{\sin \Omega \cos \nu + \cos \Omega \cos \iota \sin \nu}{\cos \Omega \cos \nu - \sin \Omega \cos \iota \sin \nu}$$

In this latter fraction we may divide all terms by the term $\cos \Omega \cos \nu$ to obtain

$$\tan \sigma = \frac{\tan \Omega + \cos \iota \tan \nu}{1 - \tan \Omega \cos \iota \tan \nu}$$

In summary we have

$$
\begin{aligned}
\sin L &= \sin \iota \sin \nu \\
\tan \alpha &= \tan \iota \cos \nu \\
\tan \sigma &= \frac{\tan \Omega + \cos \iota \tan \nu}{1 - \cos \iota \tan \nu \tan \Omega}
\end{aligned}
$$

We may now use these equations, given some point in time, to compute the latitude L to the orbiting body, as well as the angle σ. From the angle σ, given a value for λ_0 we may compute the longitude λ to the orbiting body; that is, we may compute the orbit ephemeris, which was our goal. The angle α, incidentally, is related to the path direction of the ephemeris at a given time.

4.6   Great Circle Navigation

Our last example of the application of a sequence of Euler angle rotations deals with the problem of *great circle navigation*. In this example a rotation sequence is constructed for a great circle path between two points, A and B, on the surface of the earth, as shown in Figure 4.12. We assume that the latitude and longitude for the points A and B are known, which means that in Figure 4.12 the angles λ_1, λ_2, L_1, and L_2 are all given.

Relative to this great circle path we specify three angles. The first is the angle ψ_1, which is the *heading* of the great circle path at point A, toward point B. The second is the angle θ, which is the *radian distance* along the path from point A to point B. The third angle ψ_2, is the *heading of arrival* or the *approach heading* at

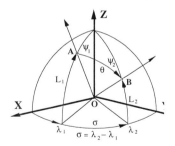

Great Circle Path

The spherical trapezoid is redrawn here in the margin for reference as we step through the sequence of rotations.

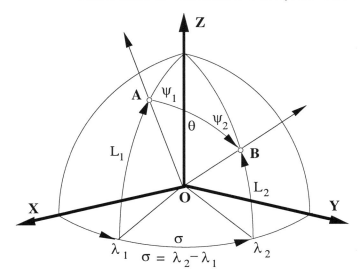

Figure 4.12: Great Circle Path

point B. Heading, by convention, is measured relative to North. In this section we derive algorithms for computing each of these angles.

Again in this example we use a sequence of Euler angle rotations. Our approach here, however, is somewhat different from our earlier examples. Here we shall construct a sequence of seven rotations, such that their product is the identity. This composite identity sequence is then partitioned into two subsequences, say, M and N, as illustrated in Section 4.2.3. From this pair (which we know are inverses of one another) we then derive the desired algorithms.

We begin the rotation sequence from an *earth-fixed*, right-handed **XYZ**-reference frame, as in Figure 4.12. The **XY**-plane lies in the earth's equatorial plane, the **X**-axis is directed positively at Greenwich zero longitude, and the **Z**-axis coincides with the Earth polar axis, directed North.

Since it is often difficult to visualize and to properly interpret the geometric effect of a sequence of several rotations, we define each rotation and accompany it with a figure which shows its geometric effect.

The first rotation in our sequence is about the **Z**-axis through the angle λ_1. As shown in Figure 4.12, the new x-axis is in the direction of the longitude of point A, measured from Greenwich zero.

The second rotation is taken about the new y-axis through an angle L_1 (the latitude of point A), such that the new x-axis contains the point A. Note, however, that the *right-hand rule*, as applied to

CAUTION

Conventionally, Longitude, λ, and Latitude, L, are positive quantities, measured either east or west from Greenwich zero, or north or south from the equator, respectively. In order to deal with these quantities algebraically, we shall define the domains for λ and L, as follows:

$$-\pi < \lambda < \pi$$

$$-\frac{\pi}{2} < L < \frac{\pi}{2}$$

For Longitude, we define east to be positive and west negative. For Latitude, we take north to be positive and south negative.

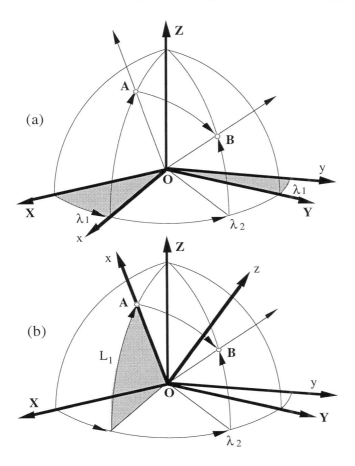

Figure 4.13:
(a) Rotation thru λ_1 about the z-axis
(b) Rotation thru L about the new y-axis.

this rotation defines a positive rotation direction which is opposite the sign of the indicated Latitude angle, L_1. Therefore, the rotation angle must be the negative of L_1, that is, $-L_1$, as indicated later in Figure 4.17.

It is very important to emphasize that rotation directions (that is, whether positive or negative) are always *frame dependent*, as determined by the right-hand rule. In this second rotation this means that if L_1 is positive, the rotation direction is negative, while if L_1 is negative, the rotation direction is positive. Hence the label $-L_1$ at this point in the sequence.

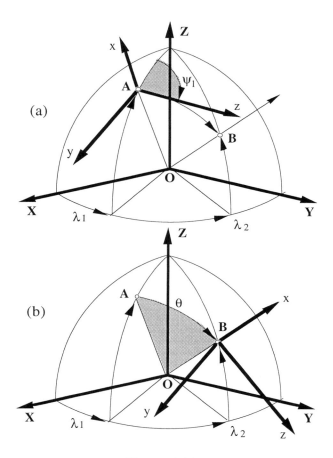

Great Circle Path

For clarity sake and to avoid pictorial conjestion, here and in some of the following figures, we show the newly rotated coordinate frame on the surface of the sphere properly related to the great circle path we seek to define.

Figure 4.14:
(a) Rotation thru ψ_1 about the x-axis
(b) Rotation thru θ about the y-axis.

The third rotation is about the new x-axis (which, incidentally, is the local vertical axis at point A) through the heading angle ψ_1. The new z-axis now lies in a direction tangent to the great circle path at point A directed toward point B. This is the heading, at point A, of the great circle path from point A to point B.

By convention, the Heading angle, ψ, is always positive ($0 \leq \psi < 2\pi$). As in the second rotation, the right-hand rule, however, dictates a negative rotation direction. The required rotation angle has the direction, $-\psi_1$, though shown as ψ_1 in Figure 4.14(a).

The fourth rotation is about the new y-axis through an angle θ. After this rotation the new x-axis contains point B and hence represents the local-vertical at point B. Note that the new z-axis now lies in a direction still tangent to the great circle path but at point B. Here again, the right-hand rule specifies the rotation angle to be $-\theta$.

Since θ is the radian distance between points A and B, the linear distance is equal to $R\theta$, where R is taken to be the radius of the Earth, in this application. See Figure 4.6.

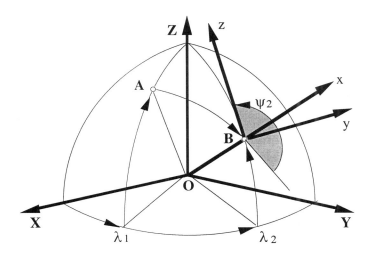

Figure 4.15: Rotation thru ψ_2 about the x-axis

The fifth rotation is about the new x-axis through the angle ψ_2. This angle is the *heading of arrival* at point B. Since the rotation is about the x-axis, it remains in the local vertical direction at point B. After rotation through the angle, ψ_2, the new z-axis points north from point B and the y-axis points east.

The sixth rotation is about the new y-axis through the angle L_2, which is the latitude of point B. After this rotation the new x-axis lies in the original **XY**-plane, and the new z-axis coincides with the original **Z**-axis of the **XYZ** Reference Frame.

The seventh and final rotation in our sequence is about the new z-axis (which is the original reference frame **Z**-axis) through the angle λ_2, the longitude of point B as measured from Greenwich zero.

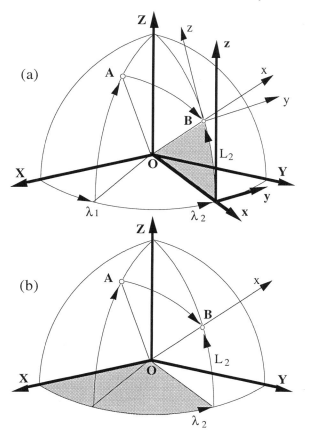

Figure 4.16:
(a) Rotation thru L_2 about the y-axis
(b) Rotation thru λ_2 about the z-axis.

This final rotation takes the x-axis back to the original **X**-axis, and therefore the y-axis back to the original **Y**-axis. That is to say, after these seven rotations the rotated coordinate frame is back to its original reference frame orientation. And, note here again, that whether the angle λ_2 is positive or negative, the right-hand rule requires that the rotation angle be -λ_2, as shown in the sequence of Figure 4.17.

This completes the detailed development of and relationship between the seven angles in the closed rotation sequence summarized Figure 4.17.

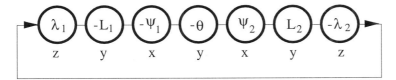

Figure 4.17: Great Circle Path Rotation Sequence

We make two useful remarks about this diagram:

First, we may begin this sequence at any point and still have the identity transformation, so long as all of the rotations are included *in the same order*. Thus the sequence of rotations represented in Figure 4.18 is also the identity, except that the three unknown an-

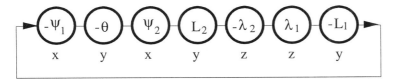

Figure 4.18: Modified Rotation Sequence

gles occur first. We notice also that two consecutive rotations in this sequence, namely, $-\lambda_2$ and λ_1, are both about the z-axis. This two rotation sequence is, of course, equivalent to a single rotation about the same axis through the angle $-\lambda_2 + \lambda_1$. Thus, we define an auxilliary angle σ by

$$\sigma = \lambda_2 - \lambda_1$$

and we replace the two rotations which are shown in the sequence of Figure 4.18 with a single rotation about the z-axis through the angle $-\sigma$ as shown in Figure 4.20.

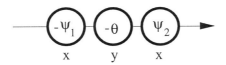

Figure 4.19: Rotation Sequence of Unknowns, M

Recall that the angles we seek to determine are the angles ψ_1, θ, and ψ_2. Note that these now occur first in the sequence and that this sequence is still the identity. The second remark we make about rotation sequences is that the entire sequence may be partitioned,

Closed Sequences

Analysis of closed rotation sequences, as introduced in Section 4.2.3, will prove to be very useful in that which follows.

as shown in Figure 4.20, into two non-overlapping segments, each of which is itself a rotation sequence. Moreover, since the product

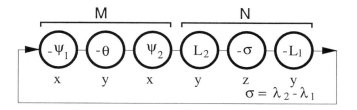

Figure 4.20: Simplified Rotation Sequence

of these two contiguous rotation sequences is the identity, it follows that these two segments are also inverses of one another.

We now apply these remarks to the sequence as shown partitioned in Figure 4.20 into subsequences, M and N. Recall that the inverse of a rotation sequence is obtained by reversing the order and changing the signs of the angles of rotation. Therefore, the inverse of the second segment of the partition is the sequence shown in Figure 4.21. It follows that this inverse sequence is now equiv-

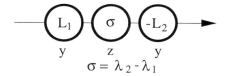

Figure 4.21: Known Rotation Sequence $N^{-1} = N^t$

alent to the first segment of the partition, namely M, as shown in Figure 4.19. We next write these rotation sequences using matrix notation. In doing so we recall that for a rotation matrix

$$R_{z,-\alpha} = R_{z,\alpha}^t$$

The equivalence of the two sequences, M and N^{-1}, represented in matrix form, gives us the important matrix equation

$$M = R_{x,\psi_2} R_{y,\theta}^t R_{x,\psi_1}^t = R_{y,L_2}^t R_{z,\sigma} R_{y,L_1} = N^{-1} \quad (4.7)$$

If we denote the elements of M by m_{ij} we have

$$M = R_{x,\psi_2} R_{y,\theta}^t R_{x,\psi_1}^t \quad (4.8)$$
$$= \begin{bmatrix} m_{11} & m_{12} & m_{13} \\ m_{21} & m_{22} & m_{23} \\ m_{31} & m_{32} & m_{33} \end{bmatrix}$$

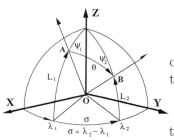

Great Circle Path

which in more detail is the product of the three matrices,

$$
\begin{bmatrix} 1 & 0 & 0 \\ 0 & \cos\psi_2 & \sin\psi_2 \\ 0 & -\sin\psi_2 & \cos\psi_2 \end{bmatrix}
\begin{bmatrix} \cos\theta & 0 & \sin\theta \\ 0 & 1 & 0 \\ -\sin\theta & 0 & \cos\theta \end{bmatrix}
\begin{bmatrix} 1 & 0 & 0 \\ 0 & \cos\psi_1 & -\sin\psi_1 \\ 0 & \sin\psi_1 & \cos\psi_1 \end{bmatrix}
$$

In like manner, we write

$$
N^{-1} \;=\; R^t_{y,L_2} R_{z,\sigma} R_{y,L_1} \tag{4.9}
$$

$$
=\; \begin{bmatrix} n_{11} & n_{12} & n_{13} \\ n_{21} & n_{22} & n_{23} \\ n_{31} & n_{32} & n_{33} \end{bmatrix}
$$

which, in more detail, is the product of the three matrices,

$$
\begin{bmatrix} \cos L_2 & 0 & \sin L_2 \\ 0 & 1 & 0 \\ -\sin L_2 & 0 & \cos L_2 \end{bmatrix}
\begin{bmatrix} \cos\sigma & \sin\sigma & 0 \\ -\sin\sigma & \cos\sigma & 0 \\ 0 & 0 & 1 \end{bmatrix}
\begin{bmatrix} \cos L_1 & 0 & -\sin L_1 \\ 0 & 1 & 0 \\ \sin L_1 & 0 & \cos L_1 \end{bmatrix}
$$

The matrices M and N^{-1} are equal, which means their corresponding elements, m_{ij} and n_{ij}, are equal. When we compute the foregoing matrix products, we obtain

$$
\begin{aligned}
m_{11} &= \cos\theta \\
&= \cos L_2 \cos\sigma \cos L_1 + \sin L_1 \sin L_2 & = n_{11} \\
m_{12} &= \sin\theta \sin\psi_1 \\
&= \cos L_2 \sin\sigma & = n_{12} \\
m_{13} &= \sin\theta \cos\psi_1 \\
&= \cos L_1 \sin L_2 - \cos L_2 \cos\sigma \sin L_1 & = n_{13} \\
m_{21} &= -\sin\theta \sin\psi_2 \\
&= -\cos L_1 \sin\sigma & = n_{21} \\
m_{22} &= \cos\psi_2 \cos\psi_1 + \sin\psi_1 \cos\theta \sin\psi_2 \\
&= \cos\sigma & = n_{22} \\
m_{23} &= \cos\psi_1 \cos\theta \sin\psi_2 - \sin\psi_1 \cos\psi_2 \\
&= \sin L_1 \sin\sigma & = n_{23} \\
m_{31} &= -\sin\theta \cos\psi_2 \\
&= -\sin L_2 \cos\sigma \cos L_1 + \sin L_1 \cos L_2 & = n_{31} \\
m_{32} &= -\cos\psi_1 \sin\psi_2 + \sin\psi_1 \cos\theta \cos\psi_2 \\
&= -\sin L_2 \sin\sigma & = n_{32} \\
m_{33} &= \cos\psi_1 \cos\theta \cos\psi_2 + \sin\psi_1 \sin\psi_2 \\
&= \cos L_1 \cos L_2 + \sin L_1 \cos\sigma \sin L_2 & = n_{33}
\end{aligned}
$$

Finally, from these nine equations we compute expressions which determine the desired unknown angles, θ, ψ_1, and ψ_2, which define the great circle path from point A to point B on the earth surface.

Equating m_{11} and n_{11} gives an expression from which we can compute θ, the radian distance from point A to point B. We get

$$\cos \theta = \cos L_2 \cos \sigma \cos L_1 + \sin L_1 \sin L_2 \qquad (4.10)$$

Dividing the equation $m_{12} = n_{12}$ by $m_{13} = n_{13}$ gives an expression for the angle ψ_1, which is the heading at point A.

$$\tan \psi_1 = \frac{\cos L_2 \sin \sigma}{\cos L_1 \sin L_2 - \cos L_2 \cos \sigma \sin L_1} \qquad (4.11)$$

Finally, dividing the equation $m_{21} = n_{21}$ by $m_{31} = n_{31}$ gives an expression for the angle ψ_2, which defines the heading of arrival at point B.

$$\tan \psi_2 = \frac{\cos L_1 \sin \sigma}{\sin L_2 \cos \sigma \cos L_1 - \sin L_1 \cos L_2} \qquad (4.12)$$

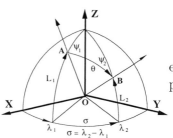

Great Circle Path

Once again we would have to specify the domain for each of these angles in order to get unique solutions for the great circle path.

This completes our analysis of sequences of rotations in R^3 using matrix representation for the rotation operators. At certain significant points in this analysis the use of quaternion operators might be more efficient. So in the next chapter we introduce quaternions, and later show how this same analysis of sequences of rotation operators in R^3 can be made from that point of view.

Exercises for Chapter 4

In Figure 4.2 the sequence $R_\beta^y R_\alpha^z$ is merely one of several alternatives to the tracking sequence. An alternative tracking sequence would be $R_\psi^z R_\theta^y$. From a common **XYZ** reference frame, these two

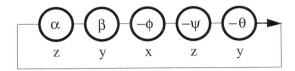

Figure 4.22: Relate two Tracking Sequences

tracking sequences both direct the x-axes of their respective tracking frames at the remote object. Use the rotation sequence illustrated in Figure 4.22.

1. Explain the need for the rotation angle ϕ about the common tracking x-axis.

2. Express the angle θ as a function of α and β.

3. Express the angle ψ as a function of α and β.

4. Express the angle ϕ as a function of α and β.

5. Illustrate the $R_\psi^z R_\theta^y$ sequence geometrically with an appropriate sketch similar to that in Figure 3.3 on page 60.

Chapter 5

Quaternion Algebra

5.1 Introduction

As we begin our consideration of quaternions and their algebra, we recall the remarks made earlier in our introductory chapter, particularly in Sections 1.1 and 1.5. There we recounted just a bit of the work done by William Rowan Hamilton in extending the notion of complex numbers to that of the quaternion. In this context we may think of the real numbers as being *hyper-complex* numbers of *rank 1*. Recall that these real numbers satisfy the field properties under ordinary addition and multiplication. Further, we may think of the *ordinary* complex numbers as being hyper-complex numbers of *rank 2*, and treat the real numbers as a subset of the complex numbers in which the imaginary part is zero. The complex numbers also satisfy the field properties, as we demonstrated earlier.

It turns out, however, that any set of hyper-complex numbers having rank greater than rank 2 does not satisfy the field properties as enumerated in Section 1.2. It was this fact that troubled and impeded those mathematicians who were seeking higher rank extensions suggested by the gradual acceptance of the complex numbers in R^2.

In 1843 Hamilton invented the so-called hyper-complex number of rank 4, to which he gave the name *quaternion*. Crucial to this invention was his celebrated rule

$$\mathbf{i}^2 = \mathbf{j}^2 = \mathbf{k}^2 = \mathbf{i}\,\mathbf{j}\,\mathbf{k} = -1$$

Be Careful!

These are *quaternion* products — not dot products. They are not ordinary products, as we shall soon see.

for dealing with the operations on the vector part of the quaternion. And finally, even though the hyper-complex numbers of rank 1 through n can be defined, few applications have been found for

103

hyper-complex numbers of rank $n > 4$. But since our concern is for graphics applications and for a variety of dynamical applications which involve rotations, we will restrict our attention from this point on to the hyper-complex numbers of rank 4, namely, *quaternions*, which as we shall see are particularly well suited for use as *rotation operators*.

The set of quaternions, along with the two operations of addition and multiplication, form a mathematical system called a *ring*, more precisely a *non-commutative division ring*. This longer title merely emphasizes the fact that the quaternion product, in general, is not commutative, and also that the multiplicative inverse exists, as usual, for every non-zero element in the set.

In summary: The set of quaternions under the operations of addition and multiplication satisfy all of the axioms of a field except for the commutative law for multiplication.

5.2 Quaternions Defined

In what follows, a quaternion will always be denoted by some lower-case letter, say p or q or r. Further, we shall continue to use bold-faced letters to denote ordinary vectors in three-dimensional space, namely R^3. In particular we use $\mathbf{i}$, $\mathbf{j}$, and $\mathbf{k}$ to denote the standard orthonormal basis for R^3. Vectors in three dimensional space may, of course, nicely be written as triplets of real numbers (that is, scalars), so that we write this orthonormal basis as

$$\begin{aligned} \mathbf{i} &= (1,0,0) \\ \mathbf{j} &= (0,1,0) \\ \mathbf{k} &= (0,0,1) \end{aligned}$$

Now a quaternion, as the name already suggests, may be regarded as a 4-tuple of real numbers, that is, as an element of R^4. In this case we would write $\quad$ *Scalar part*

$$q = (q_0, q_1, q_2, q_3)$$

where q_0, q_1, q_2, and q_3 are simply real numbers or scalars.

We shall adopt an alternative way of representing a quaternion. First, we define a *scalar part* to be some real number or scalar, say q_0. Then we define a *vector part*, say $\mathbf{q}$, which is an ordinary vector in R^3

$$\mathbf{q} = \mathbf{i}q_1 + \mathbf{j}q_2 + \mathbf{k}q_3$$

where $\mathbf{i}$, $\mathbf{j}$, and $\mathbf{k}$ are the standard orthonormal basis in R^3. We now define a *quaternion* as the sum

Quaternion Defined

Here, finally, is our definition and a representation which we adopt for the quaternion.

$$q = q_0 + \mathbf{q} = q_0 + \mathbf{i}q_1 + \mathbf{j}q_2 + \mathbf{k}q_3$$

In this sum, q_0 is called the *scalar part* of the quaternion while $\mathbf{q}$ is called the *vector part* of the quaternion. The scalars q_0, q_1, q_2, q_3 are called the *components* of the quaternion.

Defined in this way, a quaternion is a mathematically strange object – the sum of a scalar and a vector, something not defined in ordinary linear algebra. We must therefore give further meaning to this definition by showing how quaternions are to be added and multiplied.

5.3 Equality and Addition

We begin by saying two quaternions are *equal* if and only if they have exactly the same components, this is to say that if

$$p = p_0 + \mathbf{i}p_1 + \mathbf{j}p_2 + \mathbf{k}p_3$$

and

$$q = q_0 + \mathbf{i}q_1 + \mathbf{j}q_2 + \mathbf{k}q_3$$

then $p = q$ if and only if

$$
\begin{aligned}
p_0 &= q_0 \\
p_1 &= q_1 \\
p_2 &= q_2 \\
p_3 &= q_3
\end{aligned}
$$

The *sum* of the two quaternions p and q above is defined by adding the corresponding components, that is

$$p + q = (p_0 + q_0) + \mathbf{i}(p_1 + q_1) + \mathbf{j}(p_2 + q_2) + \mathbf{k}(p_3 + q_3)$$

Addition for quaternions, defined in this way, is exactly the same as that for 4-tuples of real numbers, and thus it satisfies the field properties, as these apply to addition. For example, notice that the sum of two quaternions is again a quaternion, that is, the set of quaternions is *closed* under addition. Also, there is a *zero* quaternion, in which each component of the quaternion is 0. Moreover, each quaternion q has a *negative* or an *additive inverse*, denoted $-q$, in which each component is the negative of the corresponding component of q. Further, as is easily verified, this addition of quaternions is both associative and commutative, because addition of real numbers has these properties.

5.4   Multiplication Defined

Just as in the case of vectors in R^3, the product of a scalar and a quaternion is defined in a straightforward manner. If c is a scalar and q is the quaternion

$$q = q_0 + \mathbf{i}q_1 + \mathbf{j}q_2 + \mathbf{k}q_3$$

then the product of the quaternion q and the scalar c is given by

$$cq = cq_0 + \mathbf{i}cq_1 + \mathbf{j}cq_2 + \mathbf{k}cq_3$$

Thus to multiply the quaternion by the scalar we simply multiply each component of the quaternion by the scalar. Note that the result is again a quaternion, that is, the set of quaternions is closed under multiplication by a scalar. For example, if

$$
\begin{aligned}
q &= 3 + 2\mathbf{i} - \mathbf{j} + 4\mathbf{k} \\
\text{then} \qquad 3q &= 9 + 6\mathbf{i} - 3\mathbf{j} + 12\mathbf{k}
\end{aligned}
$$

The product of two quaternions is more complicated. It must be defined so that the following fundamental *special* products are satisfied:

Again, Be Careful

These indicated products of the orthonormal basis vectors in R^3 will shortly be defined as *quaternion* products. Do not confuse these products with the familiar *dot* and *cross* products.

$$\mathbf{i}^2 = \mathbf{j}^2 = \mathbf{k}^2 = \mathbf{ijk} = -1$$

$$\mathbf{ij} = \mathbf{k} = -\mathbf{ji}$$

$$\mathbf{jk} = \mathbf{i} = -\mathbf{kj}$$

$$\mathbf{ki} = \mathbf{j} = -\mathbf{ik}$$

Recall that these products are those which Hamilton saw as being necessary for achieving his goal. Notice also that these products are not commutative, so that the product of two quaternions will, in general, not be commutative either. Now, by using ordinary rules for algebraic multiplication together with the above fundamental products, it is easy (though somewhat tedious) to verify that the product of quaternions must go as follows. If

$$p = p_0 + \mathbf{i}p_1 + \mathbf{j}p_2 + \mathbf{k}p_3$$

and

$$q = q_0 + \mathbf{i}q_1 + \mathbf{j}q_2 + \mathbf{k}q_3$$

we have

$$
\begin{aligned}
pq &= (p_0 + \mathbf{i}p_1 + \mathbf{j}p_2 + \mathbf{k}p_3)(q_0 + \mathbf{i}q_1 + \mathbf{j}q_2 + \mathbf{k}q_3) \\
&= p_0q_0 + \mathbf{i}p_0q_1 + \mathbf{j}p_0q_2 + \mathbf{k}p_0q_3 \\
&\quad + \mathbf{i}p_1q_0 + \mathbf{i}^2p_1q_1 + \mathbf{ij}p_1q_2 + \mathbf{ik}p_1q_3 \\
&\quad + \mathbf{j}p_2q_0 + \mathbf{ji}p_2q_1 + \mathbf{j}^2p_2q_2 + \mathbf{jk}p_2q_3 \\
&\quad + \mathbf{k}p_3q_0 + \mathbf{ki}p_3q_1 + \mathbf{kj}p_3q_2 + \mathbf{k}^2p_3q_3
\end{aligned}
$$

Notice that in this multiplication we have uniformly written the scalar on the right of the vector, and we have maintained the correct order in the vector products. If Hamilton's special products are to hold, we must have

$$
\begin{aligned}
pq &= p_0q_0 + \mathbf{i}p_0q_1 + \mathbf{j}p_0q_2 + \mathbf{k}p_0q_3 \\
&\quad + \mathbf{i}p_1q_0 - p_1q_1 + \mathbf{k}p_1q_2 - \mathbf{j}p_1q_3 \\
&\quad + \mathbf{j}p_2q_0 - \mathbf{k}p_2q_1 - p_2q_2 + \mathbf{i}p_2q_3 \\
&\quad + \mathbf{k}p_3q_0 + \mathbf{j}p_3q_1 - \mathbf{i}p_3q_2 - p_3q_3
\end{aligned}
$$

With some algebraic regrouping of terms, the expression may now be written as

$$
\begin{aligned}
pq &= p_0q_0 - (p_1q_1 + p_2q_2 + p_3q_3) \\
&\quad + p_0(\mathbf{i}q_1 + \mathbf{j}q_2 + \mathbf{k}q_3) + q_0(\mathbf{i}p_1 + \mathbf{j}p_2 + \mathbf{k}p_3) \\
&\quad + \mathbf{i}(p_2q_3 - p_3q_2) + \mathbf{j}(p_3q_1 - p_1q_3) + \mathbf{k}(p_1q_2 - p_2q_1)
\end{aligned}
$$

Before rewriting this expression in a more concise form, it is helpful to recall the scalar product and the cross product from the algebra of vectors in three dimensions. If we have vectors

$$
\mathbf{a} = (a_1, a_2, a_3) \quad \text{and} \quad \mathbf{b} = (b_1, b_2, b_3)
$$

then the scalar product is given by

$$
\mathbf{a} \cdot \mathbf{b} = a_1b_1 + a_2b_2 + a_3b_3
$$

and the cross product is

$$
\begin{aligned}
\mathbf{a} \times \mathbf{b} &= \begin{vmatrix} \mathbf{i} & \mathbf{j} & \mathbf{k} \\ a_1 & a_2 & a_3 \\ b_1 & b_2 & b_3 \end{vmatrix} \\
&= \mathbf{i}(a_2b_3 - a_3b_2) \\
&\quad + \mathbf{j}(a_3b_1 - a_1b_3) \\
&\quad + \mathbf{k}(a_1b_2 - a_2b_1)
\end{aligned}
$$

Using these results, we may write the above product of the two quaternions $p = p_0 + \mathbf{p}$ and $q = q_0 + \mathbf{q}$ in the more concise form

$$pq = p_0 q_0 - \mathbf{p} \cdot \mathbf{q} + p_0 \mathbf{q} + q_0 \mathbf{p} + \mathbf{p} \times \mathbf{q} \qquad (5.1)$$

Quaternion Product

Here, finally, is our definition of the quaternion product.

Although other definitions could be used, it is the expression in Equation 5.1 that we will use as our working *definition* for the product of two quaternions.

As an example of the product of two quaternions, suppose we take p and q to be

$$p = 3 + \mathbf{i} - 2\mathbf{j} + \mathbf{k}$$

and

$$q = 2 - \mathbf{i} + 2\mathbf{j} + 3\mathbf{k}$$

In order to calculate the product we note that with $p = 3 + \mathbf{p}$ and $q = 2 + \mathbf{q}$, then we have $\mathbf{p} = (1, -2, 1)$ and $\mathbf{q} = (-1, 2, 3)$.

Product of two Pure Quaternions

Note that in the special case where the real parts of the quaternions are equal to zero, then called **pure quaternions**, we get the *quaternion product* of two vectors, say, **a** and **b** as

$$\mathbf{ab} = \mathbf{a} \times \mathbf{b} - \mathbf{a} \cdot \mathbf{b}$$

Note also that applying this expression to the standard basis vectors confirms that Hamilton's special products do indeed hold.

[handwritten: $pq \neq qp$]

We calculate

$$\mathbf{p} \cdot \mathbf{q} = (1)(-1) + (-2)(2) + (1)(3) = -2$$

and $\mathbf{p} \times \mathbf{q}$ as

$$\mathbf{p} \times \mathbf{q} = \begin{vmatrix} \mathbf{i} & \mathbf{j} & \mathbf{k} \\ 1 & -2 & 1 \\ -1 & 2 & 3 \end{vmatrix}$$
$$= (-6 - 2)\mathbf{i} - (3 - (-1))\mathbf{j} + (2 - 2)\mathbf{k}$$
$$= -8\mathbf{i} - 4\mathbf{j}$$

According to Equation 5.1, the quaternion product is then given by

$$pq = 6 - (-2) + 3(-\mathbf{i} + 2\mathbf{j} + 3\mathbf{k}) + 2(\mathbf{i} - 2\mathbf{j} + \mathbf{k}) + (-8\mathbf{i} - 4\mathbf{j})$$
$$= 8 - 9\mathbf{i} - 2\mathbf{j} + 11\mathbf{k}$$

[handwritten: $2p = 8 \ 7 \ 6 \ 11$]

The product of quaternions, defined in Equation 5.1, may be written using the algebra of matrices. If we designate the product as the quaternion

$$pq = r = r_0 + \mathbf{r} = r_0 + \mathbf{i} r_1 + \mathbf{j} r_2 + \mathbf{k} r_3$$

then we have

$$r_0 = p_0 q_0 - p_1 q_1 - p_2 q_2 - p_3 q_3$$
$$r_1 = p_0 q_1 + p_1 q_0 + p_2 q_3 - p_3 q_2 \qquad (5.2)$$

$$r_2 = p_0 q_2 - p_1 q_3 + p_2 q_0 + p_3 q_1$$
$$r_3 = p_0 q_3 + p_1 q_2 - p_2 q_1 + p_3 q_0$$

or, if written in matrix notation

$$
\begin{bmatrix} r_0 \\ r_1 \\ r_2 \\ r_3 \end{bmatrix} = \begin{bmatrix} p_0 & -p_1 & -p_2 & -p_3 \\ p_1 & p_0 & -p_3 & p_2 \\ p_2 & p_3 & p_0 & -p_1 \\ p_3 & -p_2 & p_1 & p_0 \end{bmatrix} \begin{bmatrix} q_0 \\ q_1 \\ q_2 \\ q_3 \end{bmatrix} \qquad (5.3)
$$

We may now use this matrix form in order to compute the product of the two quaternions in our example above. We have

$$
\begin{bmatrix} r_0 \\ r_1 \\ r_2 \\ r_3 \end{bmatrix} = \begin{bmatrix} 3 & -1 & 2 & -1 \\ 1 & 3 & -1 & -2 \\ -2 & 1 & 3 & -1 \\ 1 & 2 & 1 & 3 \end{bmatrix} \begin{bmatrix} 2 \\ -1 \\ 2 \\ 3 \end{bmatrix}
$$

$$
= \begin{bmatrix} 8 \\ -9 \\ -2 \\ 11 \end{bmatrix}
$$

Hence the product is

$$pq = r = 8 - 9\mathbf{i} - 2\mathbf{j} + 11\mathbf{k}$$

as we obtained before.

We have remarked earlier that so far as addition is concerned, quaternions satisfy the field properties. Further, we have claimed that the set of quaternions is a non-commutative division ring, which means that all of the field properties are satisfied, except that multiplication is not commutative. To see that this is so we need to make a few more comments about the quaternion product.

Notice first that the product of two quaternions is another quaternion, with scalar part

$$p_0 q_0 - \mathbf{p} \cdot \mathbf{q}$$

and vector part

$$p_0 \mathbf{q} + q_0 \mathbf{p} + \mathbf{p} \times \mathbf{q}$$

Thus the set of quaternions is closed under multiplication as well as under addition.

Next we remark that the quaternion product is indeed associative. It is not difficult to verify that this is so, but the details are tedious, and some knowledge of the algebraic manipulation of vectors is required.

Moreover, since the cross product $\mathbf{p} \times \mathbf{q}$ is not commutative, neither is the quaternion product. And, for quaternion algebra, this is the only departure from the field properties.

We must, however, have an identity for quaternion multiplication, as well as for addition. The set of quaternions indeed does have such a multiplicative identity, namely, that quaternion with real part 1 and vector part $\mathbf{0}$. This follows directly from Equation 5.1 with

$$q = 1 + \mathbf{0}$$

It is not difficult to verify that multiplication of quaternions is also *distributive over addition*, as required.

Finally we remark that every non-zero quaternion does have a multiplicative inverse, as we show in the next three sections. This final remark completes our justification that the set of quaternions indeed is a non-commutative division ring.

Equation 5.1
Quaternion Product

$$
\begin{aligned}
pq \;=\; & p_0 q_0 - \mathbf{p} \cdot \mathbf{q} \\
& + p_0 \mathbf{q} \\
& + q_0 \mathbf{p} \\
& + \mathbf{p} \times \mathbf{q}
\end{aligned}
$$

5.5   The Complex Conjugate

An important algebraic concept relating to quaternions, as well as to ordinary complex numbers, will be useful to us in what follows, namely, that of the *complex conjugate* of a quaternion. We define the *complex conjugate* of the quaternion

$$q \;=\; q_0 + \mathbf{q} \;=\; q_0 + \mathbf{i}q_1 + \mathbf{j}q_2 + \mathbf{k}q_3$$

to be the quaternion, denoted q^*, given by

$$q^* \;=\; q_0 - \mathbf{q} = q_0 - \mathbf{i}q_1 - \mathbf{j}q_2 - \mathbf{k}q_3$$

As an example in the use of the quaternion product, defined in Equation 5.1, it is not difficult (although somewhat tedious) to show that for any two quaternions, the complex conjugate of the product of the quaternions is equal to the product of the individual complex conjugates, in reverse order. That is, given any two quaternions p and q we have

$$(pq)^* \;=\; q^* p^*$$

Note also that for any quaternion q, the *sum* of q and its complex conjugate q^* is a scalar, for we have

$$q + q^* = (q_0 + \mathbf{q}) + (q_0 - \mathbf{q}) = 2q_0$$

where $2q_0$ is the scalar. This is also true for the *product* of a quaternion q and its complex conjugate q^* as we show in the next section.

5.6   The Norm

Another important algebraic concept relating to quaternions is the *norm* of a quaternion. The *norm* of a quaternion q, denoted by $N(q)$ or $|q|$, sometimes called the *length* of q, is the scalar defined by

$$N(q) = \sqrt{q^*q}$$

Using our definition of the quaternion product, together with the fact that for any vector $\mathbf{q}$ we have $\mathbf{q} \times \mathbf{q} = \mathbf{0}$, we may calculate

$$
\begin{aligned}
N^2(q) &= (q_0 - \mathbf{q})(q_0 + \mathbf{q}) \\
&= q_0q_0 - (-\mathbf{q}) \cdot \mathbf{q} + q_0\mathbf{q} + (-\mathbf{q})q_0 + (-\mathbf{q}) \times \mathbf{q} \\
&= q_0^2 + \mathbf{q} \cdot \mathbf{q} \\
&= q_0^2 + q_1^2 + q_2^2 + q_3^2 = |q|^2
\end{aligned}
$$

As a simple example, if

$$q = 2 - \mathbf{i} + 2\mathbf{j} + 3\mathbf{k}$$

then our definition of the norm of q gives

$$N^2(q) = 2^2 + (-1)^2 + 2^2 + 3^2 = 18 \Rightarrow N(q) = \sqrt{18}$$

Notice that this definition is the same as that for the length of a vector in R^4, or for that matter it has the same meaning as any Euclidean Norm. In what follows we shall, for the most part, be working with quaternions with norm 1.

Note further that if a quaternion has norm 1 each of its components must have absolute value less than or equal to 1. Such quaternions are called *unit* or *normalized* quaternions.

Finally, it is relatively easy to show that the norm of the product of two quaternions p and q is the product of the individual norms.

Note!

It is important to note that for any quaternion q we have

$$q^*q = qq^* = |q|^2$$

as is clear from what follows.

Equation 5.1
Quaternion Product

$$
\begin{aligned}
pq = & \ p_0q_0 - \mathbf{p} \cdot \mathbf{q} + \\
& \ p_0\mathbf{q} + q_0\mathbf{p} + \mathbf{p} \times \mathbf{q}
\end{aligned}
$$

For we have

$$
\begin{aligned}
N^2(pq) &= (pq)(pq)^* \\
&= pqq^*p^* \\
&= pN^2(q)p^* \\
&= pp^*N^2(q) \\
&= N^2(p)N^2(q)
\end{aligned}
$$

It follows, of course, that the product of two *unit* quaternions is again a unit quaternion, a fact that will be important for us later on. We mention also that by mathematical induction the result extends to any product of finitely many quaternions.

5.7   Inverse of the Quaternion

Using the ideas of the complex conjugate and the norm of a quaternion, we are now able to show that every non-zero quaternion does have a multiplicative inverse, and we can develop a formula for it. If we designate the inverse q^{-1} we must, by definition of inverse, have

$$
q^{-1}q = qq^{-1} = 1
$$

Now, if we use both pre- and post-multiplication by the complex conjugate q^* we may write

$$
q^{-1}qq^* = q^*qq^{-1} = q^*
$$

Since $qq^* = N^2(q)$ we get

$$
q^{-1} = \frac{q^*}{N^2(q)} = \frac{q^*}{|q|^2} \tag{5.4}
$$

Note!

Since the quaternion product is non-commutative, we must always be mindful of the order of the factors in multiplication. Notice, however, that q always commutes with q^*.

We note here that if q is a unit or normalized quaternion, that is, $N(q) = 1$, then the inverse is simply the complex conjugate

$$
q^{-1} = q^*
$$

We remark incidentally that this result is analogous to the inverse of a rotation matrix, where $A^{-1} = A^t$.

This completes our preliminary analysis of the quaternion sum and product. We have not mentioned, however, any sort of geometric interpretation of a quaternion, and we turn to that in the next section.

5.8 Geometric Interpretations

We recall that in Section 3.4 we showed that a rotation in R^3 may be represented by a 3×3 matrix, provided that the matrix is orthogonal and has determinant $+1$. Or, alternatively, any such matrix A may be interpreted geometrically as a rotation operator in R^3. In order to find the vector $\mathbf{w}$ which is the image of a vector $\mathbf{v}$ under such a rotation, we simply represented the vector $\mathbf{v}$ by a column matrix whose entries are the components of $\mathbf{v}$, and multiplied it on the left by the rotation matrix. Thus, in matrix form, the rotation is given by the equation

$$\mathbf{w} = A\mathbf{v}$$

We also showed that the product of a sequence of rotation operators was again a rotation operator, and we developed algorithms for finding the axis and the angle of that composite rotation.

It is well known that quaternions play an important role in an *alternative form* for a rotation operator, a role that is quite different from the role played by our now familiar matrix rotation operator. Further, quaternions may be very efficient for analyzing certain situations which involve rotations in R^3. After all, a quaternion is a 4-tuple while a rotation matrix has nine elements. We must admit, however, that it is not at all clear at this point how a quaternion rotation operator can be defined, and so we must spend some time and effort in exploring and understanding just how this might go. In so doing, we hope to arrive at a definition for a quaternion operator, and we will then verify that this operator can be interpreted geometrically as a rotation in R^3. We will find that it is relatively easy to specify the axis and the angle of the rotation which this operator represents. We begin by investigating these matters from the point of view of quaternion algebra which we developed in the preceding section.

5.8.1 Algebraic Considerations

Our ultimate concern at this point is with developing mathematical methods for determining the orientation of an object in 3-space, that is, in R^3. An object in R^3 may be regarded as a set of points in R^3. We may easily identify these points as vectors in R^3, so that the orientation of the object may be studied by performing appropriate operations on these vectors. If we are to accomplish this goal by means of an operator defined in terms of quaternions, it seems reasonable to begin by asking the very important question:

> *How can a quaternion, which lives in R^4, operate on a vector, which lives in R^3?*

There is an answer to this question, which may seem obvious to some, and that is:

> *A vector $\mathbf{v} \in R^3$ can simply be treated as though it were a quaternion $q \in R^4$ whose real part is zero.*

Note!

From this point on we will always (at least, implicitly) identify a vector $\mathbf{v}$ with its corresponding pure quaternion $v = 0 + \mathbf{v}$.

Such a quaternion is called a *pure quaternion*. Consider this possibility by looking at, say Q_0, the set of all pure quaternions. It is a subset of Q, the set of all quaternions. Perhaps it is possible simply to identify vectors in R^3 with the elements of this set Q_0. At the very least, perhaps we may define a *one-to-one correspondence* between these two sets, a correspondence in which a vector $\mathbf{v} \in R^3$ corresponds to the pure quaternion $v = 0 + \mathbf{v} \in Q_0$, that is

$$\mathbf{v} \in R^3 \leftrightarrow v = 0 + \mathbf{v} \in Q_0 \subset Q$$

Isomorphism

In algebraic terms, with respect to addition, this one-to-one correspondence may be called an *isomorphism*.

See Figure 5.1. It is easy to verify that with respect to addition and multiplication by a scalar, this correspondence is quite plausible. For example, the sum of any two vectors in R^3 corresponds to the sum of each of their corresponding *pure* quaternions in Q_0.

Now, at this point it may seem reasonable to suppose that a rotation operator which is defined in terms of quaternions has the same form as the familiar matrix rotation operator. If so, this would mean that a quaternion $q \in Q$ somehow represents a rotation, and that we may find the image $\mathbf{w} \in R^3$ of some vector $\mathbf{v} \in R^3$ by using the simple product rule

$$\mathbf{w} = q\mathbf{v}$$

Such a rule, of course, would mean that the product of a quaternion q with a vector $\mathbf{v}$ must not only always be defined, but the result must always be *a vector*. It is quite possible, using the one-to-one correspondence between R^3 and Q_0 described above, to define the product of a quaternion with a vector. In the quaternion product of Equation 5.1, in place of a vector $\mathbf{v}$ we simply use its corresponding quaternion $v = 0 + \mathbf{v}$. Thus, given some quaternon $q = q_0 + \mathbf{q} \in Q$ and a vector $\mathbf{v} \in R^3$ we compute

Equation 5.1
Quaternion Product

$$
\begin{aligned}
pq &= p_0 q_0 - \mathbf{p} \cdot \mathbf{q} \\
&+ p_0 \mathbf{q} \\
&+ q_0 \mathbf{p} \\
&+ \mathbf{p} \times \mathbf{q}
\end{aligned}
$$

$$
\begin{aligned}
q\mathbf{v} &= (q_0 + \mathbf{q})(0 + \mathbf{v}) \\
&= q_0 \cdot 0 - \mathbf{q} \cdot \mathbf{v} + 0 \cdot \mathbf{q} + q_0 \mathbf{v} + \mathbf{q} \times \mathbf{v} \\
&= -\mathbf{q} \cdot \mathbf{v} + q_0 \mathbf{v} + \mathbf{q} \times \mathbf{v}
\end{aligned}
$$

This computation shows that the result is *not necessarily* in Q_0. That is, in general, the result does not correspond to a vector in R^3, except in the special case that $\mathbf{q} \cdot \mathbf{v} = 0$, which means $\mathbf{q}$ and $\mathbf{v}$ are orthogonal. Thus we cannot expect our quaternion rotation operator to consist simply of a single quaternion.

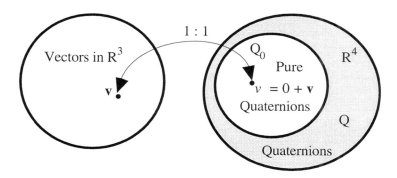

Figure 5.1: Correspondence: Vectors $\leftrightarrow$ Quaternions

Nevertheless, in spite of this difficulty with the quaternion product, in what follows we will not abandon the necessary one-to-one correspondence between vectors and pure quaternions which we have just introduced. From this point on we must understand that when we use vector notation in a quaternion product the vector is implicitly represented by its corresponding pure quaternion. In this way quaternion products involving vectors will make sense.

Since, as we have just shown, the simple product $q\mathbf{v}$ does not work, it is clear that the product $\mathbf{v}q$ will not work either, since commuting the factors does not change the real part of this product. This observation leads us to suppose that the desired operator may involve triple or perhaps even higher order products. It may then be possible to insure that the output of the operator will be a vector whenever the input is a vector. Since we wish to operate on vectors using quaternions we will let one of the factors, say p (in a triple quaternion product), be a pure quaternion, representing the vector in question.

Note!

When a vector appears in a quaternion product it must be regarded as the corresponding pure quaternion.

So we next suppose that we have two general quaternions, say q and r, from the set Q, and a third quaternion, say p, which is a pure quaternion from the set Q_0, representing some vector. There are six possible products involving these three quaternions, and for convenience we list them here

$$pqr \qquad qrp \qquad rpq$$

$$prq \qquad rqp \qquad qpr$$

Now, one of the algebraic properties of quaternions is that the set Q is closed under multiplication (although the set Q_0 is not). Thus, the products qr and rq in the first two columns in the above list are simply quaternions, and this means that these four products in this list are essentially *double* products. We have just seen that such products are not adequate for defining the operator we seek, and so we discard them as possibilities.

Equation 5.1
Quaternion Product

$$
\begin{aligned}
pq \;=\; & p_0 q_0 - \mathbf{p} \cdot \mathbf{q} \\
& + p_0 \mathbf{q} \\
& + q_0 \mathbf{p} \\
& + \mathbf{p} \times \mathbf{q}
\end{aligned}
$$

Further, although we insist that p must be a pure quaternion (thus representing a vector), we make no distinction between the quaternions q and r, so that the remaining two products in the above list need not be distinguished. Hence our last hope is the single triple product qpr.

If we let $q = q_0 + \mathbf{q}$, $p = 0 + \mathbf{p}$, and $r = r_0 + \mathbf{r}$, it is not difficult to verify (though the details are a bit tedious) that according to our quaternion product rule of Equation 5.1 the *real part* of this triple product is given by

$$-r_0(\mathbf{q} \cdot \mathbf{p}) - q_0(\mathbf{p} \cdot \mathbf{r}) - (\mathbf{q} \times \mathbf{p}) \cdot \mathbf{r}$$

Using rules of vector algebra, we may rewrite this real part in the form

$$-r_0(\mathbf{q} \cdot \mathbf{p}) - q_0(\mathbf{r} \cdot \mathbf{p}) + (\mathbf{q} \times \mathbf{r}) \cdot \mathbf{p}$$

We recall that our operator must be such that the output is a pure quaternion (that is, representing a vector) whenever the input is, and so we must require that this real part be zero. How can this be accomplished?

Well, just suppose that we had $r_0 = q_0$. This real part may then be rewritten in the form

$$-q_0(\mathbf{q} + \mathbf{r}) \cdot \mathbf{p} + (\mathbf{q} \times \mathbf{r}) \cdot \mathbf{p}$$

Clearly this real part will be zero, as required, if $\mathbf{r} = -\mathbf{q}$. But this would mean simply that

$$r = r_0 + \mathbf{r} = q_0 - \mathbf{q} = q^* \quad \Rightarrow \quad q = r^*$$

From this discussion we obtain two triple quaternion products, namely

$$qpq^* \qquad \text{and} \qquad q^*pq$$

Both of these triple products produce a pure quaternion whenever the factor p is a pure quaternion. In terms of a given input vector $\mathbf{v}$ we then have two possible *triple-product* quaternion operators, defined by

$$\mathbf{w}_1 = q\mathbf{v}q^* \qquad (5.5)$$

$$\text{and} \qquad \mathbf{w}_2 = q^*\mathbf{v}q \qquad (5.6)$$

The algebraic action of Equation 5.5 is illustrated in Figure 5.2. In summary, to this point we have found a way to handle vectors

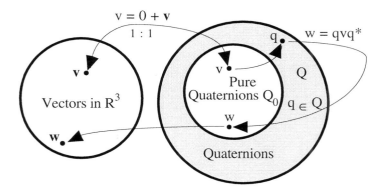

Figure 5.2: Quaternion Operations on Vectors

in quaternion space, and we have found two quaternion operators which take vectors into vectors. So, now we ask

> *What geometric interpretation can we give these operators?*

5.8.2   Geometric Considerations

The quest for geometric interpretations for the triple quaternion products in Equations 5.5 and 5.6, leads us to ask the next question:

> *Is there some way to associate an angle with a quaternion, analogous to the way we earlier associated an angle with a rotation matrix?*

If so, perhaps we can also associate an angle with the above two quaternion operators. The answer to our question is *Yes*, and in our consideration we discover an interesting and helpful fact.

From this point on, the quaternion q used to define the operator will always be a *unit* or *normalized* quaternion, that is, a quaternion

Equation 5.5

$$\mathbf{w} = q\mathbf{v}q^*$$

Equation 5.6

$$\mathbf{w} = q^*\mathbf{v}q$$

with norm 1. The reason for this will become apparent shortly.
Recall that if the quaternion

$$q = q_0 + \mathbf{q}$$

indeed has norm 1 then

$$q_0^2 + |\mathbf{q}|^2 = 1$$

However, since for any angle θ we know that

$$\cos^2 \theta + \sin^2 \theta = 1$$

there must be some angle θ such that

$$\cos^2 \theta = q_0^2$$

and

$$\sin^2 \theta = |\mathbf{q}|^2$$

This angle θ can be defined uniquely if we place the proper restriction on its domain. In general, we will ask that θ satisfy the
restriction

$$-\pi < \theta \leq \pi$$

In this way we have an angle, namely the angle θ defined above,
associated with the quaternion q. We will find it convenient to
write the unit quaternion $q = q_0 + \mathbf{q}$ in terms of this angle. Suppose
we define a unit vector $\mathbf{u}$, which represents the direction of $\mathbf{q}$ by
writing

$$\mathbf{u} = \frac{\mathbf{q}}{|\mathbf{q}|} = \frac{\mathbf{q}}{\sin \theta}$$

Then we may write the unit quaternion q in terms of the angle θ
and the unit vector $\mathbf{u}$ as

$$q = q_0 + \mathbf{q} = \cos \theta + \mathbf{u} \sin \theta \tag{5.7}$$

Note, that for a quaternion expressed in this form, substituting $-\theta$
for θ (whatever geometric meaning the angle θ might have) we get
the complex conjugate of q. That is,

$$\begin{aligned}
\cos(-\theta) + \mathbf{u} \sin(-\theta) &= \cos \theta + \mathbf{u}(-\sin \theta) \\
&= \cos \theta - \mathbf{u} \sin \theta &= q^*
\end{aligned}$$

Next, using this form for the quaternion, q, we develop an interesting
(though somewhat restricted) geometric property of the quaternion
product — one which may strengthen our impression that quaternions are somehow related to rotations in R^3.

5.9 A Special Quaternion Product

Suppose we have two unit quaternions, say p and q, with both having the *same vector*, $\mathbf{u}$. We associate an angle α with the quaternion p and an angle β with the quaternion q. Then we may write

$$p = \cos\alpha + \mathbf{u}\sin\alpha$$
$$\text{and}\quad q = \cos\beta + \mathbf{u}\sin\beta$$

The quaternion product rule in Equation 5.1 gives

$$
\begin{aligned}
r = pq &= (\cos\alpha + \mathbf{u}\sin\alpha)(\cos\beta + \mathbf{u}\sin\beta) \\
&= \cos\alpha\cos\beta - (\mathbf{u}\sin\alpha)\cdot(\mathbf{u}\sin\beta) \\
&\quad + \cos\alpha(\mathbf{u}\sin\beta) + \cos\beta(\mathbf{u}\sin\alpha) \\
&\quad + \mathbf{u}\sin\alpha \times \mathbf{u}\sin\beta \\
&= \cos\alpha\cos\beta - \sin\alpha\sin\beta \\
&\quad + \mathbf{u}(\sin\alpha\cos\beta + \cos\alpha\sin\beta) \\
&= \cos(\alpha + \beta) + \mathbf{u}\sin(\alpha + \beta) \\
&= \cos\gamma + \mathbf{u}\sin\gamma = r
\end{aligned}
$$

This is an interesting result. It says, if we multiply the quaternions, p and q, each *having the same vector*, $\mathbf{u}$, then the product is a quaternion

$$r = \cos\gamma + \mathbf{u}\sin\gamma$$

also having this same vector, $\mathbf{u}$. And, associated with the product of these two quaternions is the angle $\gamma = \alpha + \beta$, which is the sum of the angles associated with each of the factors. Now if, in fact, quaternions somehow do represent rotations, this is exactly what we would expect. Moreover, the reappearance of the vector $\mathbf{u}$ in this product quaternion, r, suggests that it also may somehow be involved in the operator action on the vector $\mathbf{v}$.

We now take a closer look at the two quaternion operators defined by Equations 5.5 and 5.6. First, it is important to note that the norm of the products of quaternions which appear in these operators is equal to the product of the individual norms. This implies that the quaternion operator of Equation 5.5, for example, does not change the length of the vector to which it is applied. For, if

$$\mathbf{w} = q\mathbf{v}q^*$$

we have, since q is a unit quaternion

$$N(\mathbf{w}) = N(q\mathbf{v}q^*)$$

Equation 5.1
Quaternion Product

$$
\begin{aligned}
pq = {}& p_0 q_0 - \mathbf{p}\cdot\mathbf{q} \\
& + p_0\mathbf{q} \\
& + q_0\mathbf{p} \\
& + \mathbf{p}\times\mathbf{q}
\end{aligned}
$$

Equation 5.5
$$\mathbf{w} = q\mathbf{v}q^*$$

Equation 5.6
$$\mathbf{w} = q^*\mathbf{v}q$$

$$\begin{aligned} &= \ N(q)N(\mathbf{v})N(q^*) \\ &= \ 1 \cdot N(\mathbf{v}) \cdot 1 \\ &= \ N(\mathbf{v}) \end{aligned} \tag{5.8}$$

It should go without saying that this same result holds, of course, for the operator $\mathbf{w} = q^*\mathbf{v}q$.

This result concerning the norm of a vector under the action of the quaternion operator is consistent with the possibility that we are dealing with a rotation operator — the length of a vector in R^3 does not change under a rotation.

Next we observe that, if in the quaternion

$$q = \cos\theta + \mathbf{u}\sin\theta$$

we replace the angle θ with the angle $-\theta$, we obtain the quaternion

$$q^* = \cos\theta - \mathbf{u}\sin\theta$$

This in turn means that, if in one of the operators of Equation 5.5 and 5.6 we replace θ by $-\theta$, we get the other operator. Hence, by the appropriate choice of the angle θ these operators may in fact represent the same geometric transformation.

At this point we have considerable evidence that the quaternion operators of Equations 5.5 and 5.6 are in some way related to rotations in R^3. We still need to investigate in more detail, however, just what geometric effect these operators have when applied to an arbitrary vector in R^3.

Equation 5.5

$$\mathbf{w} = q\mathbf{v}q^*$$

Equation 5.6

$$\mathbf{w} = q^*\mathbf{v}q$$

5.10   Incremental Test Quaternion

In order to gain more understanding of what this geometric effect is, we look at an incremental test case. First, suppose that the quaternion q in the operator of Equation 5.5 has vector

$$\mathbf{u} \ = \ 0\mathbf{i} + 0\mathbf{j} + 1\mathbf{k} \ = \ \mathbf{k}$$

where $\{\mathbf{i}, \mathbf{j}, \mathbf{k}\}$ are the standard basis vectors in R^3. And suppose further that the angle associated with this quaternion q is a *very small* positive angle, say θ. We choose a very small angle merely to gain some insight into how this incremental quaternion operator "tweaks" the input vector.

For any arbitrary angle θ, the quaternion q may, of course, be written in the form

$$q = \cos\theta + \mathbf{k}\sin\theta$$

However, it is well known that for very small values of the angle θ we may write

$$\cos\theta \approx 1$$
$$\text{and} \quad \sin\theta \approx \theta$$

and then the quaternion q may be written as

$$q \approx 1 + \mathbf{k}\theta$$

Obviously, if $\theta = 0$ then the quaternion q is the identity. We now use this incremental quaternion in the operator qvq^* in order to determine its action on, say, the basis vector $\mathbf{i}$ in R^3. Actually, of course, as we have agreed earlier, we apply the operator to the pure quaternion which corresponds to this vector, $\mathbf{i}$. In the following computation we ignore the θ^2 terms; for our purposes these terms are negligible since θ itself is taken to be very small. We then have

$$\begin{aligned} \mathbf{w} &= q\mathbf{i}q^* \\ &= (1 + \mathbf{k}\theta)(0 + \mathbf{i})(1 - \mathbf{k}\theta) \\ &= (1 + \mathbf{k}\theta)(0 + \mathbf{i} + \mathbf{j}\theta) \\ &= 0 + \mathbf{i} + 2\theta\mathbf{j} \end{aligned}$$

This result is a pure quaternion, as we knew had to be the case. We interpret this result to mean that the input vector $\mathbf{i}$ has been "tweaked" by the quaternion operator $q\mathbf{i}q^*$ to produce the output vector

$$\mathbf{w} = \mathbf{i} + 2\theta\mathbf{j}$$

Verify!

It is worth verifying that the details of this computation are correct, since several steps have been omitted here.

Notice that the *length* of this vector, given that we ignore the θ^2 term, is 1. Now, we let the angle between the input and output vectors be α. Geometrically, it should be clear, from the components of $\mathbf{w}$, that $\tan\alpha = 2\theta$. Since, for small angles $\tan\alpha \approx \alpha$ we can write $\alpha = 2\theta$. Clearly, $\mathbf{w}$ lies in the second quadrant of the $\mathbf{XY}$-plane in R^3. From this we may rightly conclude that the vector $\mathbf{i}$ seems to have been rotated *counter-clockwise* through an angle 2θ, about the vector $\mathbf{k}$ as an axis. Alternatively, we might also conclude, however, that the *frame* was rotated *clockwise* through an angle -2θ.

It is interesting to note and not difficult to check that if we apply the quaternion operator of Equation 5.6 to the vector $\mathbf{i}$, using the same quaternion q, we obtain

$$\mathbf{w} = \mathbf{i} - 2\theta\mathbf{j}$$

Comparing this with the previous result leads us to think that the difference between the two operators of Equations 5.5 and 5.6 is simply the direction of rotation. We must be somewhat careful here, however. Earlier we mentioned that we would restrict the angle θ to the domain $-\pi < \theta \leq \pi$. This restriction allows the angle θ to be negative as well as positive, which in turn means it is difficult to distinguish between these operators.

Equation 5.5

$$\mathbf{w} = q\mathbf{v}q^*$$

Equation 5.6

$$\mathbf{w} = q^*\mathbf{v}q$$

An argument for the very close relationship between the two quaternion operators goes like this. If in these equations we let

$$q = p^* \qquad \text{then, of course,} \qquad q^* = p$$

$$\text{and we may write,} \qquad \mathbf{w}_1 = q\mathbf{v}q^* = p^*\mathbf{v}p$$

$$\text{and} \qquad \mathbf{w}_2 = q^*\mathbf{v}q = p\mathbf{v}p^*$$

This says that the action these two operators have on the vector $\mathbf{v}$ is identical if, in one of the operator equations, we use the complex conjugate of q, that is, q^* instead of q. It is quite obvious that this should be the case because, if in one of these operators we use q^* instead of q we get the *other* operator equation.

Observer Perspective

From this point on, we shall not make very strong distinctions between rotations of vectors and rotations of the coordinate frame. In general, the choice depends upon the perspective of the observer — the seat of the observer is either on the frame or on the vector.

In spite of these quaternion operators being very similar in nature, there are good reasons for making distinctions between them, as we shall see. In fact, which of these two operators is appropriate, in a given application, usually depends upon the perspective adopted by the observer, as he or she formulates a mathematical model appropriate to the application. We emphasize that we do not yet know the precise geometric meaning of the angle, θ, and the vector, $\mathbf{u}$, in the expression

$$q = \cos\theta + \mathbf{u}\sin\theta$$

Nor, for that matter, do we totally understand the action these quaternion operators and their parameters produce on the vector, $\mathbf{v}$. We will, however, investigate these matters in considerable depth in what follows. The next example will lead the way.

5.11 Quaternion with Angle $\theta = \pi/6$

In this example we will associate the angle $\pi/6$ with the quaternion q, while retaining the basis vector $\mathbf{k}$ as the vector of q. Then we write the quaternion q in the form

$$q = \cos\theta + \mathbf{k}\sin\theta = \sqrt{3}/2 + 1/2\ \mathbf{k}$$

We apply the quaternion operator of Equation 5.5 to the basis vector $\mathbf{v} = 1\mathbf{i} + 0\mathbf{j} + 0\mathbf{k}$. Again, it is not difficult to verify that the following computations are correct. We have

$$
\begin{aligned}
w &= q\mathbf{v}q^* \\
&= (\sqrt{3}/2 + 1/2\ \mathbf{k})(0 + \mathbf{i})(\sqrt{3}/2 - 1/2\ \mathbf{k}) \\
&= (\sqrt{3}/2\ \mathbf{i} + 1/2\ \mathbf{j})(\sqrt{3}/2 - 1/2\ \mathbf{k}) \\
&= \frac{1}{2}\mathbf{i} + \frac{\sqrt{3}}{2}\mathbf{j}
\end{aligned}
$$

The result is again a pure quaternion, as should be the case. The angle associated with this quaternion is $\pi/3$ because $\cos(\pi/3) = 1/2$ and $\sin(\pi/3) = \sqrt{3}/2$. The vector in R^3 which corresponds to this pure quaternion is

Equation 5.5
$$w = q\mathbf{v}q^*$$

$$\mathbf{w} = \frac{1}{2}\mathbf{i} + \frac{\sqrt{3}}{2}\mathbf{j}$$

Notice again that $\mathbf{w}$ is a unit vector, that the angle between $\mathbf{w}$ and $\mathbf{i}$ is $\pi/3$ (which is *twice* $\pi/6$), and that the vector $\mathbf{w}$ lies in the *second* quadrant. We ask

> *Geometrically, how are we to view the vector-frame action of the quaternion operator, $q\mathbf{v}q^*$?*

In answering this question, an important consideration is how one views the resultant final relationship between the input vector $\mathbf{v}$, the output vector $\mathbf{w}$ and the coordinate frame with standard orthonormal basis $\{\mathbf{i}, \mathbf{j}, \mathbf{k}\}$. The answer is conditional; one may adopt either of the following two distinctly different perspectives.

First Perspective

The first perspective is that of an observer seated on and fixed with respect to the coordinate frame $\{\mathbf{i}, \mathbf{j}, \mathbf{k}\}$. To her it appears that the quaternion operator, $q\mathbf{v}q^*$, has rotated the vector $\mathbf{v}$, about $\mathbf{k}$ as an axis, through an angle, $+\pi/3$, that is, in the *counter-clockwise* direction. From this perspective it is convenient to think of the coordinate frame as being *fixed*, while the vector is rotated. This is sometimes called a *point* rotation.

Second Perspective

The second perspective is that of an observer seated on and fixed with respect to the vector $\mathbf{v}$. To him it appears that this same quaternion operator, $q\mathbf{v}q^*$, has rotated the coordinate frame $\{\mathbf{i}, \mathbf{j}, \mathbf{k}\}$, about $\mathbf{k}$ as an axis, through an angle, $-\pi/3$, that is, in the *clockwise* direction. From this second perspective we think of the vector $\mathbf{v}$ as being *fixed*, while the coordinate frame is rotated. This is sometimes called a *frame* rotation.

Choice of Perspective

Which of these two perspectives is to be preferred in a given application is usually a subjective matter — a decision made by the practitioner. In many of the applications which we consider in later chapters, we will use the operator $q^*\mathbf{v}q$, and interpret it geometrically to be a *frame rotation*. Incidentally, note that the *signs* of the angles in these two perspectives are reversed if the operator is $q^*\mathbf{v}q$.

A Remark

If the above computations are carried out using a general angle θ while still using the vector $\mathbf{k}$ as axis and applying the operator to the vector $\mathbf{i}$, then we get the expected result, which is

$$\mathbf{w} = \mathbf{i}\cos 2\theta + \mathbf{j}\sin 2\theta$$

This result confirms once again that the operator $q\mathbf{v}q^*$ seems indeed to be a rotation operator in R^3, where the angle of rotation is twice the angle associated with the quaternion used to define the rotation operator.

Equation 5.5

$$\mathbf{w} = q\mathbf{v}q^*$$

5.12   Operator Algorithm

It will be useful, for future reference, to have a general formula for the output vector when the quaternion operator of Equation 5.5 is applied to an input vector $\mathbf{v} \in R^3$. If we have the quaternion $q = q_0 + \mathbf{q}$ and the vector $\mathbf{v}$, corresponding to the pure quaternion $v = 0 + \mathbf{v}$, it is not difficult to verify that the following result is correct.

Triple Vector Product

In this verification, it is helpful to remember that for vectors $\mathbf{a}$, $\mathbf{b}$, and $\mathbf{c}$ in R^3, for the triple vector product we have the formulas

$$\mathbf{a} \times (\mathbf{b} \times \mathbf{c}) = (\mathbf{a} \cdot \mathbf{c})\mathbf{b} - (\mathbf{a} \cdot \mathbf{b})\mathbf{c}$$
$$(\mathbf{a} \times \mathbf{b}) \times \mathbf{c} = (\mathbf{a} \cdot \mathbf{c})\mathbf{b} - (\mathbf{b} \cdot \mathbf{c})\mathbf{a}$$

$$\begin{aligned} \mathbf{w} = q\mathbf{v}q^* &= (q_0 + \mathbf{q})(0 + \mathbf{v})(q_0 - \mathbf{q}) \\ &= (2q_0^2 - 1)\mathbf{v} \\ &\quad + 2(\mathbf{q} \cdot \mathbf{v})\mathbf{q} \end{aligned}$$

$$
\text{or} \quad \mathbf{w} \; = \; q\mathbf{v}q^* \; = \; \begin{aligned} &+ 2q_0(\mathbf{q} \times \mathbf{v}) \\ &(q_0^2 - |\mathbf{q}|^2)\mathbf{v} \\ &+ 2(\mathbf{q} \cdot \mathbf{v})\mathbf{q} \\ &+ 2q_0(\mathbf{q} \times \mathbf{v}) \end{aligned} \qquad (5.9)
$$

5.13 Operator action on v = kq

Equation 5.5

$$\mathbf{w} \; = \; q\mathbf{v}q^*$$

Intuitively, it is clear that any vector which lies *on the axis of rotation* must be *invariant*. In this example we verify that the quaternion operator of Equation 5.5 when applied to a vector $\mathbf{v} = k\mathbf{q}$, that is, having the same direction as the vector component of q, leaves that vector unchanged. This is not difficult to verify, for Equation 5.9 says

$$
\begin{aligned}
\mathbf{w} \; &= \; q\mathbf{v}q^* \\
&= \; q(k\mathbf{q})q^* \\
&= \; (2q_0^2 - 1)(k\mathbf{q}) \\
&\quad + 2(\mathbf{q} \cdot k\mathbf{q})\mathbf{q} + 2q_0(\mathbf{q} \times k\mathbf{q}) \\
&= \; kq_0^2\mathbf{q} - k|\mathbf{q}|^2\mathbf{q} + 2k|\mathbf{q}|^2\mathbf{q} \\
&= \; k(q_0^2 + |\mathbf{q}|^2)\mathbf{q} \\
&= \; k\mathbf{q}
\end{aligned} \qquad (5.10)
$$

In this example we learn that the vector component of q, used in the operator $q\mathbf{v}q^*$, seems to define the axis of rotation.

5.14 Quaternions to Matrices

The three terms in Equation 5.9 are now expanded to produce an algorithm which is often more convenient in applications.

$$
(2q_0^2 - 1)\mathbf{v} \; = \; \begin{bmatrix} (2q_0^2 - 1) & 0 & 0 \\ 0 & (2q_0^2 - 1) & 0 \\ 0 & 0 & (2q_0^2 - 1) \end{bmatrix} \begin{bmatrix} v_1 \\ v_2 \\ v_3 \end{bmatrix}
$$

$$
2(\mathbf{v} \cdot \mathbf{q})\mathbf{q} \; = \; \begin{bmatrix} 2q_1^2 & 2q_1q_2 & 2q_1q_3 \\ 2q_1q_2 & 2q_2^2 & 2q_2q_3 \\ 2q_1q_3 & 2q_2q_3 & 2q_3^2 \end{bmatrix} \begin{bmatrix} v_1 \\ v_2 \\ v_3 \end{bmatrix}
$$

$$
2q_0(\mathbf{q} \times \mathbf{v}) \; = \; \begin{bmatrix} 0 & -2q_0q_3 & 2q_0q_2 \\ 2q_0q_3 & 0 & -2q_0q_1 \\ -2q_0q_2 & 2q_0q_1 & 0 \end{bmatrix} \begin{bmatrix} v_1 \\ v_2 \\ v_3 \end{bmatrix}
$$

w x y z

$q = q_0 + iq_1 + jq_2 + kq_3$

The sum of these three components may be written

$$
\begin{bmatrix} w_1 \\ w_2 \\ w_3 \end{bmatrix} = \begin{bmatrix} 2q_0^2 - 1 + 2q_1^2 & 2q_1q_2 - 2q_0q_3 & 2q_1q_3 + 2q_0q_2 \\ 2q_1q_2 + 2q_0q_3 & 2q_0^2 - 1 + 2q_2^2 & 2q_2q_3 - 2q_0q_1 \\ 2q_1q_3 - 2q_0q_2 & 2q_2q_3 + 2q_0q_1 & 2q_0^2 - 1 + 2q_3^2 \end{bmatrix} \begin{bmatrix} v_1 \\ v_2 \\ v_3 \end{bmatrix}
$$

$$
\mathbf{w} = q\mathbf{v}q^* = Q\mathbf{v} \tag{5.11}
$$

In passing, we remark that a similar result is obtained for the quaternion operator of Equation 5.6

$$
\begin{aligned}
\mathbf{w} = q^*\mathbf{v}q &= (2q_0^2 - 1)\mathbf{v} \\
&+ 2(\mathbf{v} \cdot \mathbf{q})\mathbf{q} \\
&+ 2q_0(\mathbf{v} \times \mathbf{q})
\end{aligned}
$$

$$
\mathbf{w} = q^*\mathbf{v}q = Q^t\mathbf{v} \tag{5.12}
$$

Note that Equations 5.11 and 5.12 represent the two quaternion operator Equations 5.5 and 5.6 expressed in terms of the matrices, Q and Q^t, respectively. That is, in summary we have

$$
\begin{aligned}
q\mathbf{v}q^* &= Q\mathbf{v} \\
q^*\mathbf{v}q &= Q^t\mathbf{v}
\end{aligned}
$$

where $q = q_0 + \mathbf{i}q_1 + \mathbf{j}q_2 + \mathbf{k}q_3$

and
$$
Q = \begin{bmatrix} 2q_0^2 - 1 + 2q_1^2 & 2q_1q_2 - 2q_0q_3 & 2q_1q_3 + 2q_0q_2 \\ 2q_1q_2 + 2q_0q_3 & 2q_0^2 - 1 + 2q_2^2 & 2q_2q_3 - 2q_0q_1 \\ 2q_1q_3 - 2q_0q_2 & 2q_2q_3 + 2q_0q_1 & 2q_0^2 - 1 + 2q_3^2 \end{bmatrix}
$$

These algorithms offer further support that the two quaternion operators are indeed rotation operators. Their matrix representations, that is, a matrix and its transpose is exactly what we would expect.

We emphasize at this point, however, that our geometric interpretation of the preceding results are conjectural and we must, of course, prove that this interpretation is correct. This we do in the next section.

5.15 Quaternion Rotation Operator

In the preceding pages we have given considerable motivation for thinking that the triple quaternion products of Equations 5.5 and 5.6 may be interpreted as quaternion rotation operators. In order to apply these operators to a vector in R^3 we indicated that it is necessary to identify the vector $\mathbf{v}$ with the pure quaternion $v = 0 + \mathbf{v}$. With this understanding, we now use Equation 5.5 to *define* the quaternion rotation operator, L_q (associated with the quaternion q and applied to a vector $\mathbf{v} \in R^3$) by the equation

$$\mathbf{w} = L_q(\mathbf{v}) = q\mathbf{v}q^* \tag{5.13}$$

Equation 5.9 then gives us the convenient computational formula

$$\begin{aligned} L_q(\mathbf{v}) &= (q_0^2 - |\mathbf{q}|^2)\mathbf{v} \\ &\quad + 2(\mathbf{q} \cdot \mathbf{v})\mathbf{q} \\ &\quad + 2q_0(\mathbf{q} \times \mathbf{v}) \end{aligned} \tag{5.14}$$

In this section we take note of two algebraic properties of the operator L_q, defined by Equation 5.13. We also prove that this operator does in fact represent a rotation in R^3, that the axis of rotation is given by the vector part of q, and that the angle of rotation is twice the angle associated with the quaternion q.

5.15.1 $L_q(\mathbf{v}) = q\mathbf{v}q^*$ is a Linear Operator

The first algebraic property of our quaternion operator is that it is *linear*. This means that for any two vectors $\mathbf{a}$ and $\mathbf{b}$ in R^3 and for any scalar (real number) k we have

$$L_q(k\mathbf{a} + \mathbf{b}) = kL_q(\mathbf{a}) + L_q(\mathbf{b}) \tag{5.15}$$

For, using the distributive property on the quaternion product which defines this operator, we may write

$$\begin{aligned} L_q(k\mathbf{a} + \mathbf{b}) &= q(k\mathbf{a} + \mathbf{b})q^* \\ &= (kq\mathbf{a} + q\mathbf{b})q^* \\ &= kq\mathbf{a}q^* + q\mathbf{b}q^* \\ &= kL_q(\mathbf{a}) + L_q(\mathbf{b}) \qquad \textbf{QED} \end{aligned}$$

Equation 5.5

$$\mathbf{w} = q\mathbf{v}q^*$$

Equation 5.6

$$\mathbf{w} = q^*\mathbf{v}q$$

Alternative Operator

We could just as well have used Equation 5.6 to define the operator L_{q^*} by writing

$$\begin{aligned} \mathbf{w} &= L_{q^*}(\mathbf{v}) \\ &= q^*\mathbf{v}q \\ &= (q_0 - \mathbf{q})(0 + \mathbf{v})(q_0 + \mathbf{q}) \\ &= (2q_0^2 - 1)\mathbf{v} \\ &\quad + 2(\mathbf{v} \cdot \mathbf{q})\mathbf{q} \\ &\quad + 2q_0(\mathbf{v} \times \mathbf{q}) \end{aligned}$$

For now, however, we use only the operator of Equation 5.13.

5.15.2   Operator Norm

The second algebraic property is that the norm or length of a vector is invariant under the quaternion operator of Equation 5.13, that is $|L_q(\mathbf{v})| = |\mathbf{v}|$, as we have already shown in Equation 5.8. This property will be necessary, of course, if the operator is to describe a rotation.

5.15.3   Prove: Operator is a Rotation

We are finally ready to apply the foregoing results to construct a proof of the theorem (stated later) that the operator defined by Equation 5.13 indeed is the quaternion rotation operator we want.

STRATEGY for the Proof

Our argument will proceed as follows:

Equation 5.13
Quaternion Operator

$$\mathbf{w} = L_q(\mathbf{v}) = q\mathbf{v}q^*$$

We begin with a quaternion q whose vector part is $\mathbf{q}$. Then, given a vector $\mathbf{v} \in R^3$, we resolve $\mathbf{v}$ into two orthogonal components: the component $\mathbf{a}$ *along* the vector $\mathbf{q}$ and the component $\mathbf{n}$ *normal* to $\mathbf{q}$. Then we will show that under the quaternion rotation operator $q\mathbf{v}q^*$ of Equation 5.13, the first component $\mathbf{a}$ is invariant, while the second component $\mathbf{n}$ is rotated about $\mathbf{q}$ through an angle 2θ, where θ is the angle associated with the quaternion q. Since the operator is linear, and since the vector $\mathbf{v}$ is the sum of these two components, this shows that the operator $q\mathbf{v}q^*$ indeed may be interpreted as a rotation in R^3 through an angle 2θ about $\mathbf{q}$ as its axis.

Unit Quaternion

Remember, whenever a quaternion is written as a unit of the form

$$q = \cos\theta + \mathbf{u}\sin\theta$$

then the vector $\mathbf{u}$ is a unit vector, that is, $|\mathbf{u}| = 1$.

First, we recall that for any unit or normalized quaternion q, that is, $|q| = 1$, we may write

$$q = q_0 + \mathbf{q} = \cos\theta + \mathbf{u}\sin\theta$$

for some angle θ. Here $\mathbf{u} = \mathbf{q}/|\mathbf{q}|$ is the unit vector.

RESOLVE the Components of the Input Vector

We now write the vector $\mathbf{v}$ in the form

$$\mathbf{v} = \mathbf{a} + \mathbf{n}$$

where, as defined above, $\mathbf{a}$ is the component of $\mathbf{v}$ *along* the vector part of the quaternion q, and $\mathbf{n}$ is the component of $\mathbf{v}$ *normal* to the vector part of q.

INVARIANT Component

Since the vector $\mathbf{a}$ lies *along* the vector $\mathbf{q}$, $\mathbf{a}$ is simply some scalar multiple of $\mathbf{q}$, that is

$$\mathbf{a} = k\mathbf{q}$$

for some scalar k. If we invoke the results of Section 5.13 then we may write

$$L_q(\mathbf{a}) \; = \; L_q(k\mathbf{q}) \; = \; k\mathbf{q} \; = \; \mathbf{a}$$

as required.

ROTATED Component

Our proof will be complete if we now show that the operator L_q rotates the component $\mathbf{n}$ through an angle 2θ about $\mathbf{q}$ as the axis. Using Equation 5.14, with $\mathbf{v}$ replaced by $\mathbf{n}$, we may write

Equation 5.14

$$\begin{aligned} L_q(\mathbf{n}) \; &= \; (q_0^2 - |\mathbf{q}|^2)\mathbf{n} \\ &\quad + 2(\mathbf{q} \cdot \mathbf{n})\mathbf{q} + 2q_0(\mathbf{q} \times \mathbf{n}) \\ &= \; (q_0^2 - |\mathbf{q}|^2)\mathbf{n} + 2q_0(\mathbf{q} \times \mathbf{n}) \\ &= \; (q_0^2 - |\mathbf{q}|^2)\mathbf{n} + 2q_0|\mathbf{q}|(\mathbf{u} \times \mathbf{n}) \end{aligned}$$

$$\begin{aligned} L_q(\mathbf{v}) \; &= \; (q_0^2 - |\mathbf{q}|^2)\mathbf{v} \\ &\quad + 2(\mathbf{q} \cdot \mathbf{v})\mathbf{q} \\ &\quad + 2q_0(\mathbf{q} \times \mathbf{v}) \end{aligned}$$

Here we used the fact that $\mathbf{u} = \mathbf{q}/|\mathbf{q}|$. If we let $\mathbf{u} \times \mathbf{n} = \mathbf{n}_\perp$ we may write this last equation as

$$L_q(\mathbf{n}) \; = \; (q_0^2 - |\mathbf{q}|^2)\mathbf{n} + 2q_0|\mathbf{q}|\mathbf{n}_\perp \qquad (5.16)$$

ROTATION Verified

We now show that the vectors $\mathbf{n}$ and $\mathbf{n}_\perp$ have exactly the same length. First, we observe that the angle between $\mathbf{n}$ and $\mathbf{n}_\perp$ is $\pi/2$, and since $\sin \pi/2 = 1$ we can write

$$|\mathbf{n}_\perp| \; = \; |\mathbf{n} \times \mathbf{u}| \; = \; |\mathbf{n}||\mathbf{u}|\sin \pi/2 \; = \; |\mathbf{n}| \qquad (5.17)$$

Finally, using the trigonometric form for the quaternion q we may write Equation 5.16 in the form

$$\begin{aligned} L_q(\mathbf{n}) \; &= \; (\cos^2 \theta - \sin^2 \theta)\mathbf{n} \\ &\quad + (2\cos \theta \sin \theta)\mathbf{n}_\perp \\ &= \; \cos 2\theta \mathbf{n} + \sin 2\theta \mathbf{n}_\perp \end{aligned}$$

The components of $L_q(\mathbf{n})$ are illustrated in Figure 5.3.

To this point we have shown that

$$
\begin{aligned}
\mathbf{w} &= q\mathbf{v}q^* = L_q(\mathbf{v}) = L_q(\mathbf{a}+\mathbf{n}) \\
&= L_q(\mathbf{a}) + L_q(\mathbf{n}) \\
&= \mathbf{a} + \mathbf{m}
\end{aligned}
$$

$$
\text{where} \quad \mathbf{m} = L_q(\mathbf{n}) = \cos 2\theta \mathbf{n} + \sin 2\theta \mathbf{n}_\perp
$$

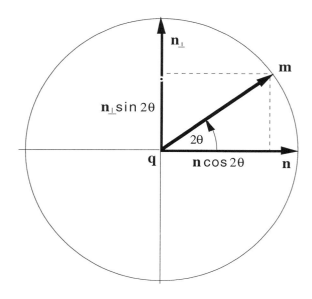

Figure 5.3: Rotated Vector Components

We verify

$$
|\mathbf{m}| = |\mathbf{n}| = |\mathbf{n}_\perp|
$$

Proof:

$$
\begin{aligned}
N^2(\mathbf{m}) &= \mathbf{m}\cdot\mathbf{m} = |\mathbf{m}|^2 \\
&= (\cos 2\theta\mathbf{n} + \sin 2\theta\mathbf{n}_\perp) \cdot \\
&\quad (\cos 2\theta\mathbf{n} + \sin 2\theta\mathbf{n}_\perp) \\
&= \cos^2 2\theta\mathbf{n}\cdot\mathbf{n} + \\
&\quad \sin^2 2\theta\mathbf{n}_\perp\cdot\mathbf{n}_\perp + \\
&\quad \cos 2\theta\sin 2\theta\mathbf{n}_\perp\cdot\mathbf{n} + \\
&\quad \sin 2\theta\cos 2\theta\mathbf{n}\cdot\mathbf{n}_\perp \\
&= (\cos^2 2\theta + \sin^2 2\theta)|\mathbf{n}|^2 \\
&= |\mathbf{n}|^2 = |\mathbf{n}_\perp|^2
\end{aligned}
$$

Therefore,

$$
|\mathbf{m}| = |\mathbf{n}| = |\mathbf{n}_\perp|
$$

SUMMARY of Proof that $q\mathbf{v}q^*$ is a Rotation

Using Equation 5.17 we get $|\mathbf{m}| = |\mathbf{n}| = |\mathbf{n}_\perp|$. Then it is clear that $\mathbf{m}$ is a rotation of $\mathbf{n}$ through 2θ, Since $\mathbf{w} = \mathbf{a} + \mathbf{m}$, it is likewise clear, $\mathbf{w} = q\mathbf{v}q^*$ may be viewed as the vector $\mathbf{v}$ rotated through 2θ about $\mathbf{q}$ as an axis.

Thus we have completed the proof of the following theorem.

Theorem 5.1:   *For any unit quaternion*

$$
q = q_0 + \mathbf{q} = \cos\theta + \mathbf{u}\sin\theta
$$

and for any vector $\mathbf{v} \in R^3$ *the action of the operator*

$$
L_q(\mathbf{v}) = q\mathbf{v}q^*
$$

on **v** *may be interpreted geometrically as a rotation of
the vector* **v** *through an angle 2θ about* **q** *as the axis of
rotation.*

Note, in Figure 5.4, that the vector **v** and and its image **w** may
be viewed as generators of the right-circular cone whose axis is the
quaternion vector **q** and whose circular base in this instance contains
the vectors, **n** and **m**. Thus, the vector **v** and its image **w** are related
by the rotation described in the above theorem.

 as shown in Figures 5.3 and 5.4.

Rotation Sense

A rotation about a directed axis
is defined as a *positive* rotation if
it has *right-handed sense*, that is,
with the thumb on the right hand
extended along the directed axis,
the fingers wrap in the direction
which defines a positive rotation.

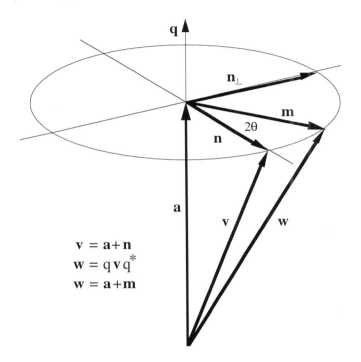

$$\mathbf{v} = \mathbf{a+n}$$
$$\mathbf{w} = q\,\mathbf{v}\,q^{*}$$
$$\mathbf{w} = \mathbf{a+m}$$

Figure 5.4: Rotation Operator Geometry

The various vectors in Figure 5.4 implicitly suggest the existence
of some reference coordinate frame. From *afar*, God sees the *frame*
within which all of this is happening. But we *local creatures*, are a
much too loosely coupled part of it all. As observer: we must more
closely relate to the *fixed axis* of rotation, the generators **v** and **w**,
the *fixed point*, and/or the *frame*.

So, the observer somehow finds a seat *fixed* in this implicit frame.
What she sees happening, under the action of the rotation operator
$q\mathbf{v}q^{*}$, is that the vector **v** goes to **w**. That is, **v** is *rotated positively*
through an angle 2θ about an axis whose *direction* is defined by

q. This is exactly the situation described earlier in Section 5.11 on page 123, and called **First Perspective**.

On the other hand, if I go find myself a seat on the vector **v**, then, for this same operator $q\mathbf{v}q^*$, what I see happening is that the coordinate frame rotates in a *negative direction* through the angle 2θ about this same axis which is defined by the quaternion vector **q**. This is what we called **Second Perspective**, also described earlier in Section 5.11 on page 124.

Equation 5.5

$$\mathbf{w} = q\mathbf{v}q^*$$

Up to this point we have used the operator of Equation 5.5. We could just as well have used the quaternion operator

$$L_{q*}(\mathbf{v}) \quad = \quad q^*\mathbf{v}q$$

Equation 5.6

$$\mathbf{w} = q^*\mathbf{v}q$$

of Equation 5.6. Keeping in mind our remarks on the observer's perspective, made above, we then would have proved the following theorem:

> **Theorem 5.2:**   *For any unit quaternion*
>
> $$q \quad = \quad q_0 + \mathbf{q} \quad = \quad \cos\theta + \mathbf{u}\sin\theta$$
>
> *and for any vector* $\mathbf{v} \in R^3$ *the action of the operator*
>
> $$L_{q*}(\mathbf{v}) \quad = \quad q^*\mathbf{v}q$$
>
> *may be interpreted geometrically as*
>
> - *a rotation of the coordinate frame with respect to the vector* **v** *through an angle* 2θ *about* **q** *as the axis, or,*
> - *an opposite rotation of the vector* **v** *with respect to the coordinate frame through an angle* 2θ *about* **q** *as the axis.*

It is important to note just how closely related these two theorems are: If in Theorem 5.1 we change the sign of the angle and therefore the direction of rotation, then Theorem 5.1 becomes Theorem 5.2, and visa versa. This tends to confirm that if the angle may be chosen from $-\pi < \theta \le \pi$ then we could get by with only one Theorem — and carefully choose the proper perspective.

On the other hand, assuming *positive* angles, it may be not only convenient but it may also be that there is precedence and therefore it would be more conventional that we properly relate these operators in the following manner:

$$q\mathbf{v}q^* \quad \rightarrow \quad \text{vector rotation}$$
$$q^*\mathbf{v}q \quad \rightarrow \quad \text{frame rotation}$$

in much the same way as we do with matrices, when we write

$$M \quad \rightarrow \quad \text{frame rotation}$$
$$\text{then} \quad M^t \quad \rightarrow \quad \text{vector rotation}$$

So we let the Theorems stand.

In summary, Theorem 5.1 states that the quaternion operator

$$L_q(\mathbf{v}) = q\mathbf{v}q^*$$

may be interpreted as a *point or vector rotation*, with respect to (or, relative to) the (fixed) coordinate frame, and Theorem 5.2 states that the quaternion operator

$$L_{q^*}(\mathbf{v}) = q^*\mathbf{v}q$$

may be interpreted as a *coordinate frame rotation*, with respect to (or, relative to) the (fixed) space of points or vectors. What is fixed may depend upon the practitioner's perspective.

Even though we stated earlier that with the appropriate choice of the angle θ (specifically whether θ is defined to be positive or defined to be negative) the two operators of Equations 5.5 and 5.6 may in fact be equivalent. However, it will still be important in many applications to maintain the distinction between the results of these two theorems. Hence from this point on we shall uniformly interpret the quaternion rotation operator of Equation 5.5 geometrically as a point rotation with respect to the frame, while we interpret the quaternion rotation operator of Equation 5.6 as a coordinate frame rotation with respect to the point or vector space. The importance of this distinction is found only in the need to decide which of these two geometric interpretations is appropriate in a given application.

We again emphasize that the choice of which of these two quaternion operators to use in a given application is actually arbitrary, and so the choice should be made on the basis of what seems natural and convenient. We can hardly emphasize too much the fact that

Equation 5.5

$$\mathbf{w} = q\mathbf{v}q^*$$

Equation 5.6

$$\mathbf{w} = q^*\mathbf{v}q$$

the important consideration is the *relative relationship* between the vectors and the coordinate frame. It is most important to realize that in a specific application we begin with an initial relative relationship between the vectors and the coordinate frame, and that these quaternion rotation operators then simply modify this relative relationship. Which rotations one uses to get to the desired final relationship, that is, whether one uses point rotations (frame fixed) or frame rotations (points fixed), is not the significant matter; these two perspectives are identical except for the *sign* on the rotation angle. Having said this, however, we must also say that in a given application the choice must be made explicitly and carefully.

5.16   Quaternion Operator Sequences

In many applications it will be important to consider, as we did in the case of matrix rotation operators, a sequence of these quaternion rotation operators. By such a sequence we mean one operator

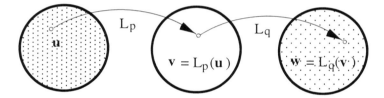

Figure 5.5: Rotation Operator Composition: $L_q{\circ}L_p$

applied after another, and so on, in some specified order. Hence we need to know that a sequence of such operators itself represents a rotation in R^3.

So, suppose that we have two operators of the sort described in Equation 5.9. Let p and q be two unit quaternions which define the quaternion rotation operators

$$L_p(\mathbf{a}) = p\mathbf{a}p^*$$
$$\text{and} \quad L_q(\mathbf{b}) = q\mathbf{b}q^*$$

Now, let $\mathbf{u}$ be a vector to which the operator L_p is applied as shown in Figure 5.5, to obtain

$$\mathbf{v} = L_p(\mathbf{u})$$
$$= p\mathbf{u}p^*$$

To this result we apply the operator L_q, thus obtaining what some-times is called the *composition* of the operators L_q and L_p, denoted $L_q \circ L_p$, we get

$$
\begin{aligned}
\mathbf{w} &= L_q(\mathbf{v}) \\
&= q\mathbf{v}q^* \\
&= q(p\mathbf{u}p^*)q^* \\
&= (qp)\mathbf{u}(qp)^* = L_{qp}(\mathbf{u}) \quad\quad (5.18)
\end{aligned}
$$

But now, since p and q are unit quaternions, so is the product qp. Hence, Equation 5.18 describes a rotation operator of the form of Equation 5.9, in which the defining quaternion is exactly the prod-uct of the two constituent quaternions p and q. The axis and the angle of this composite rotation is, of course, given by the product qp. Thus we have proved the following theorem.

> **Theorem 5.3:** *Suppose that p and q are unit quater-nions which define the quaternion rotation operators*
>
> $$ L_p(\mathbf{u}) = p\mathbf{u}p^* \quad and \quad L_q(\mathbf{v}) = q\mathbf{v}q^* $$
>
> *Then the quaternion product qp defines a quaternion ro-tation operator L_{qp} which represents a sequence of op-erators, L_p followed by L_q. The axis and the angle of rotation are those represented by the quaternion prod-uct, say, $r = qp$.*

In the proof of Theorem 5.3 we used quaternion rotation operators given by Equation 5.5. However, if we use the operator given by Equation 5.6 we obtain a result which differs distinctly from that of Theorem 5.3. That is, with $\mathbf{v} = q^*\mathbf{u}q$, a computation, similar to the one computation above, gives

$$
\begin{aligned}
\mathbf{w} &= L_q(\mathbf{v}) \\
&= q^*\mathbf{v}q \\
&= q^*(p^*\mathbf{u}p)q \\
&= (pq)^*\mathbf{u}(pq) = L_{pq}(\mathbf{u}) \quad\quad (5.19)
\end{aligned}
$$

Equation 5.19 says that, using the quaternion rotation operator of Equation 5.6, the axis and the angle of the composite rotation are those represented by the quaternion product pq. Note that the order of the quaternions is reversed, and, since the quaternion product is not commutative, the result is distinctly different. In fact, we have the theorem

Equation 5.9
Quaternion Operator

$$
\begin{aligned}
\mathbf{w} &= L_q(\mathbf{v}) \\
&= q\mathbf{v}q^* \\
&= (q_0 + \mathbf{q})(0 + \mathbf{v})(q_0 - \mathbf{q}) \\
&= (2q_0^2 - 1)\mathbf{v} + \\
&\quad 2(\mathbf{q} \cdot \mathbf{v})\mathbf{q} + \\
&\quad 2q_0(\mathbf{q} \times \mathbf{v})
\end{aligned}
$$

Equation 5.5

$$ \mathbf{w} = q\mathbf{v}q^* $$

Equation 5.6

$$ \mathbf{w} = q^*\mathbf{v}q $$

Theorem 5.4:   *Suppose that p and q are unit quaternions which define the quaternion rotation operators*

$$L_p(\mathbf{u}) = p^*\mathbf{u}p \quad and \quad L_q(\mathbf{v}) = q^*\mathbf{v}q$$

Then the quaternion product pq defines a quaternion rotation operator L_{pq} which represents a sequence of operators, L_p followed by L_q. The axis and the angle of rotation of the composite rotation operator are those represented by the quaternion product pq.

Note that in both of these theorems the sequence of operators is L_p followed by L_q. The significant difference between Theorem 5.3 and Theorem 5.4 is that if the operator of Equation 5.5 is used, the quaternion for the composite rotation is the product qp, while if the operator of Equation 5.6 is used the quaternion for the composite operator is the product pq. This distinction is important whenever an application of these ideas is made.

The four preceding theorems give us two very important results concerning the use of quaternions for defining rotations in R^3. Theorems 5.1 and 5.2 essentially define the two quaternion rotation operators which we will be using, while Theorems 5.3 and 5.4 tell how to handle sequences of such rotations.

In what follows, it usually is more convenient to express the basic defining quaternion in a rotation operator in terms of the angle $\theta/2$, so that the angle of rotation is θ. And finally, it may be well to end this section with two examples, one for each of these two important theorems.

5.16.1   Rotation Examples

Equation 5.9
Quaternion Operator

$$
\begin{aligned}
\mathbf{w} &= L_q(\mathbf{v}) \\
&= q\mathbf{v}q^* \\
&= (q_0 + \mathbf{q})(0 + \mathbf{v})(q_0 - \mathbf{q}) \\
&= (2q_0^2 - 1)\mathbf{v} + \\
&\quad 2(\mathbf{q} \cdot \mathbf{v})\mathbf{q} + \\
&\quad 2q_0(\mathbf{q} \times \mathbf{v})
\end{aligned}
$$

As an example for Theorem 5.1, consider a rotation in R^3 about an axis defined by the vector $(1,1,1)$. About this axis, the standard basis vectors $\mathbf{i}$, $\mathbf{j}$, and $\mathbf{k}$ are all generators of a *cone* — in fact, they all generate the same cone. For this axis, these standard basis vectors are equally spaced on the surface of this generated cone. If the angle of rotation about the specified axis is $2\pi/3$, then the vectors, say $a\mathbf{i}$, under this rotation becomes $a\mathbf{j}$, $b\mathbf{j}$ would become $b\mathbf{k}$, and $c\mathbf{k}$ would become $c\mathbf{i}$. We now explain exactly how this rotation goes in terms of the quaternion operator in Equation 5.9.

First, we define a unit vector $\mathbf{u}$ in the direction of the vector $(1,1,1)$ as

$$\mathbf{u} = (1/\sqrt{3}, 1/\sqrt{3}, 1/\sqrt{3})$$

Next, the appropriate choice for the angle θ in the unit quaternion q is one-half of the angle of rotation. Thus we need $\theta = \pi/3$, which gives us $\cos\theta = 1/2$ and $\sin\theta = \sqrt{3}/2$. Hence, the appropriate quaternion q to be used in defining the quaternion operator is

$$
\begin{aligned}
q &= \cos\theta + \mathbf{u}\sin\theta \\
&= \frac{1}{2} + (\mathbf{i}\frac{1}{\sqrt{3}} + \mathbf{j}\frac{1}{\sqrt{3}} + \mathbf{k}\frac{1}{\sqrt{3}})\frac{\sqrt{3}}{2} \\
&= \frac{1}{2} + \frac{1}{2}\mathbf{i} + \frac{1}{2}\mathbf{j} + \frac{1}{2}\mathbf{k}
\end{aligned}
$$

So, if we write the quaternion q in the form $q = q_0 + \mathbf{q}$ we have

$$q_0 = \frac{1}{2}$$

and

$$\mathbf{q} = \frac{1}{2}\mathbf{i} + \frac{1}{2}\mathbf{j} + \frac{1}{2}\mathbf{k}$$

We may now compute the effect of the operator qvq^* on, say, the basis vector $\mathbf{v} = \mathbf{i}$. We need (the reader should check these details)

$$\mathbf{q} \cdot \mathbf{i} = \frac{1}{2}$$

and

$$\mathbf{q} \times \mathbf{i} = \frac{1}{2}\mathbf{j} - \frac{1}{2}\mathbf{k}$$

Equation 5.9 then gives us

$$
\begin{aligned}
\mathbf{w} &= q\mathbf{i}q^* \\
&= (\frac{1}{4} - \frac{3}{4})\mathbf{i} + 2(\frac{1}{2})\mathbf{q} + 2(\frac{1}{2})(\frac{1}{2}\mathbf{j} - \frac{1}{2}\mathbf{k}) \\
&= -\frac{1}{2}\mathbf{i} + \frac{1}{2}\mathbf{i} + \frac{1}{2}\mathbf{j} + \frac{1}{2}\mathbf{k} + \frac{1}{2}\mathbf{j} - \frac{1}{2}\mathbf{k} \\
&= \mathbf{j}
\end{aligned}
$$

Equation 5.5

$$\mathbf{w} = q\mathbf{v}q^*$$

Equation 5.6

$$\mathbf{w} = q^*\mathbf{v}q$$

Our example thus shows that the operator in Equation 5.5 does have exactly the geometric effect which the theorem guarantees. It is not difficult to check that the quaternion operation indicated does, in fact, take $\mathbf{j}$ into $\mathbf{k}$ as well as $\mathbf{k}$ into $\mathbf{i}$.

We should note that the preceding result is based on a *point* or *vector* rotation. It is not difficult to verify that if instead of the quaternion rotation operator of Equation 5.5 we had used the *frame rotation* operator of Equation 5.6 we would have obtained

$$
\begin{aligned}
\mathbf{w} &= L_q(\mathbf{i}) \\
&= q^*\mathbf{i}q \\
&= \mathbf{k}
\end{aligned}
$$

A moment's reflection will tell us that this is exactly what we should expect from a frame rotation about the axis whose direction is given by the vector $(1,1,1)$ through the angle $2\pi/3$.

Rotation Sequence Example

As an example for, say, Theorem 5.4 (a frame rotation, in this case), consider first a rotation through an angle $\pi/2$ about the basis vector $\mathbf{k}$ as its axis. As we remarked before, in defining the appropriate quaternion we must use half of this angle. Since $\cos \pi/4 = \sqrt{2}/2 = \sin \pi/4$, the appropriate quaternion for this rotation is

$$
q = \frac{\sqrt{2}}{2} + \frac{\sqrt{2}}{2}\mathbf{k}
$$

Geometrically it is clear that this rotation will, for example, direct the new basis vector, $\mathbf{v} = \mathbf{i}$, along the original basis vector $\mathbf{j}$. (The reader may wish to check that new basis vector $\mathbf{j}$ then is *negatively directed* with respect to the original basis vector $\mathbf{i}$).

We follow this first rotation with a second rotation, namely, one about the *new* basis vector $\mathbf{i}$ through an angle $\pi/2$. Again, in view of our above remarks, it is clear that the quaternion appropriate for this rotation is

$$
p = \frac{\sqrt{2}}{2} + \frac{\sqrt{2}}{2}\mathbf{i}
$$

This rotation, applied after the first one, takes the basis vector $\mathbf{j}$ into the vector $\mathbf{k}$. In summary, according to our Theorem 5.4, the composite operator is $(qp)^*\mathbf{v}(qp)$ where the product qp represents the appropriate quaternion for the sequence of these two rotations. We may easily find the axis and the angle of this composite rotation by computing the product qp. We have

$$
\begin{aligned}
qp &= (\frac{\sqrt{2}}{2} + \frac{\sqrt{2}}{2}\mathbf{k})(\frac{\sqrt{2}}{2} + \frac{\sqrt{2}}{2}\mathbf{i}) \\
&= \frac{1}{2} - \frac{1}{2}\mathbf{k}\cdot\mathbf{i} + \frac{1}{2}\mathbf{i} + \frac{1}{2}\mathbf{k} + \frac{1}{2}\mathbf{k}\times\mathbf{i} \\
&= \frac{1}{2} + \frac{1}{2}\mathbf{i} + \frac{1}{2}\mathbf{j} + \frac{1}{2}\mathbf{k}
\end{aligned}
$$

From this quaternion product we note that the axis of rotation is in the direction of the vector (1,1,1) again, as in a previous example (a coincidence), and that the angle of rotation is $2\pi/3$. A moment's reflection on the geometry of the situation indeed confirms that a single rotation about this composite axis through this angle directs the new basis vector **i** along the old basis vector **j**, **j** along **k**, and **k** along **i**, if we apply the operator for each basis vector, just as Theorem 5.4 claims must happen.

It is not difficult to check that if we use the rotation operator of Equation 5.6 in the preceding example, the result may be interpreted as a negative rotation rather than a positive rotation. It should be no mystery that this is what conjugation does.

There is much more to be said about the geometry of the rotation operators we have looked at in this section, and we turn to these geometric considerations in the next chapter.

Exercises for Chapter 5

1. Write the quaternion inverse for $q = a\cos\theta - b\mathbf{u}\sin\theta$.

2. Work through the details to verify Equation 5.1.

3. Show that for any quaternion q and $\mathbf{v} \in R^3$ if q is a scalar multiple of the unit quaternion p then

$$q^{-1}\mathbf{v}q = p^*\mathbf{v}p$$

4. Work through the algebraic details to verify that the quaternion product of the two vectors **r** and **s** is given by

$$\mathbf{rs} = \mathbf{r} \times \mathbf{s} - \mathbf{r} \cdot \mathbf{s}$$

5. Geometrically, compare $q^*\mathbf{v}q$ and $q\mathbf{v}q^*$.

Chapter 6

Quaternion Geometry

6.1 Introduction

In this chapter we wish to explore in more detail certain geometric matters related to quaternions and the quaternion operator of the preceding chapter. We already know that the composition of two rotations is a rotation; this we have proved in an earlier chapter. Here, in the context of the quaternion rotation operator, we will analyze the sequence of two rotations from a geometric point of view. More specifically, we will find the axis and the angle of a composite rotation using geometric methods. In fact, we will find formulas which define this axis and angle, and these formulas will confirm the algebraic results obtained earlier.

We begin by considering a sphere, or more particularly, the set of points which constitute the surface of a sphere. We let the center of the sphere be at the origin of a fixed reference coordinate frame in R^3. Geometrically it is clear that any rotation of the sphere about a vector in this reference frame will take points on the surface of the sphere into points on that surface. The result is a mapping which takes the surface of the sphere into itself. This mapping preserves distances between points on the sphere. It also preserves the measure and the direction of the angle between any two great-circle arcs. Every such mapping has two invariant points, namely, the two points which are common to the axis of rotation and the surface of the sphere.

Suppose we choose two points, say A and B, on the surface of the sphere. These two points, in effect define two vectors $\mathbf{a}$ and $\mathbf{b}$ fixed in the reference coordinate frame. We let these vectors serve as axes of rotation of the sphere into itself. Now consider a sequence

Distance on Sphere

Distances between points on the surface of a sphere are always measured along a great-circle arc.

141

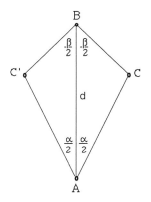

of two rotations: the first rotation about the axis **a**, followed by a second rotation about the axis **b**. We ask the following question:

Can we find a single rotation about some axis which is equivalent to a sequence of two rotations about any two given distinct axes?

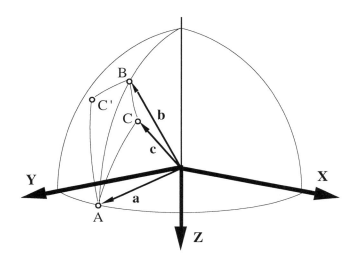

Figure 6.1: Two Rotation Construction

Points on the Sphere

The triangle ACB and its symmetrical reflection $AC'B$, on the sphere, is redrawn here in the plane to make it easier to follow the discussion in the text.

Visualize the unit vector directions **a** and **b** at points A and B, respectively, as normal to and directed out of the page.

We know that the answer to this question is affirmative. However, in what follows we will confirm this fact geometrically. We begin with a construction on the unit sphere which identifies two points which are invariant under any given sequence of two rotations. This construction relates the two given axes and their respective rotations to an equivalent composite rotation axis defined by the two invariant points. Finally, we also show how this result is related to the associated quaternion rotation operator discussed in the last chapter.

6.2   Euler Construction

So consider two distinct but arbitrary points, say A and B, on the surface of a unit sphere centered at the origin of a fixed reference frame in R^3, as illustrated in Figure 6.1. Associated with these two points are the related unit vectors, designated in the figure as **a** and **b**, respectively. For convenience we have oriented the fixed reference frame so that the vector **a** lies in the **XY** plane. Rotations of the sphere into itself are now taken, in turn, about these two vectors as axes. Let the sequence of two rotations consist of first, a rotation about the axis **a** through an angle α, followed by a second rotation about the axis **b** through an angle β.

For convenience we here introduce a new notation for a rotation. By $R[\mathbf{v}, \theta]$ we mean a rotation *about* the vector **v** through the angle θ. Further, if the rotation takes the point P into the point P' we write

$$R[\mathbf{v}, \theta]P = P'$$

In this new notation, the question asked in the introductory section of this chapter is simply this:

> *Given rotations $R[\mathbf{a}, \alpha]$ and $R[\mathbf{b}, \beta]$, is there an equivalent composite single rotation $R[\mathbf{c}, \gamma]$ such that*
>
> $$R[\mathbf{b}, \beta]R[\mathbf{a}, \alpha] = R[\mathbf{c}, \gamma] \ ?$$

In this equation the product on the left-hand side simply means that the rotation $R[\mathbf{a}, \alpha]$ is *followed by* the rotation $R[\mathbf{b}, \beta]$. We now use geometric and algebraic methods to show that the rotation $R[\mathbf{c}, \gamma]$ indeed exists, and we shall develop methods for finding formulas which define the axis **c** as well as the angle γ. The axis and angle for this composite rotation, of course, will be functions of. that is, be dependent upon the vectors **a** and **b** and the angles α and β, which are the parameters of the two constituent rotation operators.

6.2.1   Geometric Construction

Refer once again to Figure 6.1. The two points, A and B, are connected by a great circle arc. The first rotation at A, through the angle α, is about an axis whose direction is defined by the unit vector **a**; this is followed by a second rotation at B, through an angle β, about an axis whose direction is defined by the unit vector **b**. The following construction enables us to find the direction, defined by the unit vector **c**, for the axis of the composite rotation $R[\mathbf{c}, \gamma]$.

Consider points C and C' on the surface of the sphere, located so that $\angle CAB = \angle C'AB = \alpha/2$ and $\angle CBA = \angle C'BA = \beta/2$, as illustrated in the margin. If we now have a *right-handed* rotation at point A, through the angle α, it is clear, since the arc segments AC and AC' have the same length, that the point C is taken into the point C', that is,

$$R[\mathbf{a}, \alpha]C = C'$$

Note 1

Since the point B is arbitrary it may be viewed as a point in either the rotated space or the initial fixed space.

Note 2

It is important to note that, in general, if the rotation sequence is commuted, a different composite rotation results.

Right-hand-rule

All rotations about an axis, specified by a unit direction vector, follow the **right-hand rule**. That is, with the thumb of the right hand representing the specified unit direction vector, the fingers indicate the direction for a positive rotation (CCW); the direction, of course, would be CW for a negative rotation.

We follow this rotation with a similar right-handed rotation, this time about an axis defined by the vector **b** through an angle β. Geometrically, for the same reasons, it is clear that this rotation takes the point C' back into the point C, that is,

$$R[\mathbf{b}, \beta]C' = C$$

These remarks make it clear that the point C is a *fixed point* under the sequence of these two rotations. Using our new notation, we may now write

$$
\begin{aligned}
R[\mathbf{b}, \beta]R[\mathbf{a}, \alpha]C &= R[\mathbf{b}, \beta]C' \\
&= C
\end{aligned}
$$

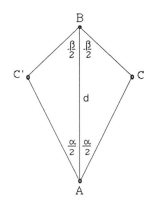

Points on the Sphere

The triangle on the sphere is redrawn here in the margin for reference as we step through the analysis.

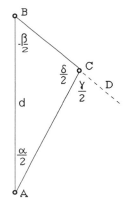

Spherical Triangle

The triangle on the sphere illustrates the relationship between the angles: α, β, δ, and γ, along with the radian distance d between points A and B.

Since we already know that the composition of two rotations in R^3 is another rotation in R^3, and since the point C is invariant under this composite rotation, it must be that the point C lies on the axis of the composite rotation. This establishes the unit vector direction **c** as the axis of the composite rotation, $R[\mathbf{c}, \gamma]$.

We next construct γ, which is the angle for the composite rotation operator, $R[\mathbf{c}, \gamma]$. This we do on the surface of the sphere by considering the rotation mapping of any arbitrary point P other than the point C. We must find the angle γ such that the equation

$$R[\mathbf{b}, \beta]R[\mathbf{a}, \alpha]P = R[\mathbf{c}, \gamma]P$$

holds for any point P (remember, the rotation axis has just been determined). Since C is invariant under the composite rotation, our only restriction on choosing the point P is that $P \neq C$; so let's take $P = A$. This choice makes the ensuing geometric and algebraic analysis much easier than if we choose an arbitrary point (although, it is easy to see we could have chosen $P = B$ with the same advantage).

So we consider the equation

$$R[\mathbf{b}, \beta]R[\mathbf{a}, \alpha]A = R[\mathbf{c}, \gamma]A$$

Since the point A is on the axis of the rotation $R[\mathbf{a}, \alpha]$ it follows that

$$R[\mathbf{a}, \alpha]A = A$$

so that our equation becomes

$$R[\mathbf{b}, \beta]A = R[\mathbf{c}, \gamma]A$$

The second rotation $R[\mathbf{b}, \beta]$ however, takes the great-circle segment BA into BA' as shown in Figure 6.2. In summary, the image of point A, under these two rotations is A', that is,

$$R[\mathbf{b}, \beta]R[\mathbf{a}, \alpha]A = R[\mathbf{c}, \gamma]A = A'$$

From the geometry in Figure 6.2 it is quite clear that the

$$\angle\gamma = \angle ACA'$$

Thus it is clear that the rotation (on the surface of the sphere)

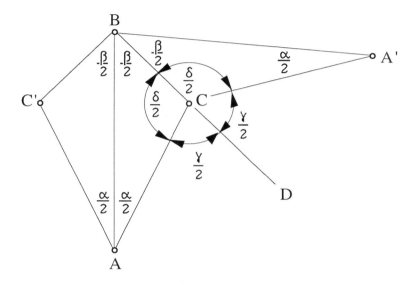

Figure 6.2: Composite rotations of Point A

about the point C through the angle γ takes the point A into the point A' just as the rotation about the point B through the angle β takes the point A into the point A'. In summary, we have constructed, on the surface of the sphere, the angle γ for the composite rotation $R[\mathbf{c}, \gamma]$.

This geometric construction does not, however, determine the magnitude of the angle γ. About all we can say for the magnitude of γ is that

$$\frac{\delta}{2} + \frac{\gamma}{2} = \pi$$

where $\delta/2$ is the interior angle of the spherical triangle shown in Figure 6.2. The angle δ is a function of the specified rotation angles α and β, and the length, say d, of the great-circle arc AB. We must now determine the magnitude of γ as a function of these given parameters, and we will do this algebraically using some identities in spherical trigonometry.

6.2.2   The Spherical Triangle

Most readers may not be familiar with the details of spherical trigonometry, by which we mean the trigonometry of triangles whose sides are great-circle arcs on the surface of a sphere. We shall not review that subject at this point; rather we shall simply appropriate the results which we need to find a formula for the rotation angle γ. The interested reader, of course, may check these results in any standard spherical trigonometry textbook. As a matter of fact, a bit later we will derive some of these formulas in the context of the quaternion rotation operator.

Consider once again the spherical triangle, ABC, introduced in Figure 6.2 and reproduced in part in the margin. It will be relatively easy to get a formula for the magnitude of the rotation angle γ if we appropriate the relevant identity from spherical trigonometry for the triange ABC. That identity is

$$\cos \frac{\delta}{2} \;=\; \sin \frac{\alpha}{2} \cos d \sin \frac{\beta}{2} - \cos \frac{\alpha}{2} \cos \frac{\beta}{2} \qquad (6.1)$$

where d is the radian distance between point A and point B, that is, $\angle d$ is the central angle subtended by the great-circle arc AB. Since, as we observed earlier,

$$\frac{\delta}{2} + \frac{\gamma}{2} = \pi$$

it follows that

$$\frac{\delta}{2} = \pi - \frac{\gamma}{2}$$

so that

$$
\begin{aligned}
\cos \frac{\delta}{2} &= \cos(\pi - \frac{\gamma}{2}) \\
&= \cos \pi \cos \frac{\gamma}{2} + \sin \pi \sin \frac{\gamma}{2} \\
&= -\cos \frac{\gamma}{2}
\end{aligned}
$$

Therefore we may write Equation 6.1 as

$$\cos \frac{\gamma}{2} = \cos \frac{\alpha}{2} \cos \frac{\beta}{2} - \sin \frac{\alpha}{2} \cos d \sin \frac{\beta}{2} \qquad (6.2)$$

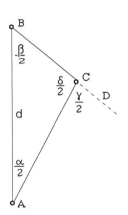

Spherical Triangle

The triangle on the sphere illustrates the relationship between the angles: α, β, δ, and γ, along with the radian distance d between points A and B.

We may then write

$$\cos\frac{\gamma}{2} = \cos\frac{(\alpha+\beta)}{2} + (1-\cos d)\sin\frac{\alpha}{2}\sin\frac{\beta}{2} \qquad (6.3)$$

It is now clear that if we are given the magnitudes for the angles α and β, and the radian length of the great-circle arc AB (that is, the magnitude of the angle d), we may use this Equation 6.3 to calculate the magnitude of the angle γ.

We remark incidentally that if triangle ABC were a *plane triangle* then we have

$$\frac{\delta}{2} + \frac{\alpha}{2} + \frac{\beta}{2} = \pi$$

and therefore

$$\frac{\gamma}{2} = \frac{\alpha}{2} + \frac{\beta}{2}$$

However, for the *spherical triangle*, Equation 6.3 merely says, if angle $d \neq 0$ then

$$\frac{\alpha}{2} + \frac{\beta}{2} + \frac{\delta}{2} > \pi$$

or, in terms of γ,

$$\frac{\gamma}{2} < \frac{\alpha}{2} + \frac{\beta}{2}$$

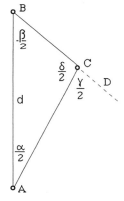

Spherical Triangle

The triangle on the sphere illustrates the relationship between the angles: α, β, δ, and γ, along with the radian distance d between points A and B.

6.3 Quaternion Geometric Analysis

In the preceding sections we described a geometric construction which identified the axis and the angle of rotation for a sequence of two given rotations. We also derived a formula from which the magnitude of the angle of rotation may be computed. In this section we accomplish essentially the same goals, but now we do this in the context of the quaternion rotation operator of Equation 5.13.

The quaternion notation for a rotation operator is algebraically more convenient and more concise, as we shall presently see. Earlier in this chapter, when we did the geometric contruction, a rotation about the point C (which defined the unit vector $\mathbf{c}$ as the axis of rotation) through the angle γ was denoted by $R[\mathbf{c}, \gamma]$. We noted that if this rotation takes the point A into the point A' we write

$$A' = R[\mathbf{c}, \gamma]A$$

Using the result of Theorem 5.1 of Section 5.8.3, we may write the

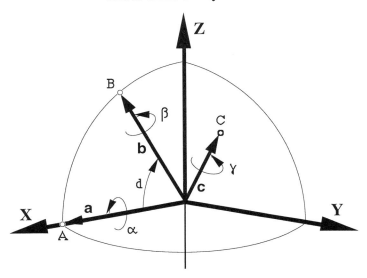

Figure 6.3: Rotation Sequence

rotation $R[\mathbf{c}, \gamma]$ in the form of the quaternion rotation operator (as applied to some vector $\mathbf{v}$)

$$\mathbf{w} = L_q(\mathbf{v}) = q_{\mathbf{c},\gamma}\mathbf{v}q_{\mathbf{c},\gamma}^*$$

In this quaternion rotation operator form the quaternions $q_{\mathbf{c},\gamma}$ and $q_{\mathbf{c},\gamma}^*$ are given by

$$\begin{aligned} q_{\mathbf{c},\gamma} &= \cos\frac{\gamma}{2} + \mathbf{c}\sin\frac{\gamma}{2} \\ \text{and} \qquad q_{\mathbf{c},\gamma}^* &= \cos\frac{\gamma}{2} - \mathbf{c}\sin\frac{\gamma}{2} \end{aligned}$$

In this quaternion notation we are using double subscripts to denote both the axis and the angle of the quaternion rotation operator. We use this notation at this point to emphasize the fact that the quaternion employed in the quaternion rotation operator exhibits *directly* both the angle of rotation and the axis about which this rotation occurs. Using this notation, we next derive expressions for the angle, γ, and the vector, $\mathbf{c}$; this will confirm the results obtained in the preceeding section.

Both the direction of the composite axis $\mathbf{c}$, and the angle of rotation γ about this axis, will be expressed as functions of the parameters which define each of the two constituent quaternions, $q_{\mathbf{a},\alpha}$ and $q_{\mathbf{b},\beta}$. Without loss of generality, for the purposes of this computation we locate the two constituent rotation axes, $\mathbf{a}$ and $\mathbf{b}$, as illustrated in Figure 6.3. Here, the points A and B are located on a unit sphere. The rotation axis $\mathbf{a}$ is directed along the reference frame $\mathbf{X}$-axis, and rotation axis $\mathbf{b}$ lies in the $\mathbf{ZX}$-plane. The points A and B are separated by a radian distance, d; that is, d is the angle between the two rotation axes. It follows that

$$\mathbf{b} = \mathbf{i} \cos d + \mathbf{k} \sin d$$

We recall that the quaternion $q_{\mathbf{c},\gamma}$ (in the composite rotation operator) equals the product of the two constituent quaternions $q_{\mathbf{a},\alpha}$ and $q_{\mathbf{b},\beta}$; that is,

$$q_{\mathbf{c},\gamma} = \cos \frac{\gamma}{2} + \mathbf{c} \sin \frac{\gamma}{2} = q_{\mathbf{b},\beta} \, q_{\mathbf{a},\alpha}$$

Using the trigonometric form for the unit quaternions $q_{\mathbf{a},\alpha}$ and $q_{\mathbf{b},\beta}$, and the above expression for the vector $\mathbf{b}$, we compute

$$
\begin{aligned}
q_{\mathbf{c},\gamma} &= \cos \frac{\gamma}{2} + \mathbf{c} \sin \frac{\gamma}{2} \qquad (6.4)\\
&= q_{\mathbf{b},\beta} \, q_{\mathbf{a},\alpha}\\
&= (\cos \frac{\beta}{2} + \mathbf{b} \sin \frac{\beta}{2})(\cos \frac{\alpha}{2} + \mathbf{i} \sin \frac{\alpha}{2})\\
&= (\cos \frac{\alpha}{2} \cos \frac{\beta}{2} - \sin \frac{\alpha}{2} \cos d \sin \frac{\beta}{2})\\
&\quad + \mathbf{i} \, (\sin \frac{\alpha}{2} \cos \frac{\beta}{2} + \cos \frac{\alpha}{2} \cos d \sin \frac{\beta}{2})\\
&\quad + \mathbf{j} \, (\sin \frac{\alpha}{2} \sin d \sin \frac{\beta}{2})\\
&\quad + \mathbf{k} \, (\cos \frac{\alpha}{2} \sin d \sin \frac{\beta}{2}) \qquad (6.5)
\end{aligned}
$$

This computation gives us another expression for the quaternion $q_{\mathbf{c},\gamma}$ and therefore the corresponding *real* and *vector* parts of the quaternions in Equation 6.4 and Equation 6.5 are equal. If we equate the real parts we have

$$\cos \frac{\gamma}{2} = \cos \frac{\alpha}{2} \cos \frac{\beta}{2} - \sin \frac{\alpha}{2} \cos d \sin \frac{\beta}{2} \qquad (6.6)$$

Thus, as was our goal, we have confirmed the earlier geometric result given in Equation 6.2.

However, this quaternion analysis also provides us with information about the direction of the axis of the composite rotation — information which was not readily available from our geometric analysis. The preceeding construction merely located a point

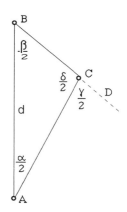

Spherical Triangle

The triangle on the sphere illustrates the relationship between the angles: α, β, δ, and γ, along with the radian distance d between points A and B.

C which was shown to be a point on the composite rotation axis. However, if we equate the vector parts of the quaternions in Equation 6.4 and Equation 6.5, we obtain

$$\mathbf{c}\sin\frac{\gamma}{2} = \mathbf{i}\,u_x + \mathbf{j}\,u_y + \mathbf{k}\,u_z$$

$$\text{where} \quad u_x = (\sin\frac{\alpha}{2}\cos\frac{\beta}{2} + \cos\frac{\alpha}{2}\cos d\sin\frac{\beta}{2})$$

$$u_y = \sin\frac{\alpha}{2}\sin d\sin\frac{\beta}{2}$$

$$u_z = \cos\frac{\alpha}{2}\sin d\sin\frac{\beta}{2}$$

Hence, the direction of the axis of the composite rotation is given by the unit vector

$$\mathbf{c} = \frac{\mathbf{i}\,u_x + \mathbf{j}\,u_y + \mathbf{k}\,u_z}{\sin\frac{\gamma}{2}} \tag{6.7}$$

Equation 6.7 thus defines the rotation axis of the composite rotation explicitly as a vector in the reference frame. In this equation we use,

$$\sin\frac{\gamma}{2} = \sqrt{1 - \cos^2\frac{\gamma}{2}}$$

where $\cos\frac{\gamma}{2}$ is defined in Equation 6.6. Note how directly this quaternion approach yields this significant result.

Note!

Note that the tracking application of Section 3.6 employed successive rotations about orthogonal axes

In the foregoing development the directions of the axes of the two successive rotations were separated by an arbitrary central angle d. This, of course, is the most general case. However, in most if not all applications, the axes for two *successive* rotations are mutually orthogonal. This orthogonality of successive rotation axes simplifies the results obtained above, because $d = \pi/2$ we have $\cos d = 0$ and $\sin d = 1$. Then the simplified expressions for both the angle and the axis are

$$\cos\frac{\gamma}{2} = \cos\frac{\alpha}{2}\cos\frac{\beta}{2} \tag{6.8}$$

and

$$\mathbf{c} = \frac{\mathbf{i}\,u_x + \mathbf{j}\,u_y + \mathbf{k}\,u_z}{\sin\frac{\gamma}{2}} \tag{6.9}$$

$$\text{where} \quad u_x = \sin\frac{\alpha}{2}\cos\frac{\beta}{2}$$

$$u_y = \sin\frac{\alpha}{2}\sin\frac{\beta}{2}$$

$$u_z = \cos\frac{\alpha}{2}\sin\frac{\beta}{2}$$

6.4   The Tracking Example Revisited

Earlier, in Sections 3.6 and 3.7 of Chapter 3, we analyzed in some detail an application called the *Tracking Transformation*. The tracking transformation employs two successive rotations about axes which are mutually orthogonal. In working out that example we used the matrix methods we had developed at that point to find formulas for the direction of the axis and for the rotation angle of the composite rotation. These are given in Equations 3.7 and 3.8

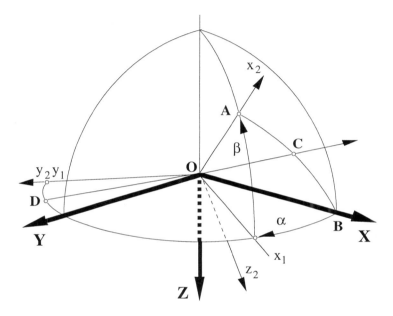

Figure 6.4: The Tracking Frame

respectively. In this section we confirm those earlier results, except that now we use the quaternion approach we have just developed.

Recall that the tracking application consists of an ordered sequence of two rotations taken about mutually orthogonal axes: the first about the **Z**-axis through an angle α, and the second about the *new* y-axis through an angle β, as illustrated in Figure 6.4. In our earlier discussion of the tracking example the angles α and β were called the *bearing* and *elevation* angles, respectively.

We now use the quaternion approach to find the axis and the angle of the composite rotation in the tracking example. Since the first rotation is about the **Z**-axis through the angle α, the appropriate quaternion for defining the required quaternion rotation operator is

$$q = \cos\frac{\alpha}{2} + \mathbf{k}\sin\frac{\alpha}{2}$$

The second rotation in the sequence is taken about the new y-axis, through the angle β. Hence the appropriate quaternion for defining the quaternion rotation operator for this second rotation is

$$p = \cos\frac{\beta}{2} + \mathbf{j}\sin\frac{\beta}{2}$$

Check it out

The reader should check the details of this computation. In so doing, she should remember that $\mathbf{k}\cdot\mathbf{j} = 0$ and that $\mathbf{k}\times\mathbf{j} = -\mathbf{i}$.

In this example we seek to direct the tracking frame $\mathbf{X}$-axis with the remote object being tracked, so it seems natural to use the frame rotation operator of Equation 5.6. Hence the quaternion associated with the composite rotation is the product qp, as required by Theorem 5.4. We compute this product as

$$
\begin{aligned}
qp &= (\cos\frac{\alpha}{2} + \mathbf{k}\sin\frac{\alpha}{2})(\cos\frac{\beta}{2} + \mathbf{j}\sin\frac{\beta}{2}) \\
&= \cos\frac{\alpha}{2}\cos\frac{\beta}{2} + \mathbf{j}\cos\frac{\alpha}{2}\sin\frac{\beta}{2} \\
&\quad + \mathbf{k}\sin\frac{\alpha}{2}\cos\frac{\beta}{2} + (\mathbf{k}\times\mathbf{j})\sin\frac{\alpha}{2}\sin\frac{\beta}{2} \\
&= \cos\frac{\alpha}{2}\cos\frac{\beta}{2} - \mathbf{i}\sin\frac{\alpha}{2}\sin\frac{\beta}{2} \\
&\quad + \mathbf{j}\cos\frac{\alpha}{2}\sin\frac{\beta}{2} + \mathbf{k}\sin\frac{\alpha}{2}\cos\frac{\beta}{2}
\end{aligned}
\tag{6.10}
$$

Quaternion Singularities?

In Section 3.8 we discussed an example of a dynamic singularity that occurs in the Euler angle-axis rotation matrix sequence, $R^y_\beta R^z_\alpha$. If a quaternion rotation operator is used to relate this *same* range of tracking frames to the reference frame, then obviously a somewhat similar dynamic singularity occurs in the vicinity of $\beta = \frac{\pi}{2}$. But the quaternion operator, in this case, is merely mimicking this particular Euler angle-axis sequence, which has a *tracking frame orientation singularity* at $\beta = \frac{\pi}{2}$.

The quaternion rotation operator is singularity-free and can relate any two independent coordinate frames in R^3. Every minimal matrix rotation operator in $\mathbf{SO}(3)$, on the other hand, will *always* have at least one singularity of the sort described earlier.

From this computation we see directly that the axis of the composite rotation is defined by the vector

$$
\begin{aligned}
\mathbf{v} &= -\mathbf{i}\sin\frac{\alpha}{2}\sin\frac{\beta}{2} \\
&\quad + \mathbf{j}\cos\frac{\alpha}{2}\sin\frac{\beta}{2} \\
&\quad + \mathbf{k}\sin\frac{\alpha}{2}\cos\frac{\beta}{2}
\end{aligned}
$$

which confirms exactly our earlier result given by Equation 3.7.

Likewise, we have a formula for the angle of rotation for the composite rotation, namely

$$\cos\frac{\phi}{2} = \cos\frac{\alpha}{2}\cos\frac{\beta}{2}$$

With an appropriate application of half-angle formulas from trigonometry, we may easily reconcile this result with that given by Equation 3.8. We point out with some joy and just a bit of satisfaction that the quaternion approach does seem to be easier and more efficient than the matrix approach.

Exercises for Chapter 6

1. Work through the details to verify Equation 6.10.

Chapter 7

Algorithm Summary

Up to this point, in the foregoing chapters we have explored some matrix algebra, and more specifically have developed the algebra of the matrix rotation operator. We have made some significant applications of the matrix rotation operator to a variety of problems, including the tracking sequence, the Aerospace sequence, determination of the orbit ephemeris for a near-earth satellite, and great circle course navigation. By way of contrast to the matrix rotation operator, we have introduced the *quaternion* rotation operator. We have explored the algebra of quaternion rotation operators, and have illustrated the application of these operators with an example or two — particularly another look, from a quaternion point of view, at the tracking sequence.

Before going on to further applications of these quaternion rotation operators, we will, in the present chapter, pull together some of the more important algebraic results developed so far. We shall also explore some of the many interesting algebraic relationships which exist between these two kinds of rotation operators. For example, we shall be a bit more explicit about how, given a rotation matrix, one finds the corresponding quaternion operator, and vice versa. Given a sequence of Euler angle rotations, we will easily represent it by means of a quaternion rotation operator. Given a set of direction cosines, we can likewise find the corresponding quaternion rotation operator. We will see how the eigenvalues and eigenvectors of a rotation matrix are related to the corresponding quaternion. Thus we seek to inter-relate the alternative transformations, algorithms, and ideas which are part of the jargon employed by practitioners of the art of rotation operators, so that terms such as *direction cosines, Euler angles, quaternions, rotation sequences, incremental rotations*, and the like, will become more familiar to the reader. We hope that in this overview the reader may gain a better perspective on these matters.

We should mention that in these notes we have been careful to distinguish between matrix and quaternion methods for studying rotations in R^3. However, this distinction will not prevent us from using the compactness of matrix notation, which can be quite convenient for expressing certain results in quaternion analysis. Now and then, in the following sections, we will use matrix notation which is not at all related to the rotation matrix.

7.1   The Quaternion Product

In Chapter 5, in a fairly detailed way, we introduced the product of two quaternions, and we will review those results here. In Equation 5.2 we wrote

$$pq = r = r_0 + \mathbf{r} = r_0 + \mathbf{i}r_1 + \mathbf{j}r_2 + \mathbf{k}r_3$$

$$\text{where} \qquad \begin{aligned} r_0 &= p_0q_0 - p_1q_1 - p_2q_2 - p_3q_3 \\ r_1 &= p_0q_1 + p_1q_0 + p_2q_3 - p_3q_2 \\ r_2 &= p_0q_2 - p_1q_3 + p_2q_0 + p_3q_1 \\ r_3 &= p_0q_3 + p_1q_2 - p_2q_1 + p_3q_0 \end{aligned} \qquad (7.1)$$

or, if written in matrix notation

$$pq = \begin{bmatrix} r_0 \\ r_1 \\ r_2 \\ r_3 \end{bmatrix} = \begin{bmatrix} p_0 & -p_1 & -p_2 & -p_3 \\ p_1 & p_0 & -p_3 & p_2 \\ p_2 & p_3 & p_0 & -p_1 \\ p_3 & -p_2 & p_1 & p_0 \end{bmatrix} \begin{bmatrix} q_0 \\ q_1 \\ q_2 \\ q_3 \end{bmatrix} \qquad (7.2)$$

$$= \begin{bmatrix} q_0 & -q_1 & -q_2 & -q_3 \\ q_1 & q_0 & q_3 & -q_2 \\ q_2 & -q_3 & q_0 & q_1 \\ q_3 & q_2 & -q_1 & q_0 \end{bmatrix} \begin{bmatrix} p_0 \\ p_1 \\ p_2 \\ p_3 \end{bmatrix} \qquad (7.3)$$

Both Equation 7.2 and Equation 7.3 are matrix representations of the quaternion product $r = pq$. Recall that because multiplication for quaternions is not commutative, the product $r = qp$ will in general be different from the product $r = pq$. For the product $r = qp$ the equations corresponding to those given just above then are

$$qp = r = r_0 + \mathbf{i}r_1 + \mathbf{j}r_2 + \mathbf{k}r_3$$

$$\text{where} \quad \begin{aligned} r_0 &= p_0q_0 - p_1q_1 - p_2q_2 - p_3q_3 \\ r_1 &= p_0q_1 + p_1q_0 - p_2q_3 + p_3q_2 \\ r_2 &= p_0q_2 + p_2q_0 + p_1q_3 - p_3q_1 \\ r_3 &= p_0q_3 + p_3q_0 - p_1q_2 + p_2q_3 \end{aligned}$$

Note that this product differs from the product in the other order simply in that the sign is changed in certain of the terms. This results, of course, from the fact that for the cross product of two vectors we have $\mathbf{q} \times \mathbf{p} = -\mathbf{p} \times \mathbf{q}$. In matrix form, this set of equations may be written in either of the two forms

$$qp = \begin{bmatrix} r_0 \\ r_1 \\ r_2 \\ r_3 \end{bmatrix} = \begin{bmatrix} p_0 & -p_1 & -p_2 & -p_3 \\ p_1 & p_0 & p_3 & -p_2 \\ p_2 & -p_3 & p_0 & p_1 \\ p_3 & p_2 & -p_1 & p_0 \end{bmatrix} \begin{bmatrix} q_0 \\ q_1 \\ q_2 \\ q_3 \end{bmatrix} \tag{7.4}$$

$$= \begin{bmatrix} q_0 & -q_1 & -q_2 & -q_3 \\ q_1 & q_0 & -q_3 & q_2 \\ q_2 & q_3 & q_0 & -q_1 \\ q_3 & -q_2 & q_1 & q_0 \end{bmatrix} \begin{bmatrix} p_0 \\ p_1 \\ p_2 \\ p_3 \end{bmatrix} \tag{7.5}$$

7.2 Quaternion Rotation Operator

We have defined the quaternion rotation operator $L_q(\mathbf{v})$, acting on a vector $\mathbf{v}$, by the equation

$$L_q(\mathbf{v}) = q^*\mathbf{v}q$$

We recall that if this operator is to represent a rotation through an angle α about a vector $\mathbf{q}$ as its axis, the quaternion q must be a unit quaternion of the form

$$q = q_0 + \mathbf{q} = \cos\frac{\alpha}{2} + \mathbf{u}\sin\frac{\alpha}{2}$$

We recall also that we have almost uniformly interpreted the quaternion rotation operator $L_q(\mathbf{v}) = q^*\mathbf{v}q$ geometrically as a *frame* rotation through a certain angle about the vector $\mathbf{q}$ as the axis. We must remark once again, however, that a frame rotation through a certain angle is entirely equivalent to a point rotation (about the same axis) but through the *negative* of that angle. Hence geometrically this operator may be viewed as either a frame rotation or a point rotation. The direction and magnitude of the rotation, in either view, is specified by the same quaternion, q.

We apply the quaternion rotation operator to vector $\mathbf{v}$, a pure quaternion defined in the reference frame, and express it as $\mathbf{w}$ in the rotated frame. That is, we write

$$
\begin{aligned}
\mathbf{w} &= L_q(\mathbf{v}) = q^*\mathbf{v}q \\
&= (q_0 - \mathbf{q})(\mathbf{v})(q_0 + \mathbf{q}) \\
&= (2q_0^2 - 1)\mathbf{v} + 2(\mathbf{v} \cdot \mathbf{q})\mathbf{q} + 2q_0(\mathbf{v} \times \mathbf{q}) \qquad (7.6)
\end{aligned}
$$

Expanding each term in Equation 7.6, in turn, gives

$$
(2q_0^2 - 1)\mathbf{v} = \begin{bmatrix} (2q_0^2 - 1) & 0 & 0 \\ 0 & (2q_0^2 - 1) & 0 \\ 0 & 0 & (2q_0^2 - 1) \end{bmatrix} \begin{bmatrix} v_1 \\ v_2 \\ v_3 \end{bmatrix}
$$

$$
2(\mathbf{v} \cdot \mathbf{q})\mathbf{q} = \begin{bmatrix} 2q_1^2 & 2q_1q_2 & 2q_1q_3 \\ 2q_1q_2 & 2q_2^2 & 2q_2q_3 \\ 2q_1q_3 & 2q_2q_3 & 2q_3^2 \end{bmatrix} \begin{bmatrix} v_1 \\ v_2 \\ v_3 \end{bmatrix}
$$

$$
2q_0(\mathbf{v} \times \mathbf{q}) = \begin{bmatrix} 0 & 2q_0q_3 & -2q_0q_2 \\ -2q_0q_3 & 0 & 2q_0q_1 \\ 2q_0q_2 & -2q_0q_1 & 0 \end{bmatrix} \begin{bmatrix} v_1 \\ v_2 \\ v_3 \end{bmatrix}
$$

Then $\mathbf{w}$ is the sum of these three matrices, and we write

$$
\mathbf{w} = \mathbf{Q}\,\mathbf{v} \qquad (7.7)
$$

$$
\begin{bmatrix} w_1 \\ w_2 \\ w_3 \end{bmatrix} = \begin{bmatrix} 2q_0^2 - 1 + 2q_1^2 & 2q_1q_2 + 2q_0q_3 & 2q_1q_3 - 2q_0q_2 \\ 2q_1q_2 - 2q_0q_3 & 2q_0^2 - 1 + 2q_2^2 & 2q_2q_3 + 2q_0q_1 \\ 2q_1q_3 + 2q_0q_2 & 2q_2q_3 - 2q_0q_1 & 2q_0^2 - 1 + 2q_3^2 \end{bmatrix} \begin{bmatrix} v_1 \\ v_2 \\ v_3 \end{bmatrix}
$$

7.3 Direction Cosines

Given a pair or vectors $\mathbf{u}$ and $\mathbf{v}$, the *direction cosine* associated with these two vectors is simply $\cos\theta$ where θ is the angle between these two vectors. We recall that the scalar or dot product of $\mathbf{u}$ and $\mathbf{v}$ is

$$
\mathbf{u} \cdot \mathbf{v} = |\mathbf{u}||\mathbf{v}| \cos\theta
$$

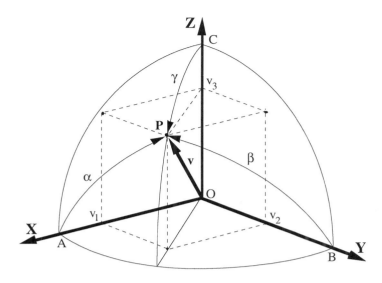

Figure 7.1: Direction Cosine Geometry

where θ is the angle between the two vectors. In particular, if $\mathbf{u}$ and $\mathbf{v}$ are *unit* vectors, that is, $|\mathbf{u}| = |\mathbf{v}| = 1$, then we have the special case

$$\mathbf{u} \cdot \mathbf{v} = \cos\theta$$

Thus for unit vectors $\mathbf{u}$ and $\mathbf{v}$, their direction cosine is simply their scalar product, $\mathbf{u} \cdot \mathbf{v}$.

Given a unit vector $\mathbf{v}$ in an orthonormal coordinate frame $\{\mathbf{X}, \mathbf{Y}, \mathbf{Z}\}$, the *direction cosines* for $\mathbf{v}$ are the cosines of the angles, say α, β, γ, between the vector $\mathbf{v}$ and the coordinate axes defined by the basis vectors $\mathbf{X}$, $\mathbf{Y}$, and $\mathbf{Z}$ respectively, as illustrated in Figure 7.1. If $\mathbf{v}$ is a unit vector, then

$$\begin{aligned}
\cos\alpha &= \mathbf{v} \cdot \mathbf{X} \\
\cos\beta &= \mathbf{v} \cdot \mathbf{Y} \\
\cos\gamma &= \mathbf{v} \cdot \mathbf{Z}
\end{aligned}$$

7.4 Frame Bases to Rotation Matrix

In this section we develop a geometric relationship between two orthonormal coordinate frames and the rotation matrix which defines the relative orientation between these two frames. As we proceed in this investigation, some interesting perspectives and a useful al-

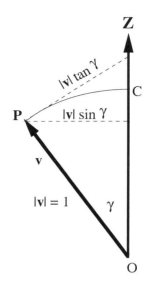

Direction Number γ

defines the orientation of the vector $\mathbf{v}$ wrt the coordinate frame $\mathbf{Z}$-axis.

Direction Angles

From the trigonometric expressions for the *direction angles* α, β, and γ, expressed in terms of the components of the unit vector $|\mathbf{v}| = 1$ it is easy to verify that

$$\cos^2 \alpha + \cos^2 \beta + \cos^2 \gamma \;=\; 1$$
$$\sin^2 \alpha + \sin^2 \beta + \sin^2 \gamma \;=\; 2$$

gorithm will emerge.

We begin with the vector

$$\mathbf{v} = v_1 \mathbf{i} + v_2 \mathbf{j} + v_3 \mathbf{k}$$

illustrated in Figure 7.1, where $\{\mathbf{i}\ \mathbf{j}\ \mathbf{k}\}$ is the standard basis in R^3. If $|\mathbf{v}| = 1$ then the direction *cosines* and *sines* for the vector $\mathbf{v}$ may be written

$$\cos \alpha \;=\; v_1 \quad \text{and} \quad \sin \alpha \;=\; \sqrt{v_2^2 + v_3^2}$$
$$\cos \beta \;=\; v_2 \quad \text{and} \quad \sin \beta \;=\; \sqrt{v_3^2 + v_1^2}$$
$$\cos \gamma \;=\; v_3 \quad \text{and} \quad \sin \gamma \;=\; \sqrt{v_1^2 + v_2^2}$$

We know the matrix A is orthogonal and has determinant $+1$. We recall that this means both the rows and columns of A form orthonormal sets of vectors. Now suppose that we regard the matrix A as representing a *frame rotation* which relates the initial reference frame $\{\mathbf{X}\ \mathbf{Y}\ \mathbf{Z}\}$ to a rotated frame $\{\mathbf{x}\ \mathbf{y}\ \mathbf{z}\}$. This means $\mathbf{x} = A\mathbf{X}, \mathbf{y} = A\mathbf{Y}$, and $\mathbf{z} = A\mathbf{Z}$. In fact, given a vector $\mathbf{v} \in R^3$, when we write $\mathbf{w} = A\mathbf{v}$ we mean, in general, that the vector $\mathbf{w}$ is simply the same vector $\mathbf{v}$, now expressed in the new frame. We will show how the rotation matrix A may be written in terms of the scalar products of the two sets of basis vectors.

In terms of ordinary linear algebra, we may think of this frame rotation simply as a *change of basis* for R^3. We begin by defining a matrix A whose *columns* are the coordinates of the final basis vectors in term of the initial basis vectors. Thus we have

$$\mathbf{x} \;=\; a_{11}\mathbf{X} + a_{21}\mathbf{Y} + a_{31}\mathbf{Z}$$
$$\mathbf{y} \;=\; a_{12}\mathbf{X} + a_{22}\mathbf{Y} + a_{32}\mathbf{Z}$$
$$\mathbf{z} \;=\; a_{13}\mathbf{X} + a_{23}\mathbf{Y} + a_{33}\mathbf{Z}$$

Consider some vector $\mathbf{v}$ in R^3, and suppose that its coordinates relative to the final basis are (v_1', v_2', v_3'). This means we may write

$$\begin{aligned}
\mathbf{v} \;&=\; v_1'\mathbf{x} + v_2'\mathbf{y} + v_3'\mathbf{z} \\
&=\; v_1'(a_{11}\mathbf{X} + a_{21}\mathbf{Y} + a_{31}\mathbf{Z}) \\
&+\; v_2'(a_{12}\mathbf{X} + a_{22}\mathbf{Y} + a_{32}\mathbf{Z}) \\
&+\; v_3'(a_{13}\mathbf{X} + a_{23}\mathbf{Y} + a_{33}\mathbf{Z})
\end{aligned}$$

Clearly this expression for $\mathbf{v}$ may be rewritten as

$$\begin{aligned}
\mathbf{v} \;&=\; (a_{11}v_1' + a_{12}v_2' + a_{13}v_3')\mathbf{X} \\
&+\; (a_{21}v_1' + a_{22}v_2' + a_{23}v_3')\mathbf{Y} \\
&+\; (a_{31}v_1' + a_{32}v_2' + a_{33}v_3')\mathbf{Z}
\end{aligned}$$

If the coordinates of $\mathbf{v}$ relative to the initial frame are (v_1, v_2, v_3), the uniqueness of coordinates in a coordinate frame tells us that we must have

$$
\begin{aligned}
v_1 &= a_{11}v_1' + a_{12}v_2' + a_{13}v_3' \\
v_2 &= a_{21}v_1' + a_{22}v_2' + a_{23}v_3' \\
v_3 &= a_{31}v_1' + a_{32}v_2' + a_{33}v_3'
\end{aligned}
$$

Since we have defined the matrix $A = [a_{ij}]$, these three equations may be represented by the single matrix equation

$$
\begin{bmatrix} v_1 \\ v_2 \\ v_3 \end{bmatrix} = A \begin{bmatrix} v_1' \\ v_2' \\ v_3' \end{bmatrix}
$$

Since bases are independent sets of vectors, it follows that the matrix A must be invertible, so we have

$$
\begin{bmatrix} v_1' \\ v_2' \\ v_3' \end{bmatrix} = A^{-1} \begin{bmatrix} v_1 \\ v_2 \\ v_3 \end{bmatrix}
$$

We now have the vector $\mathbf{v}$ expressed in the new coordinate frame, and it follows that the matrix A^{-1} is exactly the rotation matrix representing this frame rotation. But since we are here dealing with orthonormal coordinate frames, the coordinates of the old basis vectors in terms of the new basis vectors are exactly the various scalar products, as we have shown above. Hence we may write

$$
A = \begin{bmatrix} \mathbf{x} \cdot \mathbf{X} & \mathbf{y} \cdot \mathbf{X} & \mathbf{z} \cdot \mathbf{X} \\ \mathbf{x} \cdot \mathbf{Y} & \mathbf{y} \cdot \mathbf{Y} & \mathbf{z} \cdot \mathbf{Y} \\ \mathbf{x} \cdot \mathbf{Z} & \mathbf{y} \cdot \mathbf{Z} & \mathbf{z} \cdot \mathbf{Z} \end{bmatrix} \tag{7.8}
$$

The rotation matrix we seek is now the inverse of this matrix, which is simply its transpose, so in general we have

$$
A^{-1} = A^t = \begin{bmatrix} \mathbf{x} \cdot \mathbf{X} & \mathbf{x} \cdot \mathbf{Y} & \mathbf{x} \cdot \mathbf{Z} \\ \mathbf{y} \cdot \mathbf{X} & \mathbf{y} \cdot \mathbf{Y} & \mathbf{y} \cdot \mathbf{Z} \\ \mathbf{z} \cdot \mathbf{X} & \mathbf{z} \cdot \mathbf{Y} & \mathbf{z} \cdot \mathbf{Z} \end{bmatrix} \tag{7.9}
$$

7.5 Angle and Axis of Rotation

In this section we review four matters. First, given a rotation matrix, how does one determine the angle and the axis of the rotation? Second, given the angle and axis for a rotation, how does one go

about writing the corresponding rotation matrix? We then consider
these same questions, except in the context of a quaternion rotation
operator.

The first question asked above has already been answered in
Section 4 of Chapter 3. There we proved that a 3×3 matrix A is a
rotation matrix if and only if it is orthogonal and has determinant

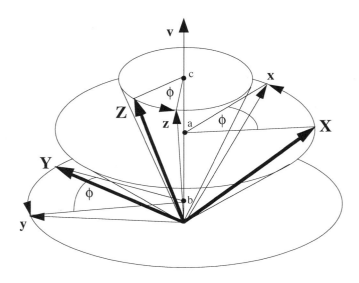

Figure 7.2: Frame Rotation about Eigenvector

$+1$. In the part of the proof which shows that a matrix with these
properties must be a rotation matrix, we proved that the rotation
matrix operator defined by the equation

$$\mathbf{u} = A\mathbf{v}$$

has a *fixed vector*, say $\mathbf{v}_0$, which has the property that

$$A\mathbf{v}_0 = \mathbf{v}_0$$

In terms of matrix algebra, if a 3×3 matrix A has the property
that

$$A\mathbf{v} = \lambda\mathbf{v}$$

then we say A has the *eigenvalue* λ, with $\mathbf{v}$ as its corresponding
eigenvector. Now it turns out that a rotation matrix always has $+1$
as an eigenvalue (although there may be other eigenvalues). Clearly,
the axis of rotation is a fixed vector for any rotation having that

vector as its axis. Hence, the axis of rotation is simply the eigenvector which corresponds to the eigenvalue $+1$. That eigenvector may be found, at least to within a scalar multiple, by solving the equation $A\mathbf{v} = \mathbf{v}$. Section 5 of Chapter 3 contains a numerical example of the procedure.

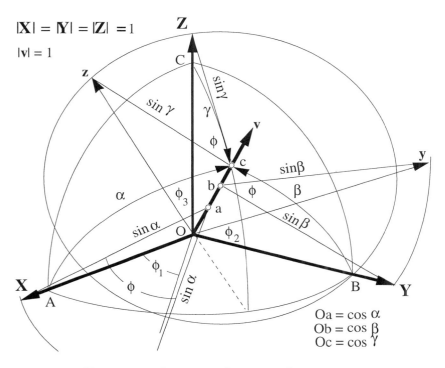

Figure 7.3: Rotation Operator Geometry

Figure 7.3

In this Figure it is helpful and important to note that the rotated frame

$$xyz$$

is related to the reference frame

$$\mathbf{XYZ}$$

by a rotation about the eigenvector $\mathbf{v}$. Note in particular,

$$
\begin{array}{rcl}
\angle \mathbf{Z}cz &=& \phi \\
\angle \mathbf{Y}by &=& \phi \\
\angle \mathbf{X}ax &=& \phi
\end{array}
$$

where the angle ϕ is the composite rotation represented by the rotation matrix, A. This geometry is discussed further in what follows.

The proof alluded to earlier also entails a formula for the angle of the rotation represented by a rotation matrix A. In Equation 3.4 we had

$$\phi = \arccos \frac{Tr(A) - 1}{2}$$

where $Tr(A)$ is the trace of the matrix A, that is, the sum of its diagonal elements. We do not review here the details of why this is so, but simply note that from this formula we easily compute the angle of rotation for any rotation matrix A.

We next write the rotation matrix in terms of its rotation axis $\mathbf{v} = (v_1, v_2, v_3)$ and its rotation angle ϕ. We do this by writing the rotated basis vectors $\{\mathbf{x}, \mathbf{y}, \mathbf{z}\}$ as linear combinations of the orthonormal reference frame vectors $\{\mathbf{X}, \mathbf{Y}, \mathbf{Z}\}$, and then use Equation 7.9 to write the rotation matrix.

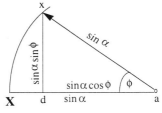

Normal Plane aXx

This is a sector of the plane base of the cone generated by the rotation of the unit basis vectors **X** and x about the eigenvector **v** as shown in Figure 7.3. This plane is normal to the eigenvector **v** and contains the endpoints of the unit basis vectors **X** and x.

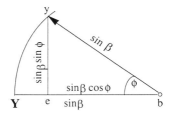

Normal Plane bYy

This is a sector of the plane base of the cone generated by the rotation of the unit basis vectors **Y** and y about the eigenvector **v** as shown in Figure 7.3. This plane is normal to the eigenvector **v** and contains the endpoints of the unit basis vectors **Y** and y.

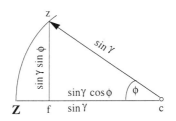

Normal Plane cZz

This is a sector of the plane base of the cone generated by the rotation of the unit basis vectors **Z** and z about the eigenvector **v** as shown in Figure 7.3. This plane is normal to the eigenvector **v** and contains the endpoints of the unit basis vectors **Z** and z.

With reference to Figure 7.3, we write expressions for the $\{\mathbf{x},\mathbf{y},\mathbf{z}\}$ frame in the $\{\mathbf{X},\mathbf{Y},\mathbf{Z}\}$ reference frame. We have

$$\mathbf{x} = \overline{Oa} + \overline{ad} + \overline{dx} \tag{7.10}$$

where $\quad \overline{Oa} = (\mathbf{i}v_1 + \mathbf{j}v_2 + \mathbf{k}v_3)\cos\alpha$

$$\overline{ad} = (\overline{OX} - \overline{Oa})\cos\phi$$

$$\overline{dx} = \mathbf{u}_1 \sin\alpha \sin\phi$$

and $\quad \mathbf{u}_1 = \mathbf{X} \times \mathbf{v}$

$$= \begin{vmatrix} \mathbf{i} & \mathbf{j} & \mathbf{k} \\ 1 & 0 & 0 \\ v_1 & v_2 & v_3 \end{vmatrix}$$

$$= \frac{-\mathbf{j}v_3 + \mathbf{k}v_2}{\sin\alpha} = \text{a unit vector}$$

$$\mathbf{y} = \overline{Ob} + \overline{be} + \overline{ey} \tag{7.11}$$

where $\quad \overline{Ob} = (\mathbf{i}v_1 + \mathbf{j}v_2 + \mathbf{k}v_3)\cos\beta$

$$\overline{be} = (\overline{OY} - \overline{Ob})\cos\phi$$

$$\overline{ey} = \mathbf{u}_2 \sin\beta \sin\phi$$

and $\quad \mathbf{u}_2 = \mathbf{Y} \times \mathbf{v}$

$$= \begin{vmatrix} \mathbf{i} & \mathbf{j} & \mathbf{k} \\ 0 & 1 & 0 \\ v_1 & v_2 & v_3 \end{vmatrix}$$

$$= \frac{\mathbf{i}v_3 - \mathbf{k}v_1}{\sin\beta} = \text{a unit vector}$$

$$\mathbf{z} = \overline{Oc} + \overline{cf} + \overline{fz} \tag{7.12}$$

where $\quad \overline{Oc} = (\mathbf{i}v_1 + \mathbf{j}v_2 + \mathbf{k}v_3)\cos\gamma$

$$\overline{cf} = (\overline{OZ} - \overline{Oc})\cos\phi$$

$$\overline{fz} = \mathbf{u}_3 \sin\gamma \sin\phi$$

and $\quad \mathbf{u}_3 = \mathbf{Z} \times \mathbf{v}$

$$= \begin{vmatrix} \mathbf{i} & \mathbf{j} & \mathbf{k} \\ 0 & 0 & 1 \\ v_1 & v_2 & v_3 \end{vmatrix}$$

$$= \frac{-\mathbf{i}v_2 + \mathbf{j}v_1}{\sin\gamma} = \text{a unit vector}$$

We expand each of these foregoing results, and gather together the terms associated with each of the rotated basis vectors $\{\mathbf{x},\mathbf{y},\mathbf{z}\}$. Each of these basis vector expressions are functions of the angle of rotation, ϕ, and of the components of the normalized eigenvector **v**, about which this rotation occurs.

Then, the rotated basis vector $\mathbf{x}$ is

$$
\begin{aligned}
\mathbf{x} = & \; \mathbf{i} \, [v_1^2 + (v_2^2 + v_3^2) \cos \phi] \\
& + \mathbf{j} \, [v_1 v_2 (1 - \cos \phi) - v_3 \sin \phi] \\
& + \mathbf{k} \, [v_1 v_3 (1 - \cos \phi) + v_2 \sin \phi]
\end{aligned}
\tag{7.13}
$$

In a similar fashion we find

$$
\begin{aligned}
\mathbf{y} = & \; \mathbf{i} \, [v_2 v_1 (1 - \cos \phi) + v_3 \sin \phi] \\
& + \mathbf{j} \, [v_2^2 + (v_3^2 + v_1^2) \cos \phi] \\
& + \mathbf{k} \, [v_2 v_3 (1 - \cos \phi) - v_1 \sin \phi]
\end{aligned}
\tag{7.14}
$$

and

$$
\begin{aligned}
\mathbf{z} = & \; \mathbf{i} \, [v_3 v_1 (1 - \cos \phi) - v_2 \sin \phi] \\
& + \mathbf{j} \, [v_3 v_2 (1 - \cos \phi) + v_1 \sin \phi] \\
& + \mathbf{k} \, [v_3^2 + (v_1^2 + v_2^2) \cos \phi]
\end{aligned}
\tag{7.15}
$$

These expressions for the rotated basis vectors $\{\mathbf{x},\mathbf{y},\mathbf{z}\}$ can nicely be written in matrix form as

$$
\mathbf{W} = A\mathbf{V}
$$

$$
\text{where} \quad \mathbf{W} = \mathrm{col}[\mathbf{x}, \mathbf{y}, \mathbf{z}]
$$

$$
\text{and} \quad \mathbf{V} = \mathrm{col}[\mathbf{i}, \mathbf{j}, \mathbf{k}]
$$

It is now easy to see that if the basis for the reference frame is the standard basis $\{\mathbf{i},\mathbf{j},\mathbf{k}\}$, then the various scalar products in Equation 7.9 give the following formula for the rotation matrix

$$
A = A(\phi, v_1, v_2, v_3)
\tag{7.16}
$$

$$
= \begin{bmatrix}
\begin{array}{c} v_1^2 + \\ (v_2^2 + v_3^2) \cos \phi \end{array} & \begin{array}{c} v_1 v_2 (1 - \cos \phi) \\ -v_3 \sin \phi \end{array} & \begin{array}{c} v_1 v_3 (1 - \cos \phi) \\ +v_2 \sin \phi \end{array} \\[1em]
\begin{array}{c} v_1 v_2 (1 - \cos \phi) \\ +v_3 \sin \phi \end{array} & \begin{array}{c} v_2^2 + \\ (v_3^2 + v_1^2) \cos \phi \end{array} & \begin{array}{c} v_2 v_3 (1 - \cos \phi) \\ -v_1 \sin \phi \end{array} \\[1em]
\begin{array}{c} v_3 v_1 (1 - \cos \phi) \\ -v_2 \sin \phi \end{array} & \begin{array}{c} v_2 v_3 (1 - \cos \phi) \\ +v_1 \sin \phi \end{array} & \begin{array}{c} v_3^2 + \\ (v_1^2 + v_2^2) \cos \phi \end{array}
\end{bmatrix}
$$

Having answered the two questions relating to axis and angle of rotation for the matrix rotation operator, we now answer these same two questions in the context of quaternions. First, suppose we are given a rotation operator in quaternion form as

$$L_q(\mathbf{v}) = q^* \mathbf{v} q$$

where the quaternion q is given by

$$q = \cos\frac{\phi}{2} + \mathbf{v}\sin\frac{\phi}{2}$$

We know that the axis and angle of rotation may be read directly from the associated quaternion. More explicitly, if the unit quaternion in terms of which the rotation operator is written is

$$q = q_0 + \mathbf{q}$$

then the rotation angle is given by

$$\phi = 2\arccos(q_0)$$

and the direction of the rotation axis is given by the vector $\mathbf{q}$.

Next, given that we have a rotation in R^3 which has the unit vector $\mathbf{v}$ as its *axis* of rotation with the angle ϕ as its *angle* of rotation we simply write the rotation operator in quaternion form as

$$L_q(\mathbf{v}) = q^* \mathbf{v} q$$

where the quaternion q is given by

$$q = \cos\frac{\phi}{2} + \mathbf{v}\sin\frac{\phi}{2}$$

Our results in this section show very clearly that these questions are much easier to answer when we use the quaternion as opposed to the matrix rotation operator. The quaternion alternative seems to offer more immediate insight into the geometric nature of rotation operators.

7.6   Euler Angles to Quaternion

Here we review the derivation of the quaternion required by an Aerospace rotation operator. This particular quaternion operator is equivalent to the rotation matrix for the Aerospace Euler angle sequence:

$$
\begin{aligned}
\psi &= \text{Heading angle} \\
\theta &= \text{Elevation angle} \\
\phi &= \text{Bank angle}
\end{aligned}
$$

The constituent quaternions are:

$$q_z = \cos\frac{\psi}{2} + \mathbf{k}\sin\frac{\psi}{2}$$

$$q_y = \cos\frac{\theta}{2} + \mathbf{j}\sin\frac{\theta}{2}$$

$$q_x = \cos\frac{\phi}{2} + \mathbf{i}\sin\frac{\phi}{2}$$

then $\quad q = q_z q_y q_x = q_0 + \mathbf{i}q_1 + \mathbf{j}q_2 + \mathbf{k}q_3$

where

$$q_0 = \cos\frac{\psi}{2}\cos\frac{\theta}{2}\cos\frac{\phi}{2} + \sin\frac{\psi}{2}\sin\frac{\theta}{2}\sin\frac{\phi}{2}$$

$$q_1 = \cos\frac{\psi}{2}\cos\frac{\theta}{2}\sin\frac{\phi}{2} - \sin\frac{\psi}{2}\sin\frac{\theta}{2}\cos\frac{\phi}{2}$$

$$q_2 = \cos\frac{\psi}{2}\sin\frac{\theta}{2}\cos\frac{\phi}{2} + \sin\frac{\psi}{2}\cos\frac{\theta}{2}\sin\frac{\phi}{2}$$

$$q_3 = \sin\frac{\psi}{2}\cos\frac{\theta}{2}\cos\frac{\phi}{2} - \cos\frac{\psi}{2}\sin\frac{\theta}{2}\sin\frac{\phi}{2}$$

Recall, the Aerospace angle/axis sequence, $(\psi\theta\phi) \rightarrow (zyx)$, is merely one of the twelve possible sequences. Other angle/axis sequences, which perhaps are more suitable for some applications, may be defined and their transformations determined as suggested above.

7.7 Quaternion to Direction Cosines

The Aerospace angle/axis sequence provides a good example of how, given the quaternion which defines such a rotation, we may write the rotation matrix in terms of direction cosines. The rotation matrix whose elements are defined in terms of the angles employed in the Aerospace angle/axis sequence, $(\psi\theta\phi) \rightarrow (zyx)$, is

$$M = [m_{ij}] = M_\phi^x M_\theta^y M_\psi^z \tag{7.17}$$

$$= \begin{bmatrix} \cos\psi\cos\theta & \sin\psi\cos\theta & -\sin\theta \\ \cos\psi\sin\theta\sin\phi & \sin\psi\sin\theta\sin\phi & \\ -\sin\psi\cos\phi & +\cos\psi\cos\phi & \cos\theta\sin\phi \\ \cos\psi\sin\theta\cos\phi & \sin\psi\sin\theta\cos\phi & \\ +\sin\psi\sin\phi & -\cos\psi\sin\phi & \cos\theta\cos\phi \end{bmatrix}$$

$$= \begin{bmatrix} 2q_0^2 - 1 \\ +2q_1^2 & 2q_1q_2 \\ +2q_0q_3 & 2q_1q_3 \\ -2q_0q_2 \\\\ 2q_1q_2 \\ -2q_0q_3 & 2q_0^2 - 1 \\ +2q_2^2 & 2q_2q_3 \\ +2q_0q_1 \\\\ 2q_1q_3 \\ +2q_0q_2 & 2q_2q_3 \\ -2q_0q_1 & 2q_0^2 - 1 \\ +2q_3^2 \end{bmatrix} = q^*(\mathbf{e}_k)q$$

Here, the quaternion q is a unit quaternion, that is, $|q| = 1$. The three column (row) vectors of M are ortho-normal and $\det |M| = +1$. Each element m_{ij} in the matrix M represents the ith direction cosine of the jth reference or basis vector. But also, as written above, $M = L_q(\mathbf{e}_k)$ for $k = 1, 2, 3$, where the $\mathbf{e}_k$'s are the standard basis vectors or reference frame.

7.8   Quaternion to Euler Angles

The results of the previous section also make it easy to determine the Euler angles in an Aerospace sequence, given the quaternion which defines the corresponding rotation operator. From the two expressions for the matrix M on the previous page we can write

$$\tan \psi = \frac{m_{12}}{m_{11}}$$
$$\sin \theta = -m_{13}$$
$$\tan \phi = \frac{m_{23}}{m_{33}}$$

where

$$m_{11} = 2q_0^2 + 2q_1^2 - 1$$
$$m_{12} = 2q_1q_2 + 2q_0q_3$$
$$m_{13} = 2q_1q_3 - 2q_0q_2$$
$$m_{23} = 2q_2q_3 + 2q_0q_1$$
$$m_{33} = 2q_0^2 + 2q_3^2 - 1$$

The trigonometric ambiguities in this Aerospace sequence can be resolved by acknowledging that by definition $\cos \theta$ must always be non-negative. Actual values for the angles can then be determined from the signs on each of the direction cosine terms involved.

7.9   Direction Cosines to Quaternion

Given a rotation matrix, say M, our previous results make is fairly easy to determine the quaternion for the corresponding quaternion rotation operator. In any proper orthogonal matrix $M = [m_{ij}] = M(q_0, q_1, q_2, q_3)$

$$
\begin{bmatrix} m_{11} & m_{12} & m_{13} \\ m_{21} & m_{22} & m_{23} \\ m_{31} & m_{32} & m_{33} \end{bmatrix} = \begin{bmatrix} 2q_0^2 - 1 & 2q_1q_2 & 2q_1q_3 \\ +2q_1^2 & +2q_0q_3 & -2q_0q_2 \\ & & \\ 2q_1q_2 & 2q_0^2 - 1 & 2q_2q_3 \\ -2q_0q_3 & +2q_2^2 & +2q_0q_1 \\ & & \\ 2q_1q_3 & 2q_2q_3 & 2q_0^2 - 1 \\ +2q_0q_2 & -2q_0q_1 & +2q_3^2 \end{bmatrix}
$$

from which we can write

$$
\begin{aligned}
4q_0q_1 &= m_{23} - m_{32} \\
4q_0q_2 &= m_{31} - m_{13} \\
4q_0q_3 &= m_{12} - m_{21} \\
\text{and} \quad tr(M) &= 4q_0^2 - 1
\end{aligned}
$$

From this last equation above it follows that

$$
q_0 = (1/2)\sqrt{m_{11} + m_{22} + m_{33} + 1}
$$

Given this expression for q_0, the remaining components of the desired quaternion are easily found. We have

$$
\begin{aligned}
q_1 &= (m_{23} - m_{32})/(4q_0) \\
q_2 &= (m_{31} - m_{13})/(4q_0) \\
q_3 &= (m_{12} - m_{21})/(4q_0)
\end{aligned}
$$

7.10 Rotation Operator Algebra

Finally, in this section we make some observations about the care that must be exercised when performing algebraic manipulations involving both matrices and quaternions.

7.10.1 Sequence of Rotation Operators

Consider the following sequence of successive rotations where P, Q, and R are distinct rotation matrices. Then writing $r^*\mathbf{v}r$ for $R\mathbf{v}$ and similar expressions for the other matrix rotation operators, we get

$$
\begin{aligned}
M\mathbf{v} &= PQR\mathbf{v} \\
&= PQr^*\mathbf{v}r \\
&= Pq^*r^*\mathbf{v}rq \\
&= p^*q^*r^*\mathbf{v}rqp \\
&= (rqp)^*\mathbf{v}(rqp)
\end{aligned}
$$

An Option

We could just as well have used the quaternion rotation operator

$$r\mathbf{v}r^*$$

depending upon the application and/or desired perspective.

For a repeated rotation, say, $R = Q = P$ then

$$
\begin{aligned}
M\mathbf{v} &= P^3\mathbf{v} \\
&= PPP\mathbf{v} \\
&= PPp^*\mathbf{v}p \\
&= Pp^*p^*\mathbf{v}pp \\
&= p^*p^*p^*\mathbf{v}ppp \\
&= (p^*)^3\mathbf{v}p^3
\end{aligned}
$$

If one of the matrices, say B, is not a rotation, then we must write

$$
\begin{aligned}
M\mathbf{v} &= PBQ\mathbf{v} \\
&= PB(q^*\mathbf{v}q) \\
&= P[B(q^*\mathbf{v}q)] \\
&= p^*[B(q^*\mathbf{v}q)]p
\end{aligned}
$$

7.10.2 Rotation of Vector Sets

In most applications it will be necessary to rotate either an entire coordinate frame which is defined by its set of three basis vectors, or an entire rigid body which is defined by a set of vectors or points. We adopt the following convenient matrix notation to represent a collection of n column vectors.

$$
V = \begin{bmatrix} | & | & & | \\ \mathbf{v}_1 & \mathbf{v}_2 & \cdots & \mathbf{v}_n \\ | & | & & | \end{bmatrix}
$$

A quaternion rotation operator which operates on this set of vectors produces a new set, which we collect in the matrix, say, W. We will *define* this operation to mean

$$
\begin{aligned}
W &= L_q(V) \\
&= qVq^*
\end{aligned}
$$

$$= q \begin{bmatrix} | & | & & | \\ \mathbf{v}_1 & \mathbf{v}_2 & \cdots & \mathbf{v}_n \\ | & | & & | \end{bmatrix} q^*$$

$$= \begin{bmatrix} | & | & & | \\ q\mathbf{v}_1 q^* & q\mathbf{v}_2 q^* & \cdots & q\mathbf{v}_n q^* \\ | & | & & | \end{bmatrix}$$

$$= \begin{bmatrix} | & | & & | \\ \mathbf{w}_1 & \mathbf{w}_2 & \cdots & \mathbf{w}_n \\ | & | & & | \end{bmatrix}$$

where each vector $\mathbf{v}_i \; \epsilon \; V$ is rotated according to the specified quaternion rotation operator L_q to give the corresponding rotated vector $\mathbf{w}_i \; \epsilon \; W$.

Note in particular that with this meaning of $L_q(V) \;=\; qVq^*$, Equation 5.11 implies that for any $3 \times n$ matrix V,

$$L_q(V) = qVq^* \;=\; QV$$

where Q is the rotation matrix which corresponds to a rotation operator using the quaternion q. But now more specifically, if we take V to be a 3×3 *identity matrix I*, whose columns represent the standard basis in R^3, we get the interesting formula

$$Q \;=\; QI \;=\; qIq^*$$

When using matrices in an expression, along with quaternion rotation operators, the algebra must be done with considerable care. In the next section we consider some examples which will confirm and emphasize this note of caution.

7.10.3 Mixing Matrices and Quaternions

At this point the reader is fairly well acquainted with the fact that there is a certain equivalence between the expressions

$$\mathbf{u} \;=\; A\mathbf{v} \qquad \text{and} \qquad \mathbf{u} \;=\; L_a(\mathbf{v}) \;=\; a^*\mathbf{v}\,a$$

Here it is understood, of course, that the matrix rotation operator A and the quaternion rotation operator L_a are alternative representations for the same rotation. We emphasize, however, that although rotation matrices may be equivalently represented using quaternions, *those matrices which are not rotations have no such quaternion representation.*

Consider first a simple operator which is the product of two matrices P and D, where D is some arbitrary 3×3 matrix, P is a rotation matrix, and the vector, $\mathbf{v}$, is an arbitrary 3-tuple. Because matrix multiplication is associative, the expression $PD\mathbf{v}$ can be grouped in different ways, in preparation for conversion to a quaternion form. We write

$$M\mathbf{v} \;=\; PD\mathbf{v} \qquad\qquad (7.18)$$

$$\text{(a)} \qquad = \; (P)D\mathbf{v} \;=\; (p^*Ip)D\mathbf{v}$$

$$\text{(b)} \qquad = \; (PD)\mathbf{v} \;=\; (p^*Dp)\mathbf{v}$$

$$\text{(c)} \qquad = \; P(D\mathbf{v}) \;=\; p^*(D\mathbf{v})p$$

The grouping in Equation 7.18a suggests first converting the rotation matrix P to p^*Ip, which is still a 3×3 matrix, but whose elements are functions of the associated quaternion elements.

The grouping in Equation 7.18b suggests first converting the matrix PD into p^*Dp, which also is still a 3×3 matrix, but whose elements are functions of the d_{ij} and the p_k elements.

The grouping in Equation 7.18c is the quaternion operation $p^*(D\mathbf{v})p$ on the matrix product $D\mathbf{v}$.

These three, of course, must all give equivalent results.

In order to illustrate a potential algebraic pitfall, we consider next a simple expression

$$M\mathbf{v} \;=\; PDP^t\mathbf{v} \qquad\qquad (7.19)$$

where D is again some arbitrary 3×3 matrix and P is a rotation matrix. This expression, rewritten using quaternions, is

$$M\mathbf{v} \;=\; p^*[D(p\mathbf{v}p^*)]p \qquad\qquad (7.20)$$

Now since quaternions as well as matrices are associative under multiplication we might be tempted to ignore the parentheses and brackets in Equation 7.20 and write

$$M\mathbf{v} = p^*[D(p\mathbf{v}\underbrace{p^*)]p}_{Collapse\ this\ ?} \;=\; (p^*Dp)\mathbf{v} \;=\; PD\mathbf{v} \;\neq\; PDP^t\mathbf{v}$$

NO. Quite obviously, something goes wrong when we perform the indicated cancellation; we lose P^t in Equation 7.19. The problem is that in the product enclosed in the square brackets, namely $D(p\mathbf{v}p^*)$

we *do not* have the required associativity. Therefore we may not write

$$[D(p\mathbf{v}p^*)] \;=\; [D(p\mathbf{v})]p^*$$

as the above algebraic manipulation would require. Notice in particular that the product $D(p\mathbf{v})$ makes no sense, because D is a 3×3 matrix while $(p\mathbf{v}$ is a 4-tuple.

We may, however, legitimately write Equation 7.19 in at least these three ways

$$
\begin{aligned}
M\mathbf{v} \;&=\; (P^t D)P\mathbf{v} \;=\; (pDp^*)(p^*\mathbf{v}p) \qquad \text{or}\\
&=\; P^t D(P\mathbf{v}) \;=\; p(Dp^*\mathbf{v}p)p^* \qquad \text{or}\\
&=\; P^t(DP\mathbf{v}) \;=\; p[D(p^*\mathbf{v}p)]p^*
\end{aligned}
$$

Whichever of the possible ways one might choose to associate the various adjacent factors in the expression, the parentheses, in general, may not be ignored.

Another way of looking at what went wrong earlier is that the matrix D in Equation 7.19 must operate on the vector

$$P^t\mathbf{v} \;=\; p\mathbf{v}p^*$$

before the final quaternion operation can be performed.

Perhaps writing the process in operator notation makes this clearer.

$$
\begin{aligned}
M\mathbf{v} \;&=\; PDP^t\mathbf{v}\\
&=\; L_p D L_{p^*}\mathbf{v} \qquad \text{where implicitly}
\end{aligned}
$$

the nesting is taken to be the most natural

$$= L_p[D(L_{p^*}\mathbf{v})]$$

As a final example which again emphasizes the caution required in the algebraic manipulation of matrices and quaternions, consider the matrix

$$M = AP^t CP$$

Here A and P are rotation matrices and C is some constant nonsingular 3×3 coupling matrix. This is called the 'normalized' signal matrix (which we will discuss in considerable detail later) in a Position and Orientation Measurement System.

We convert this matrix M to an equivalent expression using quaternions. Again, we find it helpful to have the operator M apply to some arbitrary vector, $\mathbf{v}$ — a *pure quaternion.* That is,

$$\mathbf{w} = M\mathbf{v} = a^*p[C(p^*\mathbf{v}p)]p^*a$$

Here one might be tempted to ignore the braces and brackets under the assumption that the *associativity property* will allow one to collapse those adjacent operators which are mutual inverses, namely $pp^* = p^*p = 1$, wherever they occur. For example, can we do the following

$$\mathbf{w} = M\mathbf{v} = a^*p[C(p^*\mathbf{v}\ \underbrace{p)]p^*}\ a$$
$$\text{\small\textit{May we Collapse this ?}}$$

Collapsing these terms gives

$$\mathbf{w} = M\mathbf{v} = a^*pCp^*\mathbf{v}a$$

However, note that certain associations of the remaining terms are not conformable. For example, the 3×3 coupling matrix C can not operate on the 4-tuples which, in general, are produced by the operation $p^*\mathbf{v}\,a$. For this reason, to make the operations conformable after collapsing $p)]p^*$ we also had to rearrange the parentheses as follows

$$\mathbf{u} = M\mathbf{v} = a^*\ [\underbrace{(pCp^*)}\ \mathbf{v}]a$$
$$\text{\small\textit{Do this next?}}$$

At least this next suggested operation seems possible. However, if we express this last equation in terms of equivalent matrix operations we get

$$\mathbf{u} = M\mathbf{v} = AP^tC\mathbf{v}$$

We have **lost** an operator P in the quaternion collapse.

In summary, great care must be exercised when certain matrix operators must be retained, such as the coupling operator $C = dg[2,\ -1,\ -1]$ for example, which must operate on a vector or set of vectors as we shall see in Chapter 12.

Exercises for Chapter 7

1. Prove that $\sin^2\alpha + \sin^2\beta + \sin^2\gamma = 2$, where the angles α, β, and γ are *direction angles*.

2. Find the equation of the plane tangent to the unit sphere in terms of *direction angles* of its normal vector $\mathbf{v}$.

3. Work through the necessary details (similar to Section 7.6) to get the composite quaternion components which represent the three rotation sequence for orbits represented in Figure 4.10.

Chapter 8

Quaternion Factors

8.1 Introduction

In the preceding chapters, as we developed the theory of matrix and quaternion rotation operators, we placed considerable emphasis on sequences of such operators. This in turn entailed much use of matrix and quaternion products, and up to this point, our theory and applications have been based on these products.

In this chapter we shall see whether this approach can be reversed. That is, we shall ask the question:

> *Given a certain matrix or quaternion, associated with some rotation in R^3, can we factor this matrix or quaternion into products in which the factors also represent meaningful and useful rotations in R^3?*

More specifically, suppose we have two coordinate frames related by some rotation matrix or quaternion operator.

> *Is it possible to decompose the matrix or quaternion into factors which represent a sequence of rotations about, say, principal axes?*

If so, perhaps certain computational efficiencies or advantages may result.

Our intent in this chapter is to answer these questions in the affirmative. In fact, most of the useful representations for rotations require that the matrix or quaternion be factored. We now develop the algebra which will yield a variety of factorizations, subject to certain constraints which are useful and appropriate to applications. But first, some preliminaries.

177

8.2   Factorization Criteria

At this point it is not at all obvious why one would want to factor a given rotation matrix into a product of two or more other rotation matrices. The same may be said about a quaternion which is associated with a given rotation in R^3. One conceivable reason might be that certain factorizations may afford a more time-efficient processing scheme. Another reason may be that such factorizations may be chosen so as to provide necessary geometric relationships in certain applications. For example, one or more of the factors may be assigned special geometric attributes which are meaningful to the application, in order to establish some useful relationship between the two factors for, say, computational reasons. We will encounter these matters in the applications which are discussed later.

It may be helpful for our discussion of these factorizations to list a few of these criteria which give direction to our factorization efforts — what we may call *factorization strategies*. We identify some of these as follows.

1. A specific factor may represent a rotation about a principal axis; then, the vector part of the quaternion (in the quaternion operator) has only one non-zero component.

2. The factors may be chosen so that the successive axes of the rotation sequence they represent are orthogonal.

3. Factors may be chosen so as to incorporate some combination of the foregoing.

4. We might seek three or more factors which are associated, for example, with an Euler angle-axis rotation sequence.

5. The vector defining the rotation axis associated with one of the factors is chosen in some direction especially meaningful to the application.

The above listing of possible factorization criteria surely is not exhaustive; certain applications could well impose other sorts of constraints or conditions (See, for example, the application considered in Chapter 11). In the following sections we consider some of these specific factorizations in the context of some applications which are now familiar to us. We begin with factorizations of rotation matrices and follow that with analogous factorizations of quaternions. Once again, as a result of this effort, we should be able to compare the efficacy of matrix and quaternion representations for rotations in R^3.

8.3 Transition Rotation Operators

We begin by noting that any two rotation matrices (that is, any two 3×3 orthogonal matrices with determinant $+1$), say A and B, may be related by a third such matrix, say T, which we will call a *transition matrix*. We shall say that the transition matrix T takes the rotation matrix A into the rotation matrix B if

$$B = TA$$

or equivalently if

$$T = BA^t$$

Of course we could also have factored the matrix B as

$$B = AT$$

that is

$$T = A^t B$$

where T again takes A into B. It should be noted, however, that the transition matrices, which in both cases we denote by T, are different, since we know that matrices in general do not commute under multiplication. At any rate, the idea of a transition matrix does provide a simple mechanism for representing any rotation matrix as a product of two other rotation matrices. That is to say, if B is some given rotation matrix, and if the rotation matrix A represents some significant physical attribute of the system which gives us B, then the transition matrix T is simply a rotation which relates A and B.

Which of the above two factorizations is appropriate, of course, will be dictated by the application, as we shall see. For these factorizations, however, there often are a variety of possible, as yet unspecified, constraints or attributes. These constraints and other possible factorizations will be considered in this chapter.

We next introduce *transistion quaternions* in exactly the same way. Any two *unit* quaternions, say p and q, may be related by a quaternion t, which, just as in the case of matrices, we call a *transition quaternion*. The transition quaternion t takes the quaternion p into the quaternion q if

$$q = tp$$

$$\text{or equivalently} \quad t = qp^*$$

Inverse Rotation

Recall that for a rotation matrix A, its inverse is simply its transpose, denoted A^t

Obviously, we could also have factored the quaternion q as

$$q = pt \quad \Rightarrow \quad t = p^* q$$

where t again takes p into q. Just as in the case with matrices, it should be noted that the transition quaternions, which in both of these cases we denoted as t, are different since we know quaternions, in general, also do not commute under multiplication. As before, the idea of the transition quaternion also provides a simple mechanism for representing any quaternion as a product of two quaternions. So if q is some given quaternion, and if the quaternion p represents some significant physical attribute of the system which gives us q, then the transition quaternion t simply relates p and q.

Unit Quaternions

Throughout our discussions, all quaternions are assumed to be **unit** quaternions for quite obvious reasons and unless otherwise noted — because the inverse of any unit quaternion p is simply its complex conjugate p^*. Efficiently maintaining this Euclidean norm, or any other useful norm in a dynamic environment, is a nontrivial matter, usually application dependent. Here again, the reader is given the opportunity to exploit her favorite normalizing process.

8.4   The Factorization $M = TA$

We now consider the factorization of rotation matrices, subject to the factorization criteria listed above. As we have already mentioned, there may well be other factorizations which are useful. We shall consider only two or three possibilities, hoping that the reader may learn from the details how other useful factorizations may be investigated.

8.4.1   Rotation Matrix A Specified

We assume we are given some general rotation matrix

$$M \;=\; \begin{bmatrix} m_{11} & m_{12} & m_{13} \\ m_{21} & m_{22} & m_{23} \\ m_{31} & m_{32} & m_{33} \end{bmatrix}$$

Thus M is orthogonal, that is, $M^t M = M M^t = I$, and M has determinant $+1$. This implies that both the rows and the columns of M, as vectors in R^3, form orthonormal sets. This means, of course, that the sum of the squares of the elements in any row or column is 1, and that the *scalar or dot product* of any two distinct rows or any two distinct columns is 0.

We begin, investigating the factorization of matrix

$$M = TA$$

where, in accordance with factorization criterion 1 above, the rotation matrix A is chosen to represent a rotation about a principal axis. The matrix T is an appropriate transition matrix which we wish to determine. Without loss of generality we may let A represent a rotation about, say, the **Z**-axis, through the angle ψ. Thus A has the form

$$A = \begin{bmatrix} \cos\psi & \sin\psi & 0 \\ -\sin\psi & \cos\psi & 0 \\ 0 & 0 & 1 \end{bmatrix}$$

We now define the matrix T to be

$$T = \begin{bmatrix} t_{11} & t_{12} & t_{13} \\ t_{21} & t_{22} & t_{23} \\ t_{31} & t_{32} & t_{33} \end{bmatrix}$$

Since the inverse of an orthogonal matrix is simply its transpose, it is easy to solve for the transition matrix T; we write

$$T = MA^t$$

And it follows directly that the indicated matrix product produces the following elements for the matrix T

$$
\begin{aligned}
t_{11} &= m_{11}\cos\psi + m_{12}\sin\psi \\
t_{12} &= -m_{11}\sin\psi + m_{12}\cos\psi \\
t_{13} &= m_{13} \\
t_{21} &= m_{21}\cos\psi + m_{22}\sin\psi \\
t_{22} &= -m_{21}\sin\psi + m_{22}\cos\psi \qquad\qquad (8.1) \\
t_{23} &= m_{23} \\
t_{31} &= m_{31}\cos\psi + m_{32}\sin\psi \\
t_{32} &= -m_{31}\sin\psi + m_{32}\cos\psi \\
t_{33} &= m_{33}
\end{aligned}
$$

Since the matrix M is given, the transition matrix T is determined as soon as we specify the rotation angle ψ in the matrix A. It thus appears that, given a general rotation matrix M, the rotation represented by this matrix can always be expressed as the product of an arbitrary rotation about the **Z**-axis, followed by an appropriate rotation represented by the transition matrix T.

We should remark that a rotation matrix is always invertible, so that in the above factorization the rotation represented by the matrix A could as well have been any specified rotation. We have simply specified A to represent a rotation about a principal axis, in accordance with the first factorization criterion above. Incidentally, either of the other two principal axes could have been used, with similar results.

8.4.2   Rotation Axes Orthogonal

Suppose next that we do not specify the angle ψ in the matrix A, but rather ask whether for a general rotation matrix M the second factorization criterion above can be met. More specifically, for the preceding factorization, we ask

> *With $M = TA$, is it possible that the transition matrix T may represent a rotation whose axis is orthogonal to the axis of the rotation matrix A ?*

As a specific example, suppose we wish the axis of the rotation represented by the matrix T to be the *new* y-axis. Then T must be of the form

$$T \;=\; \begin{bmatrix} \cos\theta & 0 & -\sin\theta \\ 0 & 1 & 0 \\ \sin\theta & 0 & \cos\theta \end{bmatrix}$$

where θ is the rotation angle. The question now becomes

> *Do angles θ and ψ exist for which the factorization $M = TA$ is possible for any rotation matrix M?*

If so, then we have factored the rotation represented by the rotation matrix M into a rotation about the **Z**-axis through the angle ψ, followed by a rotation about the *new* y-axis through the angle θ. We then have a factorization in which the rotation axes in the factors indeed are orthogonal, as we desired. In fact, the reader may recognize this rotation sequence as the *tracking sequence* discussed earlier.

With matrices A and T as defined, which have orthogonal rotation axes, we compute the product TA. Then we equate the elements in this matrix product with the corresponding elements in the matrix M. This correspondence produces the following system of equations

$$
\begin{aligned}
m_{11} &= \cos\theta\cos\psi \\
m_{12} &= \cos\theta\sin\psi \\
m_{13} &= -\sin\theta \\
m_{21} &= -\sin\psi \\
m_{22} &= \cos\psi \\
m_{23} &= 0 \\
m_{31} &= \sin\theta\cos\psi \\
m_{32} &= \sin\theta\sin\psi \\
m_{33} &= \cos\theta
\end{aligned}
$$

The question now is whether there are angles θ and ψ for which all of these equations are satisfied for a given rotation matrix M, and what (if any) constraints must be placed on the matrix M in order to reach this goal. We observe immediately from the sixth equation in this list that we must have

$$m_{23} = 0$$

This is a *necessary constraint* on M if the factorization we seek is possible. So from now on we suppose that M does satisfy this condition. If that is the case (that is $m_{23} = 0$), then in the second row of M we must have

$$m_{21}^2 + m_{22}^2 = 1$$

since the rows and columns of M all represent unit vectors in R^3. It is therefore possible to define an angle ψ such that

$$\sin \psi = -m_{21} \qquad \text{and} \qquad \cos \psi = m_{22}$$

For the same reason, in the third column of M we must have

$$m_{13}^2 + m_{33}^2 = 1$$

so that we may define an angle θ such that we have $\sin \theta = -m_{13}$ and $\cos \theta = m_{33}$. With these definitions for the angles θ and ψ the matrix M may be written in the form

$$M = \begin{bmatrix} m_{11} & m_{12} & -\sin \theta \\ -\sin \psi & \cos \psi & 0 \\ m_{31} & m_{32} & \cos \theta \end{bmatrix}$$

It remains to show that the elements m_{11}, m_{12}, m_{31} and m_{32} must then also be of the correct form. Consider the following set of equations

$$
\begin{aligned}
m_{11} \cos \theta \cos \psi + m_{12} \cos \theta \sin \psi \;+& \\
m_{31} \sin \theta \cos \psi + m_{32} \sin \theta \sin \psi &= 1 \\
-m_{11} \sin \psi + m_{12} \cos \psi &= 0 \\
-m_{31} \sin \psi + m_{32} \cos \psi &= 0 \\
-m_{11} \sin \theta + m_{31} \cos \theta &= 0
\end{aligned}
$$

The first equation in this array results from that fact that the matrix M has determinant $+1$. The second equation simply states that the first and second rows of M are orthogonal. Similarly, the third equation says that the second and third rows of M are orthogonal,

Restrictions

In many applications the restrictions

$$-\pi/2 < \theta < \pi/2$$

and

$$-\pi < \psi \le \pi$$

apply. With restrictions such as these, every ordered pair, (ψ, θ), defines a unique direction on the unit ball.

while the last equation results from observing that the first and third columns of M are orthogonal.

The reader may wish to check that the determinant of the coefficient matrix for the above system of equations is $-\cos\psi$. If in the matrix M we have $m_{22} \neq 0$, this determinant is non-zero, so that this system of equations has a unique solution. A little simple algebra (perhaps using Cramer's Rule) will show that this solution is

$$
\begin{aligned}
m_{11} &= \cos\theta\cos\psi \\
m_{12} &= \cos\theta\sin\psi \\
m_{31} &= \sin\theta\cos\psi \\
m_{32} &= \sin\theta\sin\psi
\end{aligned}
$$

Thus we have shown that there are angles θ and ψ such that the matrix M is exactly the product of the matrices T and A, as desired. Note that this factorization is possible for any rotation matrix M in which $m_{23} = 0$.

On the other hand, if $m_{22} = 0$, then $\cos\psi = 0$ and either $\sin\psi = +1$ or $\sin\psi = -1$. If $\sin\psi = +1$, then the matrix M must be of the form

$$
M = \begin{bmatrix} m_{11} & m_{22} & -\sin\theta \\ -1 & 0 & 0 \\ m_{31} & m_{32} & \cos\theta \end{bmatrix}
$$

Since the rows of M are othogonal, it is easily seen that $m_{11} = 0$ and $m_{31} = 0$. Hence M must be of the form

$$
M = \begin{bmatrix} 0 & m_{12} & -\sin\theta \\ -1 & 0 & 0 \\ 0 & m_{32} & \cos\theta \end{bmatrix}
$$

Since the determinant of M is $+1$, it follows immediately that we have

$$
m_{12}\cos\theta + m_{32}\sin\theta = 1
$$

Since the second and third columns of M are othogonal we also have

$$
-m_{12}\sin\theta + m_{32}\cos\theta = 0
$$

If we solve these two equations we get $m_{12} = \cos\theta$ and $m_{32} = \sin\theta$, and, since in this case $\sin\psi = 1$, our factorization is again shown to be possible.

In a similar fashion, it is not difficult to verify that the factorization is also possible in the case that $\sin\psi = -1$. Thus the factorization $M = TA$ is always possible, subject only to the constraint that $m_{23} = 0$.

Suffice it to say at this point there are five other cases in which a rotation matrix M may be factored into the product of two rotations about principal axes. It all depends on which element in M is 0. The other five cases are $m_{13} = 0$ or $m_{21} = 0$ or $m_{23} = 0$ or $m_{31} = 0$ or $m_{32} = 0$. In summary, in order for a rotation matrix, M, to be a tracking matrix, it must have a zero as one of its six off-diagonal elements. The position of the *required* 0 element in the tracking matrix M depends upon which sequence of principal rotation axes is being used. The reader should verify that the 0 element always appears in the *column* representing the *first* rotation axis and in the *row* of the *second* rotation axis. Thus if we have a z-axis rotation followed by a y-axis rotation, as above, the 0 element must appear in the m_{23} position.

8.4.3   A Slight Generalization

In the preceding section we considered a factorization of a general rotation matrix M, a factorization which was based on a rotation about the z-axis followed by a rotation about the y-axis. We were able to determine a rather simple contraint on the rotation matrix M (namely, M must have an off-diagonal zero-element) which guaranteed that such a factorization was possible. In order to illustrate the difficulties which one may experience in using this rotation matrix approach, let us consider a simple generalization of what we have just done. We consider once again a general rotation matrix M, and ask the question

> *Is it possible to factor the matrix M so that it represents a rotation about the z-axis followed by a rotation about, say, the axis* $\mathbf{v} = \mathbf{i} + \mathbf{j}$?

If the answer to this question is affirmative, then we have factored M into a product in which the rotation axes are orthogonal, but the axis of the second rotation is not a principal axis.

We know that a rotation through the angle θ about the axis $\mathbf{v} = \mathbf{i} + \mathbf{j}$ may be represented by the quaternion

$$q \;=\; q_0 + \mathbf{i}q_1 + \mathbf{j}q_2 + \mathbf{k}q_3$$

$$\text{where} \quad q_0 = \cos\frac{\theta}{2}$$

$$q_1 = \frac{1}{\sqrt{2}}\sin\frac{\theta}{2}$$

$$q_2 = \frac{1}{\sqrt{2}}\sin\frac{\theta}{2}$$

$$q_3 = 0$$

Then, using the results which we obtained in Section 5.14 for writing a rotation matrix, in terms of the components of the corresponding quaternion, we write

$$T = \frac{1}{2\sqrt{2}}\begin{bmatrix} \sqrt{2}(1+\cos\theta) & \sin\theta & -\sin\theta \\ \sin\theta & \sqrt{2}(1+\cos\theta) & \sin\theta \\ \sin\theta & -\sin\theta & 2\sqrt{2}\cos\theta \end{bmatrix}$$

The next step requires that we compute the product TA, where A is the matrix representing a rotation about the z-axis through the angle ψ, as before. This product is not exactly of a simple form. One must now equate the elements in this product with those of the matrix M, then ask whether there are angles θ and ψ for which all of these equations are satisfied. No doubt some constraints on the matrix M will emerge, but the process seems overly difficult and tedious. Only the most masochistic reader will want to pursue these details. It is (or will be) significant to note that the quaternion approach we consider later is much easier and much less tedious.

8.5   Three Principal-axis Factors

In this section we will extend the ideas studied in the preceding section by considering the possibility of expressing a general rotation in R^3 as a product of *three* rotations about principal axes. According to Euler's Theorem, such a factorization must always be possible, since the theorem states that any two coordinate frames may be related by a sequence of at most three Euler angle-axis rotations.

Again, we assume some general rotation matrix

$$M = \begin{bmatrix} m_{11} & m_{12} & m_{13} \\ m_{21} & m_{22} & m_{23} \\ m_{31} & m_{32} & m_{33} \end{bmatrix}$$

where M is orthogonal, that is, $M^t M = M M^t = I$, and M has determinant $+1$. As before, this again means that the row vectors and the column vectors of M form two orthonormal sets in R^3.

As an example of this procedure, we will determine the factorization $M = XYZ$ where XYZ is the now familiar Aerospace Euler angle/axis rotation sequence. In this sequence the matrix X represents a rotation about the x-axis, through the angle ϕ. Thus X has the form

Orthonormal Vectors

Each vector in an orthonormal set has norm equal to 1, and the scalar or dot product of any two distinct vectors in the set is equal to zero.

$$X = \begin{bmatrix} 1 & 0 & 0 \\ 0 & \cos\phi & \sin\phi \\ 0 & -\sin\phi & \cos\phi \end{bmatrix}$$

where the angle ϕ is as yet unknown. We then have a factorization of M of the form

$$M = XT$$

where the transition matrix T is a tracking matrix and the matrix X represents a rotation about the x-axis. We know that this tracking matrix, $T = YZ$, has at least one off-diagonal zero.

We now write

$$T = X^t M = \begin{bmatrix} t_{11} & t_{12} & t_{13} \\ t_{21} & t_{22} & t_{23} \\ t_{31} & t_{32} & t_{33} \end{bmatrix}$$

**Tracking Matrix
Zero Element**

Those tracking matrices which consists of only two principal-axis factors have at least one off-diagonal zero.

Since the inverse of the orthogonal matrix X is simply its transpose, it is easy to solve for the unknown elements of $T = [t_{ij}]$. The reader should compute the indicated matrix product in order to check that this matrix equation gives the following list for the elements of T

If	$T = XY$	then	$t_{12} = 0$
if	$T = XZ$	then	$t_{13} = 0$
if	$T = YX$	then	$t_{21} = 0$
if	$T = YZ$	then	$t_{23} = 0$
if	$T = ZX$	then	$t_{31} = 0$
if	$T = ZY$	then	$t_{32} = 0$

$$
\begin{aligned}
t_{11} &= m_{11} \\
t_{12} &= m_{12} \\
t_{13} &= m_{13} \\
t_{21} &= m_{21}\cos\phi - m_{31}\sin\phi \\
t_{22} &= m_{22}\cos\phi - m_{32}\sin\phi \\
t_{23} &= m_{23}\cos\phi - m_{33}\sin\phi = 0 \\
t_{31} &= m_{31}\cos\phi + m_{21}\sin\phi \\
t_{32} &= m_{32}\cos\phi + m_{22}\sin\phi \\
t_{33} &= m_{33}\cos\phi + m_{23}\sin\phi
\end{aligned}
$$

Since the elements m_{ij} of the matrix M are known, the transition matrix T is determined. For we have already specified the transition (tracking) matrix to be the Aerospace tracking sequence, $T = YZ$.

This means that $t_{23} = 0$ (as indicated in the above listing in the margin). Therefore the rotation angle ϕ in the matrix X is

$$\phi = \tan^{-1} \frac{m_{23}}{m_{33}}$$

Now that we know the angle ϕ, the remaining elements in the above list for T are readily computed. We now can use the techniques of the preceding section (with M replaced by T) to determine the angles ψ and θ. Thus the matrices Y and Z are determined and our factorization is complete.

An alternative procedure for finding the Aerospace factorization for any general rotation matrix, however, is to express

$$M = XYZ$$

where the rotations in the indicated sequence XYZ represent principal-axis rotations about the x-axis, the y-axis, and the z-axis, through angles, ϕ, θ, and ψ, respectively. This triple product of rotation matrices defines the elements of the given rotation matrix M to be

$$
\begin{aligned}
m_{11} &= \cos\psi\cos\theta \\
m_{12} &= \sin\psi\cos\theta \\
m_{13} &= -\sin\theta \\
m_{21} &= \cos\psi\sin\theta\sin\phi - \sin\psi\cos\phi \\
m_{22} &= \sin\psi\sin\theta\sin\phi + \cos\psi\cos\phi \\
m_{23} &= \cos\theta\sin\phi \\
m_{31} &= \cos\psi\sin\theta\cos\phi + \sin\psi\sin\phi \\
m_{32} &= \sin\psi\sin\theta\cos\phi - \cos\psi\sin\phi \\
m_{33} &= \cos\theta\cos\phi
\end{aligned}
$$

Rewriting this again in matrix form gives

$$
M = \begin{bmatrix} m_{11} & m_{12} & m_{13} \\ m_{21} & m_{22} & m_{23} \\ m_{31} & m_{32} & m_{33} \end{bmatrix}
$$

which term-for-term is equivalent to

$$
M = \begin{bmatrix}
\cos\psi\cos\theta & \sin\psi\cos\theta & -\sin\theta \\
\left\{ \begin{array}{l} \cos\psi\sin\theta\sin\phi \\ -\sin\psi\cos\phi \end{array} \right\} & \left\{ \begin{array}{l} \sin\psi\sin\theta\sin\phi \\ +\cos\psi\cos\phi \end{array} \right\} & \cos\theta\sin\phi \\
\left\{ \begin{array}{l} \cos\psi\sin\theta\cos\phi \\ +\sin\psi\sin\phi \end{array} \right\} & \left\{ \begin{array}{l} \sin\psi\sin\theta\cos\phi \\ -\cos\psi\sin\phi \end{array} \right\} & \cos\theta\cos\phi
\end{bmatrix}
$$

From these equations we can define the three Euler angles

$$\begin{aligned}
\psi &= \tan^{-1}\frac{m_{12}}{m_{11}} \\
\theta &= -\sin^{-1}m_{13} \\
\phi &= \tan^{-1}\frac{m_{23}}{m_{33}}
\end{aligned}$$

The above factorization procedures apply equally well to any of the other Euler *angle/axis* sequences. These are very useful factorizations of rotation matrices. Next we turn to similar factorizations of quaternion operators.

8.6 Factorization: $q = st = (s_0 + \mathbf{j}s_2)t$

We start with an arbitrary unit quaternion q as given and we wish to find two quaternion factors such that $q = st$; that is, q is equal to the quaternion product of two, as yet unspecified, quaternions, s and t. As we have already mentioned, in that which follows we will assume all of the quaternions to be unitary, that is, $|t| = |s| = |q| = 1$. One of the factors is now chosen such that it satisfies some meaningful restriction or attribute.

We begin by choosing the factor s in accordance with criterion #1 in Secion 8.2, namely, that s have only one non-zero vector component. Here we choose

$$s = s_0 + \mathbf{j}s_2$$

The motivation for this first choice is merely that such a quaternion corresponds to a simple rotation about a principal axis — in this case about the y-axis. We could just as well have chosen the factor s such that it represent a rotation about either of the other two principal axes, $\mathbf{i}$ or $\mathbf{k}$, without loss of generality. As a matter of fact, in some applications a rotation about one of these other two principal axes might well be more appropriate. But we proceed. We now take an arbitrary unit quaternion and compute the factors.

Note

Because s is a unit quaternion, the reader can easily check that in each coefficient matrix, which relates q and t, the inverse is its transpose.

$$\begin{aligned}
q &= q_0 + \mathbf{i}q_1 + \mathbf{j}q_2 + \mathbf{k}q_3 = st \\
&= (s_0 + \mathbf{j}s_2)t \\
&= (s_0 + \mathbf{j}s_2)(t_0 + \mathbf{i}t_1 + \mathbf{j}t_2 + \mathbf{k}t_3)
\end{aligned}$$

to obtain

$$\begin{aligned}
q_0 &= s_0 t_0 - s_2 t_2 \\
q_1 &= s_0 t_1 + s_2 t_3
\end{aligned}$$

$$q_2 = s_0 t_2 + s_2 t_0$$
$$q_3 = s_0 t_3 - s_2 t_1$$

These four equations can be written in matrix form as

$$\begin{bmatrix} q_0 \\ q_2 \end{bmatrix} = \begin{bmatrix} s_0 & -s_2 \\ s_2 & s_0 \end{bmatrix} \begin{bmatrix} t_0 \\ t_2 \end{bmatrix}$$

$$\begin{bmatrix} q_1 \\ q_3 \end{bmatrix} = \begin{bmatrix} s_0 & s_2 \\ -s_2 & s_0 \end{bmatrix} \begin{bmatrix} t_1 \\ t_3 \end{bmatrix}$$

The transition quaternion components then can be expressed as

$$\begin{bmatrix} t_0 \\ t_2 \end{bmatrix} = \begin{bmatrix} s_0 & s_2 \\ -s_2 & s_0 \end{bmatrix} \begin{bmatrix} q_0 \\ q_2 \end{bmatrix} \tag{8.2}$$

$$\begin{bmatrix} t_1 \\ t_3 \end{bmatrix} = \begin{bmatrix} s_0 & -s_2 \\ s_2 & s_0 \end{bmatrix} \begin{bmatrix} q_1 \\ q_3 \end{bmatrix} \tag{8.3}$$

In these last two matrix equations, we emphasize that the quaternion t is not yet determined. The reason is that we have not specified the *rotation angle*. This means that the relative magnitudes of the components s_0 and s_2 of the quaternion factor s are not known. There are, however, at least two possible options:

1. Explicitly specify s_0 and s_2, or

2. Specify some further constraint on the quaternion, t.

We now consider these two options which will, in each case, uniquely determine both factors, s and t.

8.6.1 Principal-axis Factor Specified

Note

Since s is a unit quaternion, we must have

$$s_0^2 + s_2^2 = 1$$

Here, the principal-axis factor

$$s = s_0 + \mathbf{j} s_2$$

is defined to be a rotation through a specified angle about the **y**-axis by specifying values for s_0 and s_2. Then the transition quaternion, t, is uniquely determined by the matrix equations

$$\begin{bmatrix} t_0 \\ t_2 \end{bmatrix} = \begin{bmatrix} s_0 & s_2 \\ -s_2 & s_0 \end{bmatrix} \begin{bmatrix} q_0 \\ q_2 \end{bmatrix} \tag{8.4}$$

$$\begin{bmatrix} t_1 \\ t_3 \end{bmatrix} = \begin{bmatrix} s_0 & -s_2 \\ s_2 & s_0 \end{bmatrix} \begin{bmatrix} q_1 \\ q_3 \end{bmatrix} \tag{8.5}$$

These matrix equations give us the following expressions for the components of the quaternion t in terms of the specified quaternion s and the given quaternion q.

$$
\begin{aligned}
t_0 &= s_0 q_0 + s_2 q_2 \\
t_1 &= s_0 q_1 - s_2 q_3 \\
t_2 &= -s_2 q_0 + s_0 q_2 \\
t_3 &= s_2 q_1 + s_0 q_3
\end{aligned}
$$

Thus, given some rotation in R^3 represented by the quaternion q, we may use these equations to write this rotation as a sequence of two rotations, one of which is a rotation through a specified angle about a principal axis, namely, the **y** axis. Of course, similar equations are obtained if we use a different principal axis.

8.6.2 Orthogonal Factors

As one example of Option 2, we may specify that the rotation axes represented by the two quaternion factors, s and t, be orthogonal — a constraint that may well be useful in certain applications. With one of the quaternion factors specified to be of the form

$$
s = s_0 + \mathbf{j}s_2
$$

as in the preceding section, we now place a constraint on the other quaternion factor, t, such that it be of the form

$$
t = t_0 + \mathbf{i}t_1 + \mathbf{k}t_3
$$

Note that this choice for t makes the vector parts of the two quaternion factors, s and t orthogonal. The two matrix equations which define both quaternion factors s and t then are

$$
\begin{bmatrix} t_0 \\ 0 \end{bmatrix} = \begin{bmatrix} s_0 & s_2 \\ -s_2 & s_0 \end{bmatrix} \begin{bmatrix} q_0 \\ q_2 \end{bmatrix} \tag{8.6}
$$

$$
\begin{bmatrix} t_1 \\ t_3 \end{bmatrix} = \begin{bmatrix} s_0 & -s_2 \\ s_2 & s_0 \end{bmatrix} \begin{bmatrix} q_1 \\ q_3 \end{bmatrix} \tag{8.7}
$$

The first of these two matrix equations says that

$$
\begin{aligned}
t_0 &= s_0 q_0 + s_2 q_2 \\
0 &= -s_2 q_0 + s_0 q_2
\end{aligned}
$$

Since s is a unit quaternion we must, of course, have

$$
s_0^2 + s_2^2 = 1
$$

Hence, given values for s_0 and s_2 which satisfy this requirement, there must be some angle, say α, such that $s_0 = \cos\alpha$ and $s_2 = \sin\alpha$. If this is the case, then from the second equation above we have

$$\tan\alpha = \frac{\sin\alpha}{\cos\alpha} = \frac{s_2}{s_0} = \frac{q_2}{q_0}$$

From this it appears that

$$\alpha = \arctan\frac{q_2}{q_0}$$

We conclude that the choice

$$
\begin{aligned}
s_0 &= \cos(\arctan\frac{q_2}{q_0}) \\
\text{and} \quad s_2 &= \sin(\arctan\frac{q_2}{q_0})
\end{aligned}
$$

will insure that $t_2 = 0$, and that the vector part of t will be orthogonal to the vector part of s, as we desired. Having determined values for s_0 and s_2, we may compute the remaining components of the transition quaternion t from the above matrix equations, which give

$$
\begin{aligned}
t_0 &= s_0 q_0 + s_2 q_2 \\
t_1 &= s_0 q_1 - s_2 q_3 \\
t_3 &= s_2 q_1 + s_0 q_3
\end{aligned}
$$

8.7 Euler Angle-Axis Factors

In previous chapters it was noted that, in general, a sequence of at most three angle/axis rotations (recall Euler angles) is required to relate two independent coordinate frames. In this section we will factor an arbitrary quaternion into an *ordered triple product* which must represent one of the twelve possible Euler angle/axis sequences discussed earlier. Before we embark on any specific Euler angle-axis triple factorization of an arbitrary quaternion, q, we define some helpful notation.

We introduce this new notation in Figure 8.1. Here the quaternions in the ordered sequence of rotations are designated a^i followed by b^j followed by c^k. Each of the rotations in this figure is assumed taken about one of the *principal axes* (as yet unspecified) *of the most recent frame*.

This new notation emphasizes that the quaternion a, for example, which represents the first rotation is taken about a principal axis — as we have done earlier when we first defined Euler angle/axis sequences of rotation matrices. Here, however, we define these se-

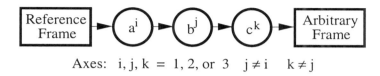

Axes: $i, j, k = 1, 2,$ or 3 $j \neq i$ $k \neq j$

Figure 8.1: The Three Rotation Angle-axis Sequence

quences in terms of the quaternion. If a represents the quaternion used for the 1st rotation, then

$$
\begin{aligned}
a &= a_0 + \mathbf{i}a_1 = a^1 \quad \text{if the rotation is about the x-axis, or} \\
a &= a_0 + \mathbf{j}a_2 = a^2 \quad \text{if the rotation is about the y-axis, or} \\
a &= a_0 + \mathbf{k}a_3 = a^3 \quad \text{if the rotation is about the z-axis}
\end{aligned}
$$

The second rotation, b, is taken about one of the principal axes of the *new* frame defined after rotation a, and so on. After these three rotations, the resulting quaternion is

$$ q = a^i b^j c^k \quad \text{where} \quad i, j, k \; \epsilon \; 1, 2, 3 \;\; j \neq i \;\; k \neq j $$

Any vector $\mathbf{u}$ in the reference frame is related to the (same) vector $\mathbf{v}$ in the new frame by the now familiar quaternion rotation operator

$$ \mathbf{v} = q^* \mathbf{u} q \tag{8.8} $$

Here the vector $\mathbf{u}$, defined in the reference frame, is expressed as the *same* vector $\mathbf{v}$, defined now in the rotated frame.

It is clear from the foregoing that a sequence of three rotations which are defined using quaternions in this fashion can properly relate any two coordinate frames using the quaternion rotation operator of Equation 8.8. The converse, however, suggests the question

> **Question 1**   *Can any unit quaternion be factored into three quaternions which represent any one of the twelve Euler angle/axis sequences?*

On the basis of the theorem cited in Section 4.3, the answer to this question must quite obviously be "yes." We will demonstrate this

Units

All quaternions are assumed to be **unit** quaternions unless otherwise noted.

later by considering the xyz (Aerospace) rotation axes sequence and the zxz (Orbit) rotation axes sequence, both of which were discussed earlier using the matrix rotation operator. First, however, we consider the simpler two rotation sequence as used in the now familiar tracking application.

8.7.1 Tracking Revisited

Earlier, we defined the sequence of two rotations illustrated in Figure 8.2 as the tracking sequence — and this tracking application has been considered in some depth. In our quaternion analysis of this sequence we used the transformation

$$\mathbf{v} = q^*\mathbf{u}q \qquad (8.9)$$

Here, the vector $\mathbf{u}$ defines the direction of North in the local tangent plane, and $\mathbf{v}$ defines the vector direction to the remote object. The quaternion rotation operator of Equation 8.9 takes $\mathbf{u}$ into $\mathbf{v}$. However, now that our immediate concern has to do with quater-

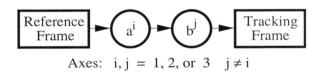

Axes: $i, j = 1, 2, \text{ or } 3 \quad j \neq i$

Figure 8.2: A Two-Rotation Angle-axis Sequence

nion *factors*, some interesting questions arise.

Question 2   *Can any arbitrary unit quaternion be factored as a two-rotation tracking quaternion?*

Obviously, the answer to this question is 'no.' If we said 'yes' it would be equivalent to saying that any two arbitrary coordinate frames can be related by an Euler angle/axis sequence of *two* rotations. This we know is not true. So we ask

Question 3   *What constraints on a unit quaternion make it a two-rotation tracking quaternion?*

To help us answer this question, we investigate the now quite familiar tracking application one more time. In this application, the tracking frame is related to the reference frame by a two rotation sequence. The first rotation is about the z-axis, followed by a second

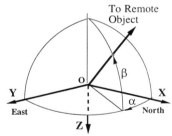

Tracking Sequence

The tracking sequence illustrated here may be represented by the quaternion

$$q = a^3 b^2$$
$$\text{where} \quad a^3 = a_0 + \mathbf{k}a_3$$
$$\text{and} \quad b^2 = b_0 + \mathbf{j}b_2$$

and where all quaternions are assumed to be **unit** quaternions unless otherwise noted.

rotation about the new y-axis. In our new notation we write this tracking quaternion as

$$q = a^3 b^2$$

If an arbitrary unit quaternion is to be a tracking quaternion of this form, we must have

$$
\begin{aligned}
q &= q_0 + \mathbf{i}q_1 + \mathbf{j}q_2 + \mathbf{k}q_3 = a^3 b^2 \\
&= (a_0 + \mathbf{k}a_3)(b_0 + \mathbf{j}b_2) \\
&= a_0 b_0 - \mathbf{i}a_3 b_2 + \mathbf{j}a_0 b_2 + \mathbf{k}a_3 b_0 \qquad (8.10)
\end{aligned}
$$

In Equation 8.10 it is easily verified, that

$$q_0 q_1 + q_2 q_3 = 0 \qquad (8.11)$$

This gives us a *necessary* constraint on the quaternion q for this particular tracking sequence. It may be verified that this constraint is also *sufficient*, and thus we have answered Question 3 for this particular tracking sequence.

However, a tracking sequence may be defined in other ways. If the two-rotation tracking sequence were defined to be, say

$$
\begin{aligned}
q &= q_0 + \mathbf{i}q_1 + \mathbf{j}q_2 + \mathbf{k}q_3 = a^1 b^2 \\
&= (a_0 + \mathbf{i}a_1)(b_0 + \mathbf{j}b_2) \\
&= a_0 b_0 + \mathbf{i}a_1 b_0 + \mathbf{j}a_0 b_2 + \mathbf{k}a_1 b_2 \qquad (8.12)
\end{aligned}
$$

then the constraint equivalent to Equation 8.11 would become

$$q_1 q_2 - q_3 q_0 = 0 \qquad (8.13)$$

We now tabulate the six possible two-rotation tracking sequences. Each of these sequences we represent as a composite quaternion, q, and each has an associated constraint equation on its elements. That is, each two-rotation tracking sequence is written in the form

$$q = q_0 + \mathbf{i}q_1 + \mathbf{j}q_2 + \mathbf{k}q_3 = a^i b^j$$

Specifically, if $\quad q = (a_0 + \mathbf{i}a_1)(b_0 + \mathbf{j}b_2) = a^1 b^2$
$$= a_0 b_0 + \mathbf{i}a_1 b_0 + \mathbf{j}a_0 b_2 + \mathbf{k}a_1 b_2$$
then the constraint equation is $\quad q_1 q_2 = q_3 q_0$

or if $\quad q = (a_0 + \mathbf{j}a_2)(b_0 + \mathbf{i}b_1) = a^2 b^1$
$$= a_0 b_0 + \mathbf{i}a_0 b_1 + \mathbf{j}a_2 b_0 - \mathbf{k}a_2 b_1$$
then the constraint equation is $\quad q_2 q_1 = -q_3 q_0$

Frame Rotation

The quaternions a and b or more specifically ab represent frame rotations with respect to the reference frame as illustrated in Figure 8.2. That is,

$$\mathbf{v} = q^* \mathbf{u} q = b^* a^* \mathbf{u} ab$$

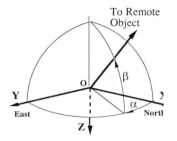

Tracking Sequence

The tracking sequence illustrated here may be represented by the quaternion

$$
\begin{aligned}
q &= a^3 b^2 \\
\text{where} \quad a^3 &= a_0 + \mathbf{k}a_3 \\
\text{and} \quad b^2 &= b_0 + \mathbf{j}b_2
\end{aligned}
$$

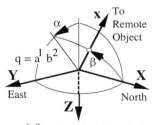

$$q = a^1 b^2 = (a_0 + \mathbf{i}a_1)(b_0 + \mathbf{j}b_2)$$

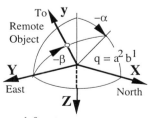

$$q = a^1 b^2 = (a_0 + \mathbf{i}a_1)(b_0 + \mathbf{j}b_2)$$

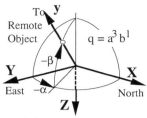

$q = a^3 b^1 = (a_0 + \mathbf{k}a_3)(b_0 + \mathbf{i}b_1)$

$$
\begin{aligned}
\text{or if} \quad q &= (a_0 + \mathbf{k}a_3)(b_0 + \mathbf{i}b_1) = a^3 b^1 \\
&= a_0 b_0 + \mathbf{i}a_0 b_1 + \mathbf{j}a_3 b_1 + \mathbf{k}a_3 b_0
\end{aligned}
$$

then the constraint equation is $\quad q_3 q_1 = q_2 q_0$

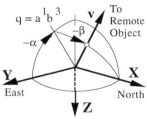

$q = a^1 b^3 = (a_0 + \mathbf{i}a_1)(b_0 + \mathbf{k}b_3)$

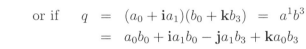

$$
\begin{aligned}
\text{or if} \quad q &= (a_0 + \mathbf{i}a_1)(b_0 + \mathbf{k}b_3) = a^1 b^3 \\
&= a_0 b_0 + \mathbf{i}a_1 b_0 - \mathbf{j}a_1 b_3 + \mathbf{k}a_0 b_3
\end{aligned}
$$

then the constraint equation is $\quad q_1 q_3 = -q_2 q_0$

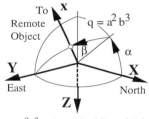

$q = a^2 b^3 = (a_0 + \mathbf{j}a_2)(b_0 + \mathbf{k}b_3)$

$$
\begin{aligned}
\text{or if} \quad q &= (a_0 + \mathbf{j}a_2)(b_0 + \mathbf{k}b_3) = a^2 b^3 \\
&= a_0 b_0 + \mathbf{i}a_2 b_3 + \mathbf{j}a_2 b_0 + \mathbf{k}a_0 b_3
\end{aligned}
$$

then the constraint equation is $\quad q_2 q_3 = q_0 q_1$

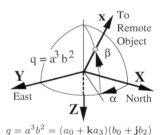

$q = a^3 b^2 = (a_0 + \mathbf{k}a_3)(b_0 + \mathbf{j}b_2)$

$$
\begin{aligned}
\text{or if} \quad q &= (a_0 + \mathbf{k}a_3)(b_0 + \mathbf{j}b_2) = a^3 b^2 \\
&= a_0 b_0 - \mathbf{i}a_3 b_2 + \mathbf{j}a_0 b_2 + \mathbf{k}a_3 b_0
\end{aligned}
$$

then the constraint equation is $\quad q_3 q_2 = -q_0 q_1$

It must be emphasized that each of these six two-quaternion sequences represents only a *tracking* rotation sequence. If a rotation sequence is to represent one of the twelve possible Euler angle/axis factorizations (for any given unit quaternion), the sequence must include a third quaternion which we denote c^k in Figure 8.1.

Note in particular that the third quaternion factor must be one of two types. This third factor must either be about a principal axis not yet employed in the first two quaternion factors, or it must be about the same axis as used in the first quaternion factor. For example, if the first two factors are represented by the familiar tracking quaternion

$$q = a^3 b^2$$

then the required third quaternion factor c^k must either be c^1 or c^3. That is, given that the first two rotations in the three rotation sequence are represented by $q = a^3 b^2$, then any unit quaternion p may be factored to give either

$$p = a^3 b^2 c^1 \quad \text{(distinct principal-axes)} \qquad (8.14)$$
$$\text{or} \quad p = a^3 b^2 c^3 \quad \text{(repeated principal-axes)} \qquad (8.15)$$

In the next two sections we develop the algorithms which will factor any arbitrary unit quaternion, p, into one of these two (of the twelve possible) factorizations. Algorithms for factorizations into any of the other ten would proceed in a similar fashion.

8.7.2 Distinct Principal Axis Factorization

Let p be any arbitrary given unit quaternion, that is

$$p = p_0 + \mathbf{i} p_1 + \mathbf{j} p_2 + \mathbf{k} p_3 \quad \text{and} \quad |p| = 1$$

The Aerospace factorization of p in Equation 8.13 is

$$
\begin{aligned}
p &= a^3 b^2 c^1 \\
&= q c^1 \\
\text{where} \quad q &= a^3 b^2 \\
\text{and} \quad c^1 &= c_0 + \mathbf{i} c_1 \\
&= \cos\frac{\phi}{2} + \mathbf{i} \sin\frac{\phi}{2} \qquad (8.16)
\end{aligned}
$$

Here ϕ is the angle of rotation (about the x-axis) associated with the quaternion c^1.

The constraint on the elements of the quaternion q in this particular Aerospace sequence (see Equation 8.11) requires that

$$q_0 q_1 + q_2 q_3 = 0 \qquad \text{or} \qquad \frac{q_0}{q_2} = -\frac{q_3}{q_1}$$

Rotation Sequences

Recall, in Section 4.3, on page 84, these twelve sequences are

xyz	yzx	zxy
xzy	yxz	zyx
xyx	yzy	zxz
xzx	yxy	zyz

Principal-axis Quaternion

By a *principal-axis quaternion* we mean a quaternion which has only two elements — the only non-zero vector component designates the principal axis about which the rotation occurs.

From the foregoing we may write

$$q = (p_0 + \mathbf{i}p_1 + \mathbf{j}p_2 + \mathbf{k}p_3)(c_0 - \mathbf{i}c_1)$$

so that
$$q_0 = p_0 c_0 + p_1 c_1$$
$$q_1 = p_1 c_0 - p_0 c_1 \qquad (8.17)$$
$$q_2 = p_2 c_0 - p_3 c_1$$
$$q_3 = p_3 c_0 + p_2 c_1$$

These four equations written in matrix form are

$$\begin{bmatrix} q_0 \\ q_2 \end{bmatrix} = \begin{bmatrix} p_0 & p_1 \\ p_2 & -p_3 \end{bmatrix} \begin{bmatrix} c_0 \\ c_1 \end{bmatrix} \qquad (8.18)$$

and
$$\begin{bmatrix} q_1 \\ q_3 \end{bmatrix} = \begin{bmatrix} p_1 & -p_0 \\ p_3 & p_2 \end{bmatrix} \begin{bmatrix} c_0 \\ c_1 \end{bmatrix} \qquad (8.19)$$

We now pre-multiply Equation 8.19 by the transpose of Equation 8.18, which gives

$$q_0 q_1 + q_2 q_3 = \begin{bmatrix} c_0 & c_1 \end{bmatrix} \begin{bmatrix} p_0 & p_2 \\ p_1 & -p_3 \end{bmatrix} \begin{bmatrix} p_1 & -p_0 \\ p_3 & p_2 \end{bmatrix} \begin{bmatrix} c_0 \\ c_1 \end{bmatrix}$$

$$= \begin{bmatrix} c_0 & c_1 \end{bmatrix} \begin{bmatrix} A & B \\ D & -A \end{bmatrix} \begin{bmatrix} c_0 \\ c_1 \end{bmatrix} \qquad (8.20)$$

where
$$A = p_0 p_1 + p_2 p_3$$
$$B = -p_0^2 + p_2^2$$
$$D = p_1^2 - p_3^2$$

Since $q_0 q_1 + q_2 q_3 = 0$, Equation 8.20 gives a quadratic form

$$0 = A(c_0^2 - c_1^2) + \frac{B + D}{2} 2 c_0 c_1 \qquad (8.21)$$

from which we can write

$$\frac{2 c_0 c_1}{c_0^2 - c_1^2} = -\frac{2A}{B + D} = \tan \phi \qquad (8.22)$$

In Equation 8.22 we have used the fact (see Equation 8.16) that

$$c_0 = \cos \frac{\phi}{2} \qquad \text{and} \qquad c_1 = \sin \frac{\phi}{2}$$

and that

$$2 c_0 c_1 = 2 \sin \frac{\phi}{2} \cos \frac{\phi}{2} = \sin \phi$$

$$c_0^2 - c_1^2 = \cos^2 \frac{\phi}{2} - \sin^2 \frac{\phi}{2} = \cos \phi$$

In Equation 8.22 the values for A, B, and D, are known; hence $\tan \phi$ is known. Therefore, we can solve for the two components c_0 and c_1 of the quaternion factor c^1. We use these values for c_0 and c_1, along with the known values of the components of the given quaternion p, in Equations 8.17, to solve for the elements of q. Then, because $q = a^3 b^2$, it follows that

Note

Remember, if

$$\tan \theta = \frac{y}{x}$$

then $\sin \theta = \dfrac{y}{\sqrt{x^2 + y^2}}$

and $\cos \theta = \dfrac{x}{\sqrt{x^2 + y^2}}$

$$q_0 = a_0 b_0 \qquad (8.23)$$
$$q_1 = -a_3 b_2 \qquad (8.24)$$
$$q_2 = a_0 b_2 \qquad (8.25)$$
$$q_3 = a_3 b_0 \qquad (8.26)$$

And, since a^3 and b^2 are unit quaternions, we may write

$$
\begin{aligned}
a^3 &= a_0 + \mathbf{k} a_3 \quad \text{and} \quad |a^3| = 1 \\
&= \cos \frac{\psi}{2} + \mathbf{k} \sin \frac{\psi}{2}
\end{aligned}
$$

$$
\begin{aligned}
\text{and} \quad b^2 &= b_0 + \mathbf{j} b_2 \quad \text{and} \quad |b^2| = 1 \\
&= \cos \frac{\theta}{2} + \mathbf{j} \sin \frac{\theta}{2}
\end{aligned}
$$

Then their respective components can be determined in the following manner. We divide Equation 8.26 by Equation 8.23 to give

$$\frac{q_3}{q_0} = \frac{a_3}{a_0} = \tan \frac{\psi}{2}$$

The reader may then easily verify that

$$a_0 = \frac{q_0}{\sqrt{q_0^2 + q_3^2}}$$

$$\text{and} \quad a_3 = \frac{q_3}{\sqrt{q_0^2 + q_3^2}}$$

In a similar manner, divide Equation 8.25 by Equation 8.23 to give

$$\frac{q_2}{q_0} = \frac{b_2}{b_0} = \tan \frac{\theta}{2}$$

Again, from this we can write

$$b_0 = \frac{q_0}{\sqrt{q_0^2 + q_2^2}}$$

$$\text{and} \quad b_2 = \frac{q_2}{\sqrt{q_0^2 + q_2^2}}$$

With these computations we have shown that any arbitrary unit quaternion p can be factored into

$$p = a^3 b^2 c^1 = (a_0 + \mathbf{k}a_3)(b_0 + \mathbf{j}b_2)(c_0 + \mathbf{i}c_1)$$

in which the rotations in the sequence are about distinct principle axes. There are, of course, five other such sequences (about distinct principal axes) and for each of these the computations would be similar.

In the next section we consider the factorization of p into

$$p = a^3 b^2 c^3 = (a_0 + \mathbf{k}a_3)(b_0 + \mathbf{j}b_2)(c_0 + \mathbf{k}c_3)$$

where the final rotation c is again about the $\mathbf{k}$-axis — the only other choice, given that the first two rotations are specified about the $\mathbf{k}$ and $\mathbf{j}$ axes, respectively.

8.7.3   Repeated Principal Axis Factorization

Again, let p be any arbitrary given unit quaternion, that is

$$p = p_0 + \mathbf{i}p_1 + \mathbf{j}p_2 + \mathbf{k}p_3 \quad \text{and} \quad |p| = 1$$

In this section, however, the desired factorization of the given quaternion p is

$$p = a^3 b^2 c^3 = qc^3 \quad \text{where again we let} \quad q = a^3 b^2$$

The constraint on the elements of the quaternion q in this sequence (which is the same as that specified above in the Aerospace factorization) again requires that

$$q_0 q_1 + q_2 q_3 = 0 \qquad \text{or} \qquad \frac{q_0}{q_2} + \frac{q_3}{q_1} = 0 \qquad (8.27)$$

Just as before, we can write

$$
\begin{aligned}
q &= (p_0 + \mathbf{i}p_1 + \mathbf{j}p_2 + \mathbf{k}p_3)(c_0 - \mathbf{k}c_3) \\
\text{so that} \quad q_0 &= p_0 c_0 + p_3 c_3 \\
q_1 &= p_1 c_0 - p_2 c_3 \\
q_2 &= p_2 c_0 + p_1 c_3 \\
q_3 &= p_3 c_0 - p_0 c_3
\end{aligned}
\qquad (8.28)
$$

In matrix form these equations may be written as

$$
\begin{bmatrix} q_0 \\ q_2 \end{bmatrix} = \begin{bmatrix} p_0 & p_3 \\ p_2 & p_1 \end{bmatrix} \begin{bmatrix} c_0 \\ c_3 \end{bmatrix}
\qquad (8.29)
$$

$$
\begin{bmatrix} q_1 \\ q_3 \end{bmatrix} = \begin{bmatrix} p_1 & -p_2 \\ p_3 & -p_0 \end{bmatrix} \begin{bmatrix} c_0 \\ c_3 \end{bmatrix}
\qquad (8.30)
$$

We again pre-multiply Equation 8.30 by the transpose of Equation 8.29, to obtain

$$
q_0 q_1 + q_2 q_3 = \begin{bmatrix} c_0 & c_3 \end{bmatrix} \begin{bmatrix} p_0 & p_1 \\ p_3 & p_2 \end{bmatrix} \begin{bmatrix} p_1 & -p_2 \\ p_3 & -p_0 \end{bmatrix} \begin{bmatrix} c_0 \\ c_3 \end{bmatrix}
$$

$$
= \begin{bmatrix} c_0 & c_3 \end{bmatrix} \begin{bmatrix} A & B \\ D & E \end{bmatrix} \begin{bmatrix} c_0 \\ c_3 \end{bmatrix} \qquad (8.31)
$$

$$
\text{where} \quad
\begin{aligned}
A &= p_0 p_1 + p_1 p_3 \\
B &= -p_0 p_1 - p_0 p_2 \\
D &= p_1 p_3 + p_2 p_3 \\
E &= -p_2 p_3 - p_2 p_0
\end{aligned}
$$

Since $q_0 q_1 + q_2 q_3 = 0$, Equation 8.31 gives the quadratic form

$$
0 = A c_0^2 + (B + D) c_0 c_3 + E c_3^2
$$

whose solutions can be written

$$
\frac{c_3}{c_0} = \frac{B + D}{2E} \left[-1 \pm \sqrt{1 - \frac{4AE}{(B+D)^2}} \right] \qquad (8.32)
$$

From Equation 8.32 we note that there are two possible solutions for this ratio. Since c is a unit quaternion, we may solve for the two components of

$$
c^3 = c_0 + \mathbf{k} c_3
$$

We then use these values for c_0 and c_3, along with the known values of the components of the given quaternion p, in Equations 8.28, to solve for the elements of q, just as in the previous section. Because $q = a^3 b^2$ it follows that

$$
\begin{aligned}
q_0 &= a_0 b_0 \\
q_1 &= -a_3 b_2 \\
q_2 &= a_0 b_2 \\
q_3 &= a_3 b_0
\end{aligned}
$$

And, since a^3 and b^2 also are unit quaternions, which means that

$$
a^3 = a_0 + \mathbf{k} a_3 \qquad b^2 = b_0 + \mathbf{j} b_2 \qquad |a^3| = |b^2| = 1
$$

their respective components can be readily determined, just as before.

There are five other factorizations of this sort (repeated principal-axes), and once again the computation for each of these would be similar to the foregoing.

8.8   Some Geometric Insight

Rather than forming the vector Equations 8.29 and 8.30 from the four equations listed in 8.28, we could form the following

$$\mathbf{U} = \begin{bmatrix} q_0 \\ q_3 \end{bmatrix} = \begin{bmatrix} c_0 & c_3 \\ -c_3 & c_0 \end{bmatrix} \begin{bmatrix} p_0 \\ p_3 \end{bmatrix} = M\mathbf{u} \qquad (8.33)$$

$$\mathbf{V} = \begin{bmatrix} q_1 \\ q_2 \end{bmatrix} = \begin{bmatrix} c_0 & -c_3 \\ c_3 & c_0 \end{bmatrix} \begin{bmatrix} p_1 \\ p_2 \end{bmatrix} = M^t\mathbf{v} \qquad (8.34)$$

for the geometric insight it affords.

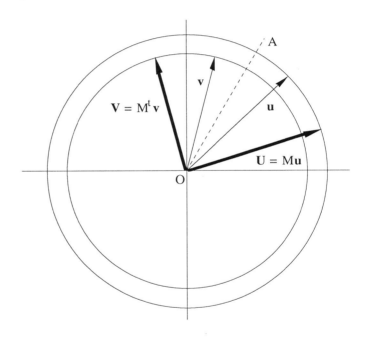

Figure 8.3: Roll Rotation

In accordance with Equations 8.33 and 8.34 the known vectors **u** and **v** are subjected to equal and opposite rotations, until the vectors **U** and **V** are mutually orthogonal, which is required by Equation 8.27. Figure 8.3 illustrates the geometric relationships.

The angle, say θ, between **u** and **v** is given by

$$\theta = \arccos \frac{\mathbf{u} \cdot \mathbf{v}}{|\mathbf{u}||\mathbf{v}|}$$

and if we let $$\frac{\sigma}{2} = \frac{1}{2}(\frac{\pi}{2} - \theta)$$

then the elements in the rotation matrices in Equations 8.33 and 8.34 may be defined

$$c_0 = \cos\frac{\sigma}{2}$$
$$c_3 = \sin\frac{\sigma}{2}$$

It seems this particular geometry applies only in those cases with *repeated* principal-axis factorizations.

Exercises for Chapter 8

1. Find the t_{ij}'s of Equations 8.1 listed in Subsection 8.4.1 if the the matrix A is a rotation about the y-axis.

2. Subsection 8.7.2 described the factorization of $p = a^3b^2c^1$ as indicated in Equation 8.13. Work through the distinct princple-axes quaternion factors for $p = a^2b^3c^1$.

3. Work through the repeated principal axis factorization for $p = a^1b^3c^1$ in a fashion similar to that done in Subsection 8.7.3.

Chapter 9

More Quaternion Applications

9.1   Introduction

In the preceding three chapters we considered at some length the algebra of quaternions and how rotation operators may be defined in terms of quaternions. Our claim in these notes is that defining rotation operators in terms of appropriate quaternions is a very useful alternative to the rotation matrix method. We ended Chapter 6 on Quaternion Geometry with an application of the quaternion rotation operator method to the tracking example of Chapter 4. We were able to show that the quaternion method easily and more efficiently produces the same results as the rotation matrix method.

In this present chapter we wish to do the same thing with the other examples considered in Chapter 4. We begin with an application of the quaternion operator method to the Aerospace Sequence example.

9.2   The Aerospace Sequence

We have already considered the *Aerospace Sequence* in some detail in Section 4.4 of Chapter 4, using matrix methods. There we introduced this well-known sequence, which consists of the three successive *coordinate frame* rotations illustrated in Figure 9.1. The first is a rotation about the **Z**-axis through a *heading* angle ψ. The second is a rotation about the new y-axis through an *elevation* angle θ. And the third is a rotation about the resulting x-axis through the *bank* angle ϕ.

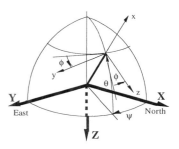

Figure 9.1
Aerospace Rotation Sequence

We emphasize again, the orientation angles illustrated in Figure 9.1 are too often called yaw, pitch and roll. Yaw, pitch and roll are actually angular perturbations relative to a given state. The angles Heading, Elevation, and Bank angle, define the given state. Yaw, pitch and roll represent incremental angular motions about the respective principal axes of the body, at this state.

It is this **Aerospace sequence** which is displayed by the **Heading and Attitude Indicator** — one of the primary instruments in the cockpit of every aircraft.

205

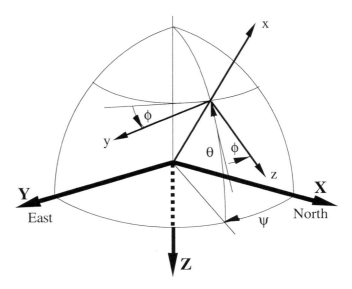

Figure 9.1: Aerospace Rotation Sequence

Proceeding with the quaternion rotation operator approach, we first define

$$\alpha = \frac{\psi}{2} \qquad \beta = \frac{\theta}{2} \qquad \gamma = \frac{\phi}{2}$$

These half-angle relationships are convenient in order to avoid writing the many half-angle expressions in what follows. In terms of the angles α, β, and γ, the three quaternions required to define the three appropriate rotation operators are then

$$
\begin{aligned}
q_{z,\psi} &= \cos\alpha + \mathbf{k}\sin\alpha \\
q_{y,\theta} &= \cos\beta + \mathbf{j}\sin\beta \\
q_{x,\phi} &= \cos\gamma + \mathbf{i}\sin\gamma
\end{aligned}
$$

Further, since we have a sequence of *frame* rotations, the appropriate quaternion product for representing this composite rotation is given by

Order

The *order* in this quaternion product is based upon the results which lead to Equation 5.19 on page 135.

$$q = q_{z,\psi}q_{y,\theta}q_{x,\phi}$$

The computation of this quaternion product, although somewhat tedious, produces the following results. First,

$$
\begin{aligned}
q_{y,\theta}q_{x,\phi} &= \cos\beta\cos\gamma + \mathbf{i}\cos\beta\sin\gamma \\
&\quad + \mathbf{j}\sin\beta\cos\gamma - \mathbf{k}\sin\beta\sin\gamma
\end{aligned}
$$

and then finally

$$q = q_{z,\psi}q_{y,\theta}q_{x,\phi} = q_0 + \mathbf{i}q_1 + \mathbf{j}q_2 + \mathbf{k}q_3$$

where

$$
\begin{aligned}
q_0 &= \cos\alpha\cos\beta\cos\gamma + \sin\alpha\sin\beta\sin\gamma \\
q_1 &= \cos\alpha\cos\beta\sin\gamma - \sin\alpha\sin\beta\cos\gamma \\
q_2 &= \cos\alpha\sin\beta\cos\gamma + \sin\alpha\cos\beta\sin\gamma \\
q_3 &= \sin\alpha\cos\beta\cos\gamma - \cos\alpha\sin\beta\sin\gamma
\end{aligned}
$$

From this quaternion product we are now able to read directly an expression for the composite rotation angle and for the composite rotation axis. If the rotation angle is δ, then we have

$$
\cos\delta/2 = q_0 = \cos\alpha\cos\beta\cos\gamma + \sin\alpha\sin\beta\sin\gamma
$$

Further, the rotation axis may be defined as

$$
\begin{aligned}
\mathbf{v} &= (v_1, v_2, v_3) \\
\text{where} \quad v_1 = q_1 &= \cos\alpha\cos\beta\sin\gamma - \sin\alpha\sin\beta\cos\gamma \\
v_2 = q_2 &= \cos\alpha\sin\beta\cos\gamma + \sin\alpha\cos\beta\sin\gamma \\
v_3 = q_3 &= \sin\alpha\cos\beta\cos\gamma - \cos\alpha\sin\beta\sin\gamma
\end{aligned} \tag{9.1}
$$

We thus have expressions, obtained rather easily using quaternions, for both the angle and the axis of the composite rotation — a single rotation equivalent to the Aerospace Sequence.

We next reconcile these results with the comparable results obtained earlier using matrix rotation operators. We recall that Equation 4.4 gave us the rotation matrix which represents the composite rotation for the aerospace sequence. That matrix, say A, is reproduced here for convenience.

$$
A = \begin{bmatrix}
\cos\psi\cos\theta & \sin\psi\cos\theta & -\sin\theta \\
\begin{pmatrix} \cos\psi\sin\theta\sin\phi \\ -\sin\psi\cos\phi \\ \cos\psi\sin\theta\cos\phi \\ +\sin\psi\sin\phi \end{pmatrix} & \begin{pmatrix} \sin\psi\sin\theta\sin\phi \\ +\cos\psi\cos\phi \\ \sin\psi\sin\theta\cos\phi \\ -\cos\psi\sin\phi \end{pmatrix} & \begin{matrix} \cos\theta\sin\phi \\ \\ \cos\theta\cos\phi \end{matrix}
\end{bmatrix}
$$

9.2.1 The Rotation Angle

We recall that when we found this matrix A, we mentioned that we could compute formulas for the angle and the axis of the composite rotation, but we did not do so then. In order to compare results with that of the quaternion rotation operator method, as well as to illustrate the relative difficulty involved, we shall attempt to do so now.

A formula for the composite rotation angle δ is fairly easy to compute, given the matrix A. From Equation 3.4 we have

Equation 3.4

The composite angle of rotation for any rotation matrix A was derived in Section 3.5.1 on pages 56 and 57 resulting in Equation 3.4.

$$\cos \delta = \frac{Tr(A) - 1}{2}$$

where $\mathrm{Tr}(A)$ denotes the trace of A, that is, the sum of the diagonal elements of A. Hence for the Aerospace Sequence rotation matrix given above, the rotation angle δ may be computed from the formula

$$\cos \delta = \frac{\cos \psi \cos \theta + \sin \psi \sin \theta \sin \phi + \cos \psi \cos \phi + \cos \theta \cos \phi - 1}{2}$$

This matrix result may be reconciled with the corresponding quaternion rotation operator result by the appropriate use of half-angle formulas from trigonometry. The algebraic details, however, are somewhat tedious. So here we simply note that one might begin with the identity

$$\cos \delta = 2 \cos^2 \frac{\delta}{2} - 1$$

As a matter of fact, if we now replace the factor $\cos \delta/2$ by the quaternion rotation operator result given earlier, we indeed do obtain the result given by the matrix rotation operator method.

9.2.2   The Rotation Axis

We have already obtained, fairly easily, an expression which gives the axis of rotation for the composite rotation sequence in the Aerospace example. In this section we find a corresponding expression from the rotation matrix. This approach entails a great deal more work than required by the quaternion method. But, let's do it just this once, to confirm that quaternions indeed do offer an attractive alternative to the matrix approach. Even a cursory glance at the computational details which follow supports this contention.

We begin by observing that the rotation axis, say $\mathbf{v}$, for any given rotation is invariant under that rotation. Thus we need to find this *fixed vector*, $\mathbf{v} = (v_1, v_2, v_3)$, such that if A is the rotation matrix for the Aerospace sequence, then

$$A\mathbf{v} = \mathbf{v}$$
$$\text{or} \quad [A - I]\mathbf{v} = \mathbf{0}$$

This matrix equation is expanded to give the following system of homogeneous equations

$$(\cos\psi\cos\theta - 1)v_1 + \sin\psi\cos\theta v_2 - \sin\theta v_3 = 0$$
$$(\cos\psi\sin\theta\sin\phi - \sin\psi\cos\phi)v_1 +$$
$$(\sin\psi\sin\theta\sin\phi + \cos\psi\cos\phi - 1)v_2 + \cos\theta\sin\phi v_3 = 0$$
$$(\cos\psi\sin\theta\cos\phi + \sin\psi\sin\phi)v_1 +$$
$$(\sin\psi\sin\theta\cos\phi - \cos\psi\sin\phi)v_2 + (\cos\theta\cos\phi - 1)v_3 = 0$$

We will find a non-trivial solution for this homogeneous system by setting , say, $v_3 = 1$. The first two equations in the above system then become

$$(\cos\psi\cos\theta - 1)v_1 + \sin\psi\cos\theta v_2 = \sin\theta$$
$$(\cos\psi\sin\theta\sin\phi - \sin\psi\cos\phi)v_1 +$$
$$(\sin\psi\sin\theta\sin\phi + \cos\psi\cos\phi - 1)v_2 = -\cos\theta\sin\phi$$

We may use the well-known Cramer's Rule for solving these two equations for v_1 and v_2. Then we get

$$\mathbf{v} = (v_1, v_2, v_3) = (\frac{D_1}{D}, \frac{D_2}{D}, 1)$$

where
$$\begin{aligned} D_1 &= \sin\psi\sin\phi + \cos\psi\sin\theta\cos\phi - \sin\theta \\ D_2 &= -\cos\psi\sin\phi + \cos\theta\sin\phi + \sin\psi\sin\theta\cos\phi \\ D &= \cos\theta\cos\phi - \cos\psi\cos\theta - \sin\psi\sin\theta\sin\phi \\ &\quad - \cos\psi\cos\phi + 1 \end{aligned}$$

Since the rotation axis may be expressed as any scalar multiple of this vector, we may write

$$\mathbf{v} = (D_1, D_2, D) \tag{9.2}$$

where the elements D_1, D_2, and D are as given above.

Substitution of numerical values for the various angles in both Equation 9.1 and Equation 9.2 does in fact yield vectors which are scalar multiples of each other.

A Cruel Challenge

A cruel challenge to the reader: Show *analytically* that the axis of rotation in Equation 9.2 has the same direction as does the vector in Equation 9.1 found using quaternion methods.

Perhaps, for most of us, such a demonstration is best left to the powerful algebraic manipulations performed by programs such as *Maple©* or *Mathematica©*.

9.3 Computing the Orbit Ephemeris

In our second example in Chapter 4 we derived algorithms for computing the *orbit ephemeris* for an orbiting body, using the rotation matrix method. Here we use quaternions to accomplish the same objective. At this point the reader might wish to review the details of the rotation matrix operator approach as given in Section 4.5 of Chapter 4.

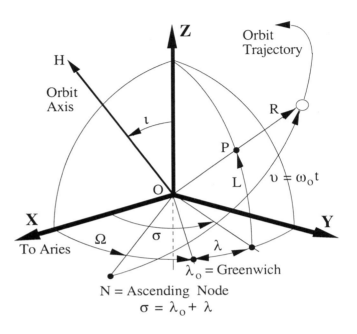

Figure 9.2: Orbit Ephemeris Rotation Sequence

The *orbit ephemeris* for an orbiting body is a tabulation of the latitude and longitude of point P (on the earth's surface) over time, as shown in Figure 9.2. We again derive algorithms for computing the orbit ephemeris parameters, σ, L, and α, of a near-earth satellite in terms of the orbit parameters, Ω, ι, and ν. As in Section 4.5.2, we equate these two *Euler angle* sequences of three frame rotations, illustrated in Figure 9.2. Both of these 3-rotation sequences locate the orbiting body in the fixed reference frame **XYZ**.

9.3.1   The Orbit Euler Angle Sequence

The first of the three-rotation sequences mentioned above is called the *Euler Angle Sequence for Orbits*. It comprises a sequence of three rotations, the first of which is a rotation about the reference frame **Z**-axis through the angle Ω (Omega). The second rotation is about the new x-axis through the angle ι (iota), while the third is a rotation about the resulting z-axis through the angle ν (nu). For convenience, we reproduce here the matrix (obtained earlier) which represents the composite rotation equivalent to the orbit Euler angle sequence.

$$
\begin{bmatrix}
\left(\begin{array}{c}\cos\Omega\cos\nu- \\ \sin\Omega\cos\iota\sin\nu\end{array}\right) & \left(\begin{array}{c}\sin\Omega\cos\nu+ \\ \cos\Omega\cos\iota\sin\nu\end{array}\right) & \sin\iota\sin\nu \\
\left(\begin{array}{c}-\cos\Omega\sin\nu- \\ \sin\Omega\cos\iota\cos\nu\end{array}\right) & \left(\begin{array}{c}-\sin\Omega\sin\nu+ \\ \cos\Omega\cos\iota\cos\nu\end{array}\right) & \sin\iota\cos\nu \\
\sin\Omega\sin\iota & -\cos\Omega\sin\iota & \cos\iota
\end{bmatrix}
$$

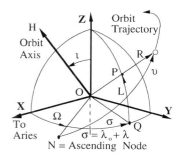

Notice that although we remarked at the time that from this matrix we could compute expressions for both the angle and the axis of rotation, we did not do so.

The appropriate quaternions for representing the Euler Angle Sequence for Orbits are given by

$$
\begin{aligned}
q_{z,\Omega} &= \cos\omega + \mathbf{k}\sin\omega \\
q_{x,\iota} &= \cos\beta + \mathbf{i}\sin\beta \\
q_{z,\nu} &= \cos\gamma + \mathbf{k}\sin\gamma
\end{aligned}
$$

Orbit Euler Angles

$$
\begin{aligned}
\Omega &= \text{to Ascending node} \\
\iota &= \text{Orbit inclination} \\
\nu &= \text{Argument of Latitude}
\end{aligned}
$$

Since the quaternion analysis proceeds in term of half-angle rotations, we here have defined

$$
\omega = \frac{1}{2}\Omega \qquad \beta = \frac{1}{2}\iota \qquad \gamma = \frac{1}{2}\nu
$$

The quaternion product needed for representing the composite rotation operator is then given by

$$
q = q_{z,\Omega}q_{x,\iota}q_{z,\nu}
$$

We compute this product in two steps. First (the reader may wish to check out the details of computing this product)

$$
\begin{aligned}
q_{x,\iota}q_{z,\nu} &= (\cos\beta + \mathbf{i}\sin\beta)(\cos\gamma + \mathbf{k}\sin\gamma) \\
&= \cos\beta\cos\gamma + \mathbf{i}\sin\beta\cos\gamma - \mathbf{j}\sin\beta\sin\gamma + \mathbf{k}\cos\beta\sin\gamma
\end{aligned}
$$

It follows that

$$
\begin{aligned}
q_{z,\Omega}q_{x,\iota}q_{z,\nu} &= (\cos\omega + \mathbf{k}\sin\omega)(\cos\beta\cos\gamma + \mathbf{i}\sin\beta\cos\gamma \\
&\quad - \mathbf{j}\sin\beta\sin\gamma + \mathbf{k}\cos\beta\sin\gamma) \\
&= q_0 + q_1\mathbf{i} + q_2\mathbf{j} + q_3\mathbf{k}
\end{aligned}
$$

where

$$
\begin{aligned}
q_0 &= \cos\omega\cos\beta\cos\gamma - \sin\omega\cos\beta\sin\gamma \\
q_1 &= \cos\omega\sin\beta\cos\gamma + \sin\omega\sin\beta\sin\gamma \\
q_2 &= -\cos\omega\sin\beta\sin\gamma + \sin\omega\sin\beta\cos\gamma \\
q_3 &= \cos\omega\cos\beta\sin\gamma + \sin\omega\cos\beta\cos\gamma
\end{aligned}
$$

From this computation we learn immediately that if δ is the angle of this composite rotation, then

$$\begin{aligned}
\cos \frac{1}{2}\delta &= q_0 \\
&= \cos\omega \cos\beta \cos\gamma - \sin\omega \cos\beta \sin\gamma
\end{aligned}$$

Further, the axis of the rotation is given by the vector

$$\mathbf{v} = (q_1, q_2, q_3)$$

where the components q_1, q_2, and q_3 are given above.

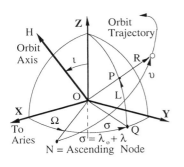

Orbit Euler Angles

Ω = to Ascending node

ι = Orbit inclination

ν = Argument of Latitude

It is relatively easy to reconcile the matrix rotation operator result for the angle of the composite rotation with that just obtained for the quaternion rotation operator. One needs only to observe that if δ is the angle of rotation, our matrix rotation operator analysis says

$$\cos\delta = \frac{Trace(A) - 1}{2}$$

$$\begin{aligned}
\text{where} \quad Trace(A) &= \cos\Omega \cos\nu - \sin\Omega \cos\iota \sin\nu \\
&\quad + \cos\Omega \cos\iota \cos\nu - \sin\Omega \sin\nu + \cos\iota
\end{aligned}$$

A familiar trigonometric identity gives

$$\cos\delta = 2\cos^2 \frac{1}{2}\delta - 1$$

If we use the above expression for $\cos\delta/2$ as given by the quaternion analysis, we indeed do obtain the expression for $\cos\delta$ as given by the matrix analysis.

Notice again that the quaternion analysis immediately provides an expression for the rotation axis. To obtain such an expression from the matrix which represents this rotation, as we did for the Aerospace Sequence in the preceding section, is somewhat more difficult. And, to analytically reconcile the quaternion and the matrix results is again a very difficult matter indeed. We shall not pursue those tedious details.

9.3.2   Orbit Ephemeris

The second of the three-rotation sequences mentioned above is the *Orbit Ephemeris* sequence, described in detail in Section 4.5.2. In that section we computed its matrix representation to be

$$
\begin{bmatrix}
\cos\sigma\cos L &
\sin\sigma\cos L &
\sin L \\
\left\{\begin{array}{l} -\cos\sigma\sin L\sin\alpha \\ -\sin\sigma\cos\alpha \end{array}\right\} &
\left\{\begin{array}{l} -\sin\sigma\sin L\sin\alpha \\ +\cos\sigma\cos\alpha \end{array}\right\} &
\cos L\sin\alpha \\
\left\{\begin{array}{l} -\cos\sigma\sin L\cos\alpha \\ +\sin\sigma\sin\alpha \end{array}\right\} &
\left\{\begin{array}{l} -\sin\sigma\sin L\cos\alpha \\ -\cos\sigma\sin\alpha \end{array}\right\} &
\cos L\cos\alpha
\end{bmatrix}
$$

Recall, in this rotation sequence, the first rotation is about the **Z**-axis through the angle σ (longitude), as illustrated. We follow this by a rotation about the new y-axis through the angle $-L$ (latitude). The third rotation is about the new x-axis through the angle α (not shown). The reader should once again review the details as given in Section 4.5.2, to assure herself that this sequence also relates the fixed reference frame to the orbit frame.

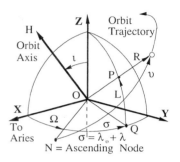

Since the quaternions used in rotation operators are conveniently defined in terms of half-angle rotations, we again introduce symbols for the half-angles

$$
\mu = \frac{\sigma}{2} \qquad \epsilon = \frac{L}{2} \qquad \rho = \frac{\alpha}{2}
$$

The appropriate quaternions for representing the three rotations in the composite rotation operator then are

$$
\begin{aligned}
q_{z,\sigma} &= \cos\mu + \mathbf{k}\sin\mu \\
q_{y,L}^{*} &= \cos\epsilon - \mathbf{j}\sin\epsilon \\
q_{x,\alpha} &= \cos\rho + \mathbf{i}\sin\rho
\end{aligned}
$$

Orbit & Ephemeris Geometry & Angles

Ω	=	to Ascending node
ι	=	Orbit inclination
ν	=	Argument of Latitude
L	=	Ephemeris Latitude
λ	=	Ephemeris Longitude
α	=	Track Direction

The Orbit Ephemeris sequence is represented by the following quaternion product.

$$
\begin{aligned}
q_{z,\sigma}q_{y,L}^{*}q_{x,\alpha} &= (\cos\mu + \mathbf{k}\sin\mu)(\cos\epsilon - \mathbf{j}\sin\epsilon)(\cos\rho + \mathbf{i}\sin\rho) \\
&= \cos\mu\cos\epsilon\cos\rho - \sin\mu\sin\epsilon\sin\rho \\
&\quad + \mathbf{i}(\cos\mu\cos\epsilon\sin\rho + \sin\mu\sin\epsilon\cos\rho) \\
&\quad + \mathbf{j}(\sin\mu\cos\epsilon\sin\rho - \cos\mu\sin\epsilon\cos\rho) \\
&\quad + \mathbf{k}(\cos\mu\sin\epsilon\sin\rho + \sin\mu\cos\epsilon\cos\rho)
\end{aligned}
$$

To Avoid Half-angles

Ω	=		2ω
ι	=		2β
ν	=		2γ
λ	=		2μ
L	=		2ϵ
α	=		2ρ
σ	=		$\lambda + \lambda_0$
$\Omega - \sigma$	=	ψ =	2τ

As we have already observed, the Orbit Sequence and the Ephemeris Sequence are equivalent. Therefore, the two composite quaternion representations for these sequences are equal, that is,

$$
q_{z,\sigma}q_{y,L}^{*}q_{x,\alpha} = q_{z,\Omega}q_{x,\iota}q_{z,\nu}
$$

Once again, in order to avoid the cumbersome half-angle notation in the Orbit sequence, we write

$$\omega = \frac{\Omega}{2} \qquad \beta = \frac{\iota}{2} \qquad \gamma = \frac{\nu}{2}$$

Using these relationships we will derive a set of algorithms for calculating the ephemeris parameters, namely, μ, ϵ, and ρ, each as functions of the orbit parameters, ω, β, and γ.

Should we so desire, we might now equate the components of the quaternion for the composite Orbit Sequence with the corresponding components of the quaternion for the composite Ephemeris Sequence. This would give us a set of equations from which we might be able to derive algorithms for calculating the orbit ephemeris parameters, namely, μ, ϵ, and ρ as functions of the Orbit parameters, ω, β, and γ. However, this direct approach is rather difficult, so we make a slight simplification.

q-inverse

Here we recall that the inverse of $q_{z,\sigma}$ is just its conjugate $q_{z,\sigma}^*$. Recall also that if we multiply quaternions having the same vector part, the angles simply add algebraically.

We know that the Orbit Sequence and the Ephemeris Sequence are equivalent. As stated before, this says that

$$q_{z,\sigma} q_{y,L}^* q_{x,\alpha} = q_{z,\Omega} q_{x,\iota} q_{z,\nu}$$

If now we multiply both sides of this equation on the left by $q_{z,\sigma}^*$, then with $\psi = 2\tau = \Omega - \sigma$ we get

$$q_{y,L}^* q_{x,\alpha} = q_{z,\psi} q_{x,\iota} q_{z,\nu}$$

The quaternions in these indicated products are

$$
\begin{aligned}
q_{z,\Omega} &= \cos\omega + \mathbf{k}\sin\omega \\
q_{x,\iota} &= \cos\beta + \mathbf{i}\sin\beta \\
q_{z,\nu} &= \cos\gamma + \mathbf{k}\sin\gamma \\
q_{y,L}^* &= \cos\epsilon - \mathbf{j}\sin\epsilon \\
q_{z,\psi} &= \cos\tau + \mathbf{k}\sin\tau \qquad \tau = \psi/2
\end{aligned}
$$

If we let $p = q_{y,L}^* q_{x,\alpha}$ and $r = q_{z,\psi} q_{x,\iota} q_{z,\nu}$, we may write

$$
\begin{aligned}
p &= p_0 + \mathbf{i}p_1 + \mathbf{j}p_2 + \mathbf{k}p_3 \\
&= (\cos\epsilon - \mathbf{j}\sin\epsilon)(\cos\rho + \mathbf{i}\sin\rho) \\
r &= r_0 + \mathbf{i}r_1 + \mathbf{j}r_2 + \mathbf{k}r_3 \\
&= (\cos\tau + \mathbf{k}\sin\tau)(\cos\beta + \mathbf{i}\sin\beta)(\cos\gamma + \mathbf{k}\sin\gamma)
\end{aligned}
$$

If we now compute the quaternions p and r and equate their corresponding components we get the following relationships

$$
\begin{aligned}
p_0 &= \cos\epsilon\cos\rho \\
&= \cos\tau\cos\beta\cos\gamma - \sin\tau\cos\beta\sin\gamma &= r_0 \\
p_1 &= \cos\epsilon\sin\rho \\
&= \cos\tau\sin\beta\cos\gamma + \sin\tau\sin\beta\sin\gamma &= r_1 \\
p_2 &= -\sin\epsilon\cos\rho \\
&= -\cos\tau\sin\beta\sin\gamma + \sin\tau\sin\beta\cos\gamma &= r_2 \\
p_3 &= \sin\epsilon\sin\rho \\
&= \cos\tau\cos\beta\sin\gamma + \sin\tau\cos\beta\cos\gamma &= r_3
\end{aligned}
$$

We now simplify these expressions in the following way. Equating p_0 and r_0 gives

$$\cos\epsilon\cos\rho = \cos\beta\cos(\gamma+\tau) \tag{9.3}$$

Equating p_1 and r_1 gives

$$\cos\epsilon\sin\rho = \sin\beta\cos(\gamma-\tau) \tag{9.4}$$

Equating p_2 and r_2 gives

$$\sin\epsilon\cos\rho = \sin\beta\sin(\gamma-\tau) \tag{9.5}$$

Equating p_3 and r_3 gives

$$\sin\epsilon\sin\rho = \cos\beta\sin(\gamma+\tau) \tag{9.6}$$

We shall now use these four equations to obtain expressions from which the orbit ephemeris parameters λ, L, and α, may be determined. We notice that dividing Equation 9.6 by Equation 9.5 gives

$$\tan\rho\tan\beta = \frac{\sin(\gamma+\tau)}{\sin(\gamma-\tau)} \tag{9.7}$$

Dividing Equation 9.4 by Equation 9.3 gives

$$\frac{\tan\rho}{\tan\beta} = \frac{\cos(\gamma-\tau)}{\cos(\gamma+\tau)} \tag{9.8}$$

We now have two equations involving the unknowns, ρ and τ. We eliminate ρ by dividing Equation 9.7 by Equation 9.8 which gives

$$\tan^2\beta = \frac{\sin 2(\gamma+\tau)}{\sin 2(\gamma-\tau)} \tag{9.9}$$

To Avoid Half-angles

$$
\begin{aligned}
\Omega &= 2\omega \\
\iota &= 2\beta \\
\nu &= 2\gamma \\
\lambda &= 2\mu \\
L &= 2\epsilon \\
\alpha &= 2\rho \\
\sigma &= \lambda + \lambda_0 \\
\Omega - \sigma &= \psi = 2\tau
\end{aligned}
$$

With $2\tau = \Omega - \sigma$, some algebraic manipulations on Equation 9.9, using appropriate half-angle trigonometric identities, give

$$\tan\sigma = \frac{\tan\Omega + \tan\nu\cos\iota}{1 - \tan\Omega\tan\nu\cos\iota} \qquad (9.10)$$

This expression for σ agrees with our earlier result obtained in Section 4.5 using rotation matrices. Since $\sigma = \lambda + \lambda_0$, the orbit ephemeris parameter λ is now determined.

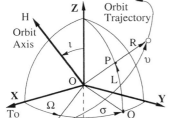

Orbit & Ephemeris

Next, dividing Equation 9.6 by Equation 9.3 gives

$$\tan\rho\tan\epsilon = \tan(\gamma + \tau) \qquad (9.11)$$

and dividing Equation 9.5 by Equation 9.4 gives

$$\frac{\tan\epsilon}{\tan\rho} = \tan(\gamma - \tau) \qquad (9.12)$$

Multipying Equation 9.11 and Equation 9.12 gives

$$\tan^2\epsilon = \tan(\gamma - \tau)\tan(\gamma + \tau) \qquad (9.13)$$

Since τ is already determined, this equation determines ϵ. Since $L = 2\epsilon$, the orbit ephemeris parameter L is determined.

Finally, dividing Equation 9.11 by Equation 9.12 we get

$$\tan^2\rho = \frac{\tan(\gamma + \tau)}{\tan(\gamma - \tau)} \qquad (9.14)$$

Once again, since τ has been determined, we have now also determined ρ, or rather, the orbit ephemeris parameter, $\alpha = 2\rho$. This parameter α, incidentally, is related to the direction of the orbit ephemeris path at any given time. With the derivation of these algorithms we may now plot the points, (λ, L), which represent the ephemeris path.

We note that Equations 9.10, 9.13, and 9.14, do not determine unique values for the angles, λ, L, and α. The ambiguities in these equations must be resolved from information peculiar to the application.

9.4   Great Circle Navigation

The final example considered in Chapter 4 was the Great Circle Navigation problem. We urge the reader to review that example as we analyzed it from the matrix rotation operator point of view in Section 4.6 of Chapter 4. Recall that the longitude and latitude for the two points A and B on the surface of the earth are assumed to be known; that is, the angles λ_1, L_1, and λ_2, L_2, as shown in Figure 9.3, are given. The problem is to determine expressions for the great circle path connecting points A and B, as specified by three angles. The first is the angle ψ_1, which is the *heading* of the great

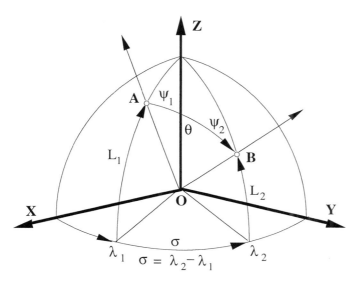

Figure 9.3: Great Circle Path

circle path at point A toward B. The second is the angle θ, which is the *radian distance* from point A to point B along the path. The third angle is ψ_2, which is the angle of *arrival heading* or *approach* at point B.

Our approach in Section 4.6 was to specify a sequence of seven rotations, such that their product yields the identity. It was from this sequence that the algorithms we seek were derived. The se-

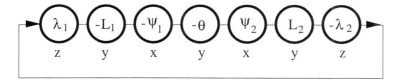

Figure 9.4: Great Circle Rotation Sequence

quence is illustrated in Figure 4.17 which we reproduce here as Figure 9.4. From this sequence we derived two equivalent sequences, one of which involves only the known angles λ_1, L_1, and λ_2, L_2, the other of which involves only the unknown angles ψ_1, θ and ψ_2. By finding matrix representations for each of these sequences and equating corresponding elements we were finally able to obtain the expressions given in Equations 4.8, 4.9, and 4.10.

The geometry of the situation is illustrated in Figure 9.3. We now solve this same problem using the quaternion rotation operator approach.

9.5   Quaternion Method

The ordered product of the quaternions associated with the rotation sequence of Figure 9.4 must also be equivalent to the identity, as was the case in the matrix analysis. Therefore, we write

$$q_{z,\lambda_1} q_{y,L_1}^* q_{x,\psi_1}^* q_{y,\theta}^* q_{x,\psi_2} q_{y,L_2} q_{z,-\lambda_2} = \text{identity}$$

As we have observed before, we may begin this sequence at any point and still have the identity, so long as the order of the factors is maintained. Hence, if we define $\sigma = \lambda_2 - \lambda_1$ we may write

$$q_{y,L_1}^* q_{x,\psi_1}^* q_{y,\theta}^* q_{x,\psi_2} q_{y,L_2} q_{z,\sigma}^* = e = 1$$

Next, if we multiply this equation on the left by q_{y,L_1} and on the right by $q_{z,\sigma} q_{y,L_2}^*$ we obtain the equation

$$p = q_{x,\psi_1}^* q_{y,\theta}^* q_{x,\psi_2} \quad = \quad q_{y,L_1} q_{z,\sigma} q_{y,L_2}^* = r \qquad (9.15)$$

Notice that in this last equation, the unknown angles ψ_1, θ, and ψ_2 are on the left side, while the given angles L_1, $\sigma = \lambda_2 - \lambda_1$, and L_2 appear only on the right side.

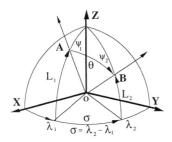

Great Circle Path

The quaternions for the rotations in these products are

$$q_{x,\psi_1} = \cos\frac{\psi_1}{2} + \mathbf{i}\sin\frac{\psi_1}{2}$$

$$q_{y,\theta} = \cos\frac{\theta}{2} + \mathbf{j}\sin\frac{\theta}{2}$$

$$q_{x,\psi_2} = \cos\frac{\psi_2}{2} + \mathbf{i}\sin\frac{\psi_2}{2}$$

$$q_{y,L_1} = \cos\frac{L_1}{2} + \mathbf{j}\sin\frac{L_1}{2}$$

$$q_{z,\sigma} = \cos\frac{\sigma}{2} + \mathbf{k}\sin\frac{\sigma}{2}$$

$$q_{y,L_2} = \cos\frac{L_2}{2} + \mathbf{j}\sin\frac{L_2}{2}$$

We now compute the quaternion products indicated in Equation 9.15 for p and r and write

$$p = p_0 + \mathbf{i}p_1 + \mathbf{j}p_2 + \mathbf{k}p_3 \tag{9.16}$$

$$= (\cos\frac{\psi_1}{2} - \mathbf{i}\sin\frac{\psi_1}{2})(\cos\frac{\theta}{2} - \mathbf{j}\sin\frac{\theta}{2})(\cos\frac{\psi_2}{2} + \mathbf{i}\sin\frac{\psi_2}{2})$$

$$r = r_0 + \mathbf{i}r_1 + \mathbf{j}r_2 + \mathbf{k}r_3 \tag{9.17}$$

$$= (\cos\frac{L_1}{2} + \mathbf{j}\sin\frac{L_1}{2})(\cos\frac{\sigma}{2} + \mathbf{k}\sin\frac{\sigma}{2})(\cos\frac{L_2}{2} - \mathbf{j}\sin\frac{L_2}{2})$$

Since the two composite quaternions p and r are equal we may equate their corresponding components. Notice that the four components of the composite quaternion p are functions of the rotation angles ψ_1, θ and ψ_2; and the components of the composite quaternion r are functions of L_1, σ and L_2.

Equating p_0 and r_0 gives

$$p_0 = \cos\frac{\psi_1}{2}\cos\frac{\theta}{2}\cos\frac{\psi_2}{2} + \sin\frac{\psi_1}{2}\cos\frac{\theta}{2}\sin\frac{\psi_2}{2}$$

$$= \cos\frac{\theta}{2}\cos(\frac{\psi_2}{2} - \frac{\psi_1}{2})$$

$$= \cos\frac{L_1}{2}\cos\frac{\sigma}{2}\cos\frac{L_2}{2} + \sin\frac{L_1}{2}\cos\frac{\sigma}{2}\sin\frac{L_2}{2} =$$

$$\cos\frac{\sigma}{2}\cos(\frac{L_2}{2} - \frac{L_1}{2}) = r_0$$

Equating p_1 and r_1 gives

$$p_1 = \cos\frac{\psi_1}{2}\cos\frac{\theta}{2}\sin\frac{\psi_2}{2} - \sin\frac{\psi_1}{2}\cos\frac{\theta}{2}\cos\frac{\psi_2}{2}$$

$$= \cos\frac{\theta}{2}\sin(\frac{\psi_2}{2} - \frac{\psi_1}{2})$$

$$= \sin\frac{L_1}{2}\sin\frac{\sigma}{2}\cos\frac{L_2}{2} + \cos\frac{L_1}{2}\sin\frac{\sigma}{2}\sin\frac{L_2}{2} =$$

$$\sin\frac{\sigma}{2}\sin(\frac{L_2}{2} + \frac{L_1}{2}) = r_1$$

Equating p_2 and r_2 gives

$$p_2 = \sin\frac{\psi_1}{2}\sin\frac{\theta}{2}\sin\frac{\psi_2}{2} - \cos\frac{\psi_1}{2}\sin\frac{\theta}{2}\cos\frac{\psi_2}{2}$$

$$= -\sin\frac{\theta}{2}\cos(\frac{\psi_2}{2} + \frac{\psi_1}{2})$$

$$= \sin\frac{L_1}{2}\cos\frac{\sigma}{2}\cos\frac{L_2}{2} - \cos\frac{L_1}{2}\cos\frac{\sigma}{2}\sin\frac{L_2}{2} =$$

$$-\cos\frac{\sigma}{2}\sin(\frac{L_2}{2} - \frac{L_1}{2}) = r_2$$

Half-Angles

It might have been less cumbersome if we had avoided writing all these half-angles, as we had done before. However, we press on!

Verify

Verify these equations. It may help one recall some long forgotten 'sum and difference' trigonometric identities.

Equating p_3 and r_3 gives

$$
\begin{aligned}
p_3 &= \sin\tfrac{\psi_1}{2}\sin\tfrac{\theta}{2}\cos\tfrac{\psi_2}{2} + \cos\tfrac{\psi_1}{2}\sin\tfrac{\theta}{2}\sin\tfrac{\psi_2}{2} \\[2mm]
&= \sin\tfrac{\theta}{2}\sin(\tfrac{\psi_2}{2} + \tfrac{\psi_1}{2}) \\[2mm]
&= \cos\tfrac{L_1}{2}\sin\tfrac{\sigma}{2}\cos\tfrac{L_2}{2} - \sin\tfrac{L_1}{2}\sin\tfrac{\sigma}{2}\sin\tfrac{L_2}{2} \quad = \\[2mm]
&\qquad\qquad\qquad \sin\tfrac{\sigma}{2}\cos(\tfrac{L_2}{2} + \tfrac{L_1}{2}) \quad = \quad r_3
\end{aligned}
$$

In summary we have

$$
\begin{aligned}
\cos\frac{\theta}{2}\cos\frac{\psi_2 - \psi_1}{2} &= \cos\frac{\sigma}{2}\cos\frac{L_2 - L_1}{2} \\[2mm]
\cos\frac{\theta}{2}\sin\frac{\psi_2 - \psi_1}{2} &= \sin\frac{\sigma}{2}\sin\frac{L_2 + L_1}{2} \\[2mm]
\sin\frac{\theta}{2}\cos\frac{\psi_2 + \psi_1}{2} &= \cos\frac{\sigma}{2}\sin\frac{L_2 - L_1}{2} \\[2mm]
\sin\frac{\theta}{2}\sin\frac{\psi_2 + \psi_1}{2} &= \sin\frac{\sigma}{2}\sin\frac{L_2 + L_1}{2}
\end{aligned}
$$

It is from these equations that we may now derive expressions for the unknown parameters ψ_1, ψ_2, and θ, as functions of the known parameters L_1, L_2, and σ. The algebraic details are quite tedious, and here we give only hints as to how this may be done.

First, if we square the first two equations and add the results, and remember that for any angle, say A, we have

$$
\cos^2 A + \sin^2 A = 1
$$

we obtain the equation

$$
\cos^2\frac{\theta}{2} = \cos^2\frac{\sigma}{2}\cos^2\frac{L_2 - L_1}{2} + \sin^2\frac{\sigma}{2}\sin^2\frac{L_2 + L_1}{2}
$$

We now make liberal use of the half-angle formulas

$$
\cos^2\frac{A}{2} = \frac{1 + \cos A}{2}
$$

and

$$
\sin^2\frac{A}{2} = \frac{1 - \cos A}{2}
$$

to simplify the preceding equation. We then solve the resulting equation for $\cos\theta$, which gives

$$
\cos\theta = \cos\sigma\cos L_1\cos L_2 + \sin L_1\sin L_2
$$

This confirms Equation 4.10 which we obtained in Chapter 4 using the rotation matrix method.

Second, it also is possible to confirm our earlier results for angles ψ_1 and ψ_2. To do this it will be helpful to define auxiliary angles A and B

$$A = \psi_2 - \psi_1$$
$$B = \psi_2 + \psi_1$$

It is easy to verify that with these definitions we have

$$\psi_1 = \frac{B - A}{2}$$
$$\psi_2 = \frac{B + A}{2}$$

With this notation, the equations we summarized above become

$$\cos\frac{\theta}{2}\cos\frac{A}{2} = \cos\frac{\sigma}{2}\cos\frac{L_2 - L_1}{2} \tag{9.18}$$

$$\cos\frac{\theta}{2}\sin\frac{A}{2} = \sin\frac{\sigma}{2}\sin\frac{L_2 + L_1}{2} \tag{9.19}$$

$$\sin\frac{\theta}{2}\cos\frac{B}{2} = \cos\frac{\sigma}{2}\sin\frac{L_2 - L_1}{2} \tag{9.20}$$

$$\sin\frac{\theta}{2}\sin\frac{B}{2} = \sin\frac{\sigma}{2}\sin\frac{L_2 + L_1}{2} \tag{9.21}$$

Now, dividing Equation 9.19 by Equation 9.18 gives

$$\tan\frac{A}{2} = \frac{\tan\frac{\sigma}{2}\sin\frac{L_2+L_1}{2}}{\cos\frac{L_2-L_1}{2}}$$

while if we divide Equation 9.21 by Equation 9.20 we obtain

$$\tan\frac{B}{2} = \frac{\tan\frac{\sigma}{2}\sin\frac{L_2+L_1}{2}}{\sin\frac{L_2-L_1}{2}}$$

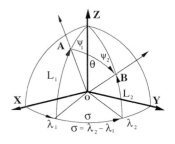

Great Circle Path

Finally, we need only to recognize that

$$\tan\psi_1 = \tan\frac{B - A}{2}$$
$$= \frac{\tan\frac{B}{2} - \tan\frac{A}{2}}{1 + \tan\frac{B}{2}\tan\frac{A}{2}}$$
$$\tan\psi_2 = \tan\frac{B + A}{2}$$
$$= \frac{\tan\frac{B}{2} + \tan\frac{A}{2}}{1 - \tan\frac{B}{2}\tan\frac{A}{2}}$$

With use of the appropriate trigonometric identities, and the above expressions for $\tan \frac{A}{2}$ and $\tan \frac{B}{2}$, it is possible to simplify the expressions for $\tan \psi_1$ and $\tan \psi_2$ to obtain

$$\tan \psi_1 = \frac{\sin \sigma \cos L_2}{\sin L_2 \cos L_1 - \cos L_2 \cos \sigma \sin L_1}$$
$$\tan \psi_2 = \frac{\sin \sigma \cos L_1}{\sin L_2 \cos L_1 \cos \sigma - \cos L_2 \sin L_1}$$

Note that these are exactly the expressions which we obtained earlier in Equations 4.11 and 4.12 from the rotation matrix method.

9.6   Reasons for the Seasons

In the preceding sections we have explored the use of the quaternion rotation operator in analyzing certain problems which mostly are of interest to those engaged in the aerospace industry. In this final section we turn to a problem of quite a different nature, one which

Raison d'être

The seasons are fixed by wisdom divine,
The slow changing moon shows forth God's design;
The sun in its orbit his Maker obeys,
And running his journey hastes not nor delays.

1959 CRC Psalter Hymnal #207

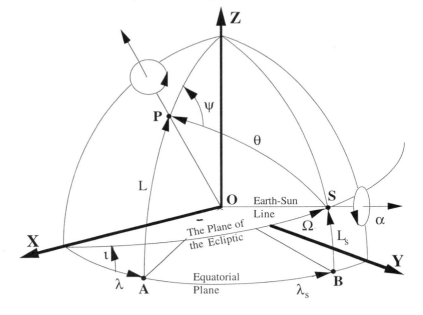

Figure 9.5: Earth Equatorial and Ecliptic Planes

all of us experience day by day. The problem is: how the earth's orbit, and the earth's orientation in that orbit relative to the sun, determine the change of seasons as we experience them — another interesting application of rotation operators. We shall assume at least some reader familiarity with terms such as, say, the *equatorial plane* and the *vernal equinox*, as represented in Figure 9.5. Note specifically, that the fixed reference frame, **XYZ**, is earth-centered with the **X**-axis (the intersection of the Equatorial and Ecliptic planes) is directed 'toward' the constellation Aries.

We will construct a rotation operator sequence which relates the earth equatorial and ecliptic planes. We will determine the number of daylight hours and the directions to the sunrise and sunset for any point $P(\lambda, L)$ on the earth surface, for any season of the year defined by the angle Ω. The angle Ω defines the direction to the Sun in the ecliptic plane from the Vernal equinox. In Figure 9.5 we begin the rotation sequence from the **X**-axis in the fixed reference frame **XYZ**. The **XY**-plane contains the earth equatorial plane. Recall, the **X**-axis lies in the intersection of the Equatorial and Ecliptic planes and is directed positively toward Aries (the direction of the Vernal equinox). The **Z**-axis of the reference frame is normal to the equatorial plane and is directed positively North.

We begin by defining three rotation sequences which relate the various parameters shown in Figure 9.5. These parameters govern the geometry of the situation, and how this geometry changes with the Seasons. These sequences are based on the following three spherical n-gons:

1. Sequence #1: Spherical Polygon - **XSPAX**

2. Sequence #2: Spherical Trapezoid - (X)**APSBA**(X)

3. Sequence #3: Spherical Triangle - **XSBAX**

We will define the rotation sequence for each of these n-gons, and following this we will derive the equations which properly relate the various parameters of interest. The reader may wish to follow each rotation in these sequences as we did earlier in Section 4.6.

9.6.1   Sequence #1: Polygon - XSPAX

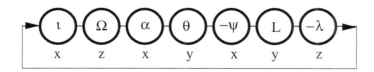

Figure 9.6: Earth-Sun Rotation Sequence #1

Aries

Over the centuries, the line of nodes has 'drifted' from the direction to Aries due to complex gravitational perturbations which are present in any n-body system or environment.

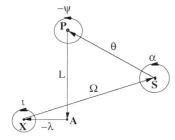

Figure 9.5a
Spherical Polygon

The spherical polygon is redrawn here in the margin for reference as we step through the sequence of rotations.

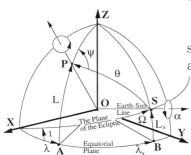

Seasons Model

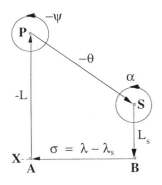

Figure 9.5b
Spherical Trapezoid

The spherical trapezoid is redrawn here in the margin for reference as we step through the sequence of rotations.

The **1**st rotation in this sequence is about the **X**-axis through the angle ι (approximately $23\frac{1}{2}$ degrees), which is the angle between the equatorial and ecliptic planes.

The **2**nd rotation is then about the new z-axis through the angle between the **X**-axis (Vernal equinox) and the Earth-Sun line, denoted by Ω. This angle specifies the Season of the year.

The **3**rd rotation follows about the new x-axis through an angle α, such that the resulting y-axis is normal to the plane POS. Further, if we view the new frame as being centered at S, the new z-axis points away from the observation point P.

The **4**th rotation is then about the new y-axis through the angle θ, such that the new x-axis is the local vertical at P.

The **5**th rotation is about the new x-axis through an angle ψ, such that the new z-axis is directed North. Note, heading, ψ, is always positive — increasing clockwise, thus the rotation is -ψ.

The **6**th is a rotation about the new y-axis through the angle L — the latitude of the observation point P.

The **7**th and final rotation about the z-axis, is opposite in sense to that of λ, the longitude of the observation point P. The new frame then coincides with the original reference frame.

9.6.2 Sequence #2: Trapezoid - APSBA

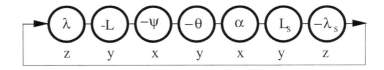

Figure 9.7: Earth-Sun Rotation Sequence #2

The **1**st rotation in this trapezoidal sequence is taken about the **Z**-axis through the angle λ, from the Vernal equinox to the longitude of the observation point **P**.

The **2**nd rotation is then taken about the new y-axis through the angle L — the latitude of the point **P**. Note the opposite sense here. The new x-axis is the local vertical at point P.

The **3**rd rotation is about the new x-axis through an angle ψ such that the new zx-plane contains the Earth-Sun line, with the z-axis pointing toward S. Note again the opposite sense.

The **4**th rotation is about the new y-axis through the angle, $-\theta$, so that the new x-axis is directed along the Earth-Sun line.

The **5**th rotation is then taken about this new x-axis through an angle, α, such that the new z-axis at S again points North.

The **6**th rotation is then about the new y-axis (which is now parallel to the Equatorial plane) through the angle L_s — the latitude of the point S (on the Earth-Sun line).

The **7**th and final rotation is about the new z-axis through the angle λ_s (opposite sense) which is the longitude of S measured from the Vernal equinox. The new frame is now coincident with the original reference **XYZ** frame — and the loop is closed!

9.6.3   Sequence #3: Triangle - XSBAX

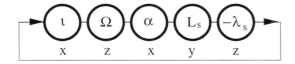

Figure 9.8: Earth-Sun Rotation Sequence #3

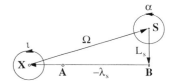

Figure 9.5c
Spherical Triangle

The spherical triangle is redrawn here in the margin for reference as we step through the sequence of rotations.

The **1**st rotation in this triangular sequence is about the **X**-axis through the angle ι such that the new xy-plane is the plane of the ecliptic.

The **2**nd rotation is then about the new z-axis through the angle Ω. This angle (in the plane of the ecliptic) defines the Season by specifying the direction of the Earth-Sun line with respect to the reference frame **X**-axis (the Vernal Equinox).

The **3**rd rotation is about the new x-axis through an angle, α such that the new y-axis is parallel to the Equatorial plane.

The **4**th rotation is taken about this new y-axis through the angle L_s, which represents the Latitude of the Earth-Sun line.

The **5**th and final rotation is about the new z-axis through an angle λ_s, which is the longitude of S (the Earth-Sun line) measured from the Vernal equinox. Note that here the right-hand-rule sense is opposite the conventional sense of the measured longitude of S. As before, we indicate this fact by writing $-\lambda_s$ in Figure 9.8. This makes the final xyz-frame again coincident with the original **XYZ** reference frame.

9.6.4   Summary of Rotation Angles

In our three foregoing closed-loop sequences we have used several angles, some of which are longitudes. It will be convenient to define some of these longitude angles with respect to the Vernal equinox (**X**-axis), rather than Greenwich zero. We must however, carefully establish the relationship between these longitude angles. In summary we have

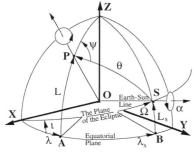

Figure 9.5
Seasons Model

$$
\begin{aligned}
\lambda &= \text{Longitude of P wrt Vernal equinox (}\mathbf{X}\text{-axis)} \\
L &= \text{Latitude of point P} \\
\Omega &= \text{Argument of Latitude of Earth-Sun line} \\
\lambda_P &= \text{Longitude of P wrt Greenwich} \\
\lambda_0 &= \text{Longitude of Greenwich wrt the Vernal equinox} \\
L_s &= \text{Latitude of the Earth-Sun line} \\
\lambda_s &= \text{Longitude of Earth-Sun line wrt Vernal equinox} \\
\theta &= \text{Co-elevation of Earth-Sun line} \\
\iota &= \text{Inclination of equatorial plane (}23\tfrac{1}{2}\text{ degrees)} \\
\psi &= \text{Heading of Earth-Sun line, viewed from P}
\end{aligned}
$$

In general, many of the above angles are functions of time. For example, if ω_e is the Earth rotation rate, then

$$\lambda = \lambda_P - \lambda_0 + \omega_e t$$

In like manner, if ω_s is the Earth orbital revolution rate, and Ω_0 is the initial value of Ω then

$$\Omega = \Omega_0 + \omega_s t$$

The foregoing model may well serve as the starting point for seeking the answer to many challenging questions in local celestial mechanics. In our present application, however, we choose to determine only the number of seasonal daylight hours for a specified location on the earth surface. That is, we specify values for the angles Ω,

λ_P and L. Then, for $\theta = \frac{\pi}{2}$ (the condition for sunrise and sunset), we determine two values for the angles, λ and ψ. The difference between the two λ values (corresponding to sunrise and sunset) determines the number of daylight hours. The two values for ψ give the directions relative to North, for sunrise and sunset. We shall now analyze Sequence #1 in considerable detail, for pedagogical reasons. The reader will then note the results obtained for sequences #2 and #3 come from a similar analytical methodology.

9.6.5 Matrix Method on Sequence #1

Use Quaternions

In this application we use matrix rotation operators. We leave as a worth-while exercise to confirm the results obtained here, using quaternion rotation operators.

It is important to note that the sequence in Figure 9.6, reproduced here for convenience, includes a rotation through an angle α. This angle α, in general, has little or no interesting geometric signifi-

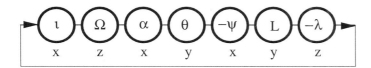

Figure 9.9: Earth-Sun Rotation Sequence #1

cance, but it is required to close the sequence. The result obtained in Section 4.2.3 allows us to begin this loop with the α-rotation, as

Closed Sequence

We again emphasize that the product of a closed ordered sequence of rotation operators is equivalent to the identity transformation.

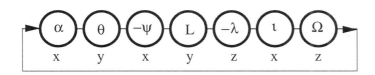

Figure 9.10: Begin Sequence #1 with α-Rotation

Identity Sequence

A sequence of rotations may be *closed* iff it represents an identity. The operations represented by such a sequence, opened at any point, takes any vector introduced at that point, into itself.

illustrated in Figure 9.10. We notice that the α-rotation is about the local x-axis. It follows, if we input the vector $\mathbf{i} = col[1, 0, 0]$, we may eliminate the α-transformation from the sequence, as shown in Figure 9.11.

The product of rotations, ordered according to the sequence of Figure 9.11, takes the vector $\mathbf{i} = col[1, 0, 0]$ into itself, so we may write

$$T_{\Omega,z}T_{\iota,x}T_{\lambda,z}^t T_{L,y}T_{\psi,x}^t T_{\theta,y}\mathbf{i} = \mathbf{i}$$

If we pre-multiply, in turn, both sides of this matrix equation by $T^t_{\Omega,z}$, $T^t_{\iota,x}$, $T_{\lambda,z}$, and finally by $T^t_{L,y}$, we get the matrix equation

$$T^t_{\psi,x} T_{\theta,y}\mathbf{i} \;=\; T^t_{L,y} T^t_{\lambda,z} T^t_{\iota,x} T^t_{\Omega,z}\mathbf{i} \tag{9.22}$$

We solve this equation, where the indicated sequence operates on

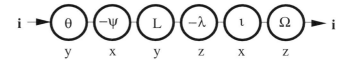

Figure 9.11: α Eliminated in Opened Sequence #1

Matrix Product

Here we use Matrix Equations 3.1, 3.2, and 3.3 on pages 49 to 51 to write Equation 9.22 in matrix form. Computing the matrix product gives the indicated result.

the vector $\mathbf{i}$, using matrix methods. Equation 9.22 produces equivalent column matrices, that is

$$\begin{bmatrix} \cos\theta \\ -\sin\psi\sin\theta \\ \cos\psi\sin\theta \end{bmatrix} =$$

$$\begin{bmatrix} \cos L \cos\lambda \cos\Omega + \cos L \sin\lambda \sin\Omega \cos\iota + \sin L \sin\Omega \sin\iota \\ -\sin\lambda\cos\Omega + \cos\lambda\sin\Omega\cos\iota \\ \cos L \sin\Omega\sin\iota - \sin L \sin\lambda\sin\Omega\cos\iota - \sin L \cos\lambda\cos\Omega \end{bmatrix}$$

If we equate the two first elements we get

$$\cos\theta \;=\; \cos\Omega\cos\lambda\cos L + \sin\Omega\sin\iota\sin L +$$
$$\sin\Omega\cos\iota\sin\lambda\cos L$$

Dividing their 2nd elements by their 3rd gives (9.23)

$$\tan\psi \;=\; \frac{\tan\Omega\cos\iota\cos\lambda - \sin\lambda}{\tan\Omega(\cos\iota\sin\lambda\sin L - \sin\iota\cos L) + \cos\lambda\sin L}$$

For $|L| \leq \frac{\pi}{2} - \iota$, a necessary condition for *sunrise* or *sunset* is that $\theta = \frac{\pi}{2}$, that is, $\cos\theta = 0$. We may then write

$$\tan L \;=\; -\frac{\sin\Omega\cos\iota\sin\lambda + \cos\Omega\cos\lambda}{\sin\Omega\sin\iota} \tag{9.24}$$

The angle ι is $23\frac{1}{2}$ degrees and the angle Ω is chosen to specify the season. Then for some specified latitude $|L| \leq \frac{\pi}{2} - i$ on the earth, we may obtain sunrise and sunset values for λ. This we do by rewriting Equation 9.24 as a quadratic equation in $\sin\lambda$, which gives

$$\sin^2\lambda + 2B\sin\lambda - C \;=\; 0$$
$$\text{where} \qquad B \;=\; \frac{\tan^2\Omega\sin\iota\cos\iota\tan L}{1 + \tan^2\Omega\cos^2\iota}$$

$$\text{and} \qquad C = \frac{1 - \tan^2 \Omega \sin^2 \iota \tan^2 L}{1 + \tan^2 \Omega \cos^2 \iota}$$

from which we may write

$$\sin \lambda = -B \pm \sqrt{B^2 + C}$$

$$\text{that is} \qquad \lambda = \arcsin(-B \pm \sqrt{B^2 + C})$$

These values for λ are then used in Equation 9.23 to compute the corresponding values for heading, ψ, which are the directions to sunrise and sunset.

9.6.6 Matrix Method on Sequence #2

In the analysis of Sequence #2 we will again solve for $\cos \theta$ and $tan\psi$ as was done in Sequence #1. This time, however, the results are expressed in terms of the meaningful latitude/longitude parameters. The product of the rotation matrices, ordered according to the rotation sequence of Figure 9.12, takes the vector $\mathbf{i} = col[1, 0, 0]$

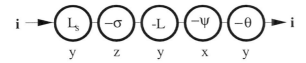

Figure 9.12: Season Sequence #2 Simplified

into itself so we write

Notation

By the notation $T_{\theta,y}$ we mean a rotation through an angle θ about the y-axis — and

$$T_{\theta,y}^t = T_{-\theta,y}$$

$$T_{\theta,y}^t T_{\psi,x}^t T_{L,y}^t T_{\sigma,z}^t T_{L_s,y} \mathbf{i} = \mathbf{i} \qquad (9.25)$$

where the angle $\sigma = \lambda_s - \lambda$. Then pre-multiplying both sides of the matrix Equation 9.25 by $T_{\psi,x} T_{\theta,y}$ we get

$$T_{L,y}^t T_{\sigma,z}^t T_{L_s,y} \mathbf{i} = T_{\psi,x} T_{\theta,y} \mathbf{i} \qquad (9.26)$$

Solving matrix Equation 9.26 yields the following relationships

$$\cos \theta = \cos L \cos \sigma \cos L_s + \sin L \sin L_s \qquad (9.27)$$

$$\sin \theta \sin \psi = \cos L_s \sin \sigma \qquad (9.28)$$

$$\sin \theta \cos \psi = \sin L_s \cos L - \cos L_s \cos \sigma \sin L \qquad (9.29)$$

Dividing Equation 9.28 by Equation 9.29 we get

$$\tan \psi = \frac{\cos L_s \sin \sigma}{\sin L_s \cos L - \cos L_s \cos \sigma \sin L} \qquad (9.30)$$

Once again we have solved for $\cos \theta$ and $\tan \psi$, except this time in terms of Latitude and Logitude parameters.

9.6.7   Matrix Method on Sequence #3

The seasons are conveniently defined in terms of the direction of the Earth-Sun line in the ecliptic plane. The direction of the Earth-Sun line in Figure 9.5 is defined by the angle Ω in the Ecliptic Plane relative to the *Vernal* equinox (the positive Reference **X**-axis). The angle ι, which relates the Ecliptic and Equatorial planes, measures $23\frac{1}{2}$ degrees. In Figure 9.5 we consider the following sequence of rotations which relate these two planes in the spherical triangle $XSBAX$. As before, we open the closed sequence of Figure 9.8 at the rotation α, and input the vector $\mathbf{i} = col[1, 0, 0]$ at this point.

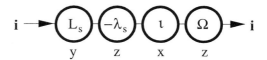

Figure 9.13: Sequence #3 Simplified

This results in the sequence shown in Figure 9.13. From the matrix equation indicated by this sequence we get

$$\cos \Omega \;=\; \cos \lambda_s \cos L_s \qquad (9.31)$$
$$\sin L_s \;=\; \sin \iota \sin \Omega \qquad (9.32)$$
$$\tan \lambda_s \;=\; \cos \iota \tan \Omega \qquad (9.33)$$
$$\tan L_s \;=\; \tan \iota \sin \lambda_s \qquad (9.34)$$

These equations suggest several interesting investigations. For example, the third equation allows us to write

$$\tan \mu = \tan(\Omega - \lambda_s) = \frac{\tan \Omega - \tan \lambda_s}{1 + \tan \lambda_s \tan \Omega} = \frac{(1 - \cos \iota) \tan \Omega}{1 + \cos \iota \tan^2 \Omega}$$

The parametric plot of the points (L_s, μ), as the angle $\Omega = 0 \to 2\pi$, produces the figure called an *analemma*, shown in the margin. It is closely related to the 'figure-eight' which appears on most globes representing the earth.

In summary, the equations from the three sequences are:

For sequence #1 we have:

$$\cos \theta \;=\; \cos \Omega \cos \lambda \cos L + \sin \Omega \sin \iota \sin L +$$
$$\sin \Omega \cos \iota \sin \lambda \cos L$$
$$\tan \psi \;=\; \frac{\tan \Omega \cos \iota \cos \lambda - \sin \lambda}{\cos \lambda \sin L + \tan \Omega (\cos \iota \sin \lambda \sin L - \sin \iota \cos L)}$$

L radians

$$0.4$$
$$0.2$$
$$-0.03 \qquad 0.03 \quad \mu$$
$$-0.2$$
$$-0.4$$

Seasonal Analemma

Plotted using Mathematica©

For sequence #2 we have:

$$
\begin{aligned}
\cos\theta &= \cos L_s \cos\sigma \cos L + \sin L_s \sin L \\
\sin\theta\sin\psi &= \cos L_s \sin\sigma \\
\sin\theta\cos\psi &= \sin L_s \cos L - \cos L_s \cos\sigma \sin L \\
\tan\psi &= \frac{\cos L_s \sin\sigma}{\sin L_s \cos L - \cos L_s \cos\sigma \sin L} \\
&= \frac{\sin\sigma}{\tan L_s \cos L - \cos\sigma \sin L}
\end{aligned}
$$

For sequence #3 we have:

$$
\begin{aligned}
\cos\Omega &= \cos\lambda_s \cos L_s \\
\sin L_s &= \sin\iota \sin\Omega \\
\tan\lambda_s &= \cos\iota \tan\Omega \\
\tan L_s &= \tan\iota \sin\lambda_s
\end{aligned}
$$

The parametric plot of the points (L_s, μ) for $\Omega = 0 \to 2\pi$ produces a figure-eight type figure called an *analemma*.

$$
\begin{aligned}
\tan\mu &= \tan(\Omega - \lambda_s) \\
&= \frac{\tan\Omega - \tan\lambda_s}{1 + \tan\lambda_s \tan\Omega} \\
&= \frac{(1 - \cos\iota)\tan\Omega}{1 + \cos\iota \tan^2\Omega}
\end{aligned}
$$

9.7 Seasonal Daylight Hours

In this section we will derive an expression which defines the number of hours of daylight as a function of the day of the year. In sequence #2 we derived the expression

$$
\cos\theta = \cos L_s \cos\sigma \cos L + \sin L_s \sin L
$$

We now replace the angle $\sigma = \lambda_s - \lambda$ by $\sigma - t$, where t is the term which accounts for the *earth rotation rate*. We include this in the expression for σ and write

$$
\begin{aligned}
\cos\theta &= \cos L_s \cos(\sigma - t)\cos L + \sin L_s \sin L \\
&= (\cos\sigma\cos t + \sin\sigma\sin t)\cos L_s \cos L \\
&\quad + \sin L_s \sin L
\end{aligned}
$$

It is convenient to let $t = 0$ at sunrise ($\theta = \frac{\pi}{2}$), which means

$$
\cos\sigma = -\frac{\sin L_s \sin L}{\cos L_s \cos L} \qquad \text{and}
$$

Sunrise/Sunset

The angle

$$
\theta = \pi/2 \Rightarrow \cos\theta = 0
$$

defines the necessary condition for *sunrise* and *sunset*. For this condition the line-of-sight to the Sun is tangent to the earth's surface for the observation point P, and

$$
\begin{aligned}
\cos\sigma &= -\tan L \tan L_s \\
\text{and} \quad L_s &= \arcsin(\sin\iota \sin\Omega)
\end{aligned}
$$

$$\sin \sigma \;=\; \frac{\sqrt{\cos^2 L_s \cos^2 L - \sin^2 L_s \sin^2 L}}{\cos L_s \cos L}$$

$$=\; \frac{\sqrt{(\mathbf{Sum})(\mathbf{Diff})}}{\cos L_s \cos L}$$

where $\mathbf{Sum} \;=\; \cos L_s \cos L + \sin L_s \sin L$

and $\mathbf{Diff} \;=\; \cos L_s \cos L - \sin L_s \sin L$

Then $\cos \theta \;=\; \sin t \sqrt{\cos(L + L_s) \cos(L - L_s)}$
$$+(1 - \cos t) \sin L \sin L_s$$

Finally, we may write

$$\theta(t) \;=\; \frac{180}{\pi} \arccos[\sin t \sqrt{\cos(L + L_s) \cos(L - L_s)}$$
$$+(1 - \cos t) \sin L \sin L_s]$$

and $\beta \;=\; 90 - \theta \;=\;$ Elevation angle to the Sun

Here the following constants are defined:

$L \;=\;$ Latitude of point of interest on Earth

$L_s \;=\; \arcsin[\sin \iota \sin \Omega]$

$\iota \;=\;$ angle between equatorial & ecliptic planes

$\;=\;$ 23.5   degrees

$\Omega \;=\;$ Earth-Sun line wrt Vernal Equinox

A plot of $\beta(t)$, which is the Sun elevation angle over the hours in one day at the peak of winter, is shown in Figure 9.14 for an observer at latitude 45 degrees.

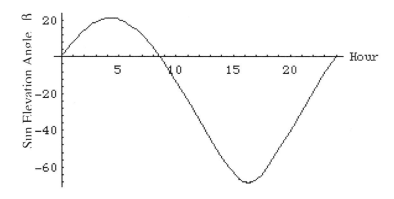

Figure 9.14: Daylight Hours in Peak of Winter

Plotted using Mathematica©

Figure 9.15 shows how the daylight hours vary over the days and seasons of the year.

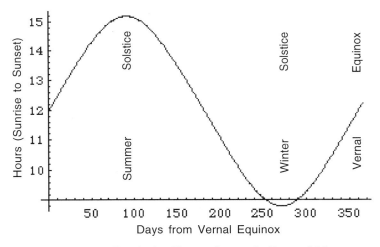

Figure 9.15: Daylight Hours for each Day of Year

Plotted using Mathematica©

From this and the previous work we have done so far it should be clear that rotations, whether we use rotation matrices or quaternions, should also be quite useful in exploring the relationships which hold in spherical trigonometry. This is indeed the case, and we turn to this substantial subject in the next chapter.

Exercises for Chapter 9

1. Show that the vector part of the quaternion rotation operator for the Aerospace application is consistent with the axis computed from the matrix operator using Equation 3.9.

2. Make a plot of the analemma for an earth synchronous circular orbit whose angle of inclination is 45 degrees with respect to the equatorial plane.

3. Make a plot of the analemma for a synchronous earth orbit whose eccentricity $\epsilon = 0$ and whose inclination $\iota = \frac{\pi}{4}$.

4. Make a plot of the analemma generated by a satellite in a synchronous earth orbit with eccentricity $\epsilon = 0.25$ and with inclination $\iota = \frac{\pi}{4}$.

Chapter 10

Spherical Trigonometry

10.1   Introduction

As an additional application of the theory of the quaternion rotation operator, as we have thus far developed it, we turn in this chapter to the development of some results in spherical trigonometry. In particular, using certain sequences of the now familiar quaternion rotation operator, we analyze certain relationships which hold in spherical triangles, that is, triangles which lie on the surface of a sphere. In so doing we shall derive some of the well-known relationships and identities in spherical trigonometry.

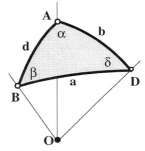

Spherical Triangle

In this analysis we shall use sequences of both matrix rotation operators and quaternion rotation operators. The two different rotation operators may then be compared and evaluated, particularly with respect to their relative computational efficiency. Use of both operators often allows us also to determine where and in what way, in some sense, the two operators complement each other.

10.2   Spherical Triangles

Consider three distinct points A, B and D which lie on the surface of a *unit sphere*. We specify a *unit* sphere, that is, a sphere of radius 1, merely so that lengths of great circle arcs on the surface of the unit sphere are numerically the same as the radian measure of the central angle subtended by that arc.

We may, without loss of generality, choose an **XYZ** coordinate frame (right-handed, as usual) such that the point A is on the **Z**-axis, and the point D lies in the **YZ** plane. In order to have a spherical triangle, on the surface of the sphere, with the points A,

Note!

We use the letter D here rather than the letter C simply because the upper and lower cases of the letter C have the same form, and may cause some confusion; not so with the letter D.

235

B and D as its vertices, the point B, of course, should *not* lie in the **YZ** plane.

Let each of the pairs of these points be connected by great circle arcs, as shown in Figure 10.1. We note that great circle arcs on the surface of a sphere are equivalent to straight line segments in a plane, in the sense that the shortest distance between any two points on the sphere is measured along the *great circle arc* joining the two points. Further, it is helpful to note that the plane which contains any great circle arc also *always* contains the center of the sphere.

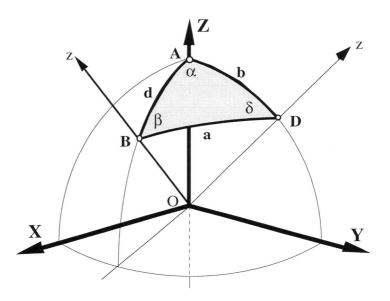

Figure 10.1: Spherical Triangle

We designate the great circle arc which connects the points B and D by a, which is equal to the radian length of that arc on our unit sphere. Similarly, b is the great circle arc connecting the points A and D, while d is the great circle arc connecting points A and B. These three arcs, a, b, and d, are the sides of the spherical triangle we wish to consider.

Each of these arcs lies on the intersection of the sphere with the plane which contains the two endpoints of that arc and the center of the sphere. The central angles $\angle AOB$, $\angle BOD$ and $\angle DOA$ correspond respectively to the sides d, a and b. Because we have chosen to work on the surface of a unit sphere, the lengths d, a and

b of the sides of the triangle are the same as the radian measure of the central angles $\angle AOB$, $\angle BOD$ and $\angle DOA$, respectively. The three *interior angles* of the spherical triangle at each vertex, A, B and D, we designate α, β and δ, respectively.

10.3 Closed-loop Rotation Sequences

Our analysis of the spherical triangle ABD proceeds in terms of a sequence of rotations in R^3, about appropriate axes, through the three interior angles α, β and δ, and the opposite three sides or central angles a, b and d. These angles, properly ordered, define the six rotations in an *identity* sequence.

The concatenation of these six rotations in the sequence is *closed*, in the sense that the initial coordinate frame, subjected to this sequence of rotations, begins and ends with the same frame   orien-

Rotations Only

We emphasize here that *all* transformations are rotations; that is, linear translations of a frame are *not relevant* in our present consideration. Therefore, that we show the origin of a frame at the origin of the reference frame, or at a point on the surface of the sphere is not important. What is important is the *orientation* of the frames.

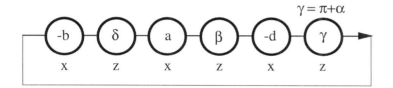

Figure 10.2: Rotation Sequence

To Simplify

In the analysis which follows it is convenient that we define

$$\gamma = \pi + \alpha$$

as we shall see.

tation. This simply says that the product of the rotations in this sequence is the identity, an idea we have used before. Hence the term *closed* or *closed-loop* rotation sequence.

In what follows, we will analyze in considerable detail the six-rotation closed-loop sequence defined in Figure 10.9. Clearly, this sequence involves the parameters we have assigned to the spherical triangle ABD. Further, all of the parameters (the three interior angles α, β and δ, and the central angles a, b and d) which characterize the spherical triangle ABD are positive. The signed symbol in each circle represents the magnitude and direction of the frame rotation. The axis about which this particular rotation occurs is indicated below the circle.

A coordinate frame, always designated xyz, which is initially coincident with the reference frame $\mathbf{XYZ}$, is now subjected to the ordered sequence of rotations involving these six angles, in turn.

Notation

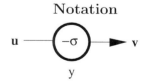

As in Chapter 4, this symbol represents a rotation about the y-axis, through an angle σ. The *minus sign* means the rotation direction, as determined by the right-hand rule, is opposite the sign of the angle σ, whatever the convention be that determined this sign.

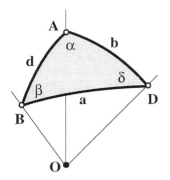

Spherical Triangle

We now describe successive rotations in the sequence, and illustrate each in an accompanying figure. The reader should take care to note that the rotating coordinate frame *xyz* indeed *begins* and *ends* with an orientation cooincident with that of the fixed reference frame **XYZ**. It is for this reason we tie the two ends together, that is, we *close-the-loop* on this sequence. This rotation sequence

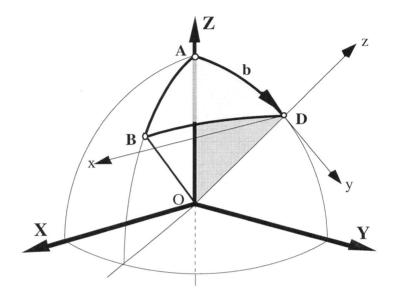

Figure 10.3: First Rotation — radian-length of Side **b**

Remember!

Remember, all rotations are taken in the "right-hand-rule" sense

is easy to follow, since it simply goes around the spherical triangle ABD in an orderly fashion. Note, however, that we have chosen the direction ADB, around the triangle — merely because most of the rotations are then positive.

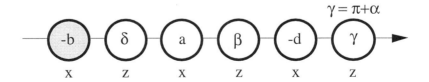

Figure 10.4: First Rotation in the Sequence

The first rotation in the sequence, denoted $R^t_{x,b}$, is about the x-axis through the central angle $\angle AOD$, which represents the side AD, labelled b on the triangle. Notice that the rotation direction, in the right-hand rule sense, is negative. Hence, the transpose. The magnitude of the rotation is such that the z-axis of the rotation frame, which initially contained vertex A, now contains the vertex D, as shown in Figure 10.3. We illustrate this first rotation in the sequence of Figure 10.4.

The second rotation in the sequence, denoted $R_{z,\delta}$, is about the new z-axis through the angle δ, such that the new yz plane contains

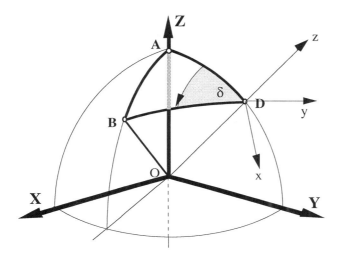

Figure 10.5: Second Rotation — interior angle δ

the side a of the triangle. This rotation is illustrated in Figure 10.5. Notice that the rotation direction, in the right-hand rule sense, is positive. This δ rotation and its place in the overall sequence of rotations is illustrated in Figure 10.6.

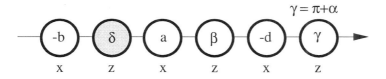

Figure 10.6: Second Rotation in the Sequence

The third rotation, $R_{x,a}$, is about the new x-axis through the angle a, such that the newly rotated z-axis now contains the vertex B of the triangle, as shown in Figure 10.7(a). Figure 10.7(b) shows where this rotation occurs in the sequence.

The fourth rotation in the sequence, denoted $R_{z,\beta}$, is about the new z-axis through the angle β, such that the new yz plane now contains the side d of the triangle ABD. Figure 10.7(c) illustrates this fourth rotation and Figure 10.7(d) its place in the sequence.

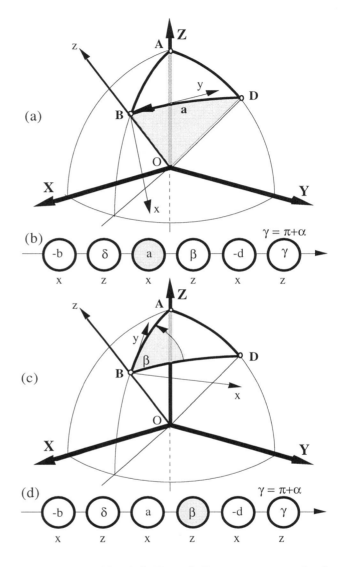

Figure 10.7: Third & Fourth Rotations — **a** & β

The fifth rotation, denoted $R_{x,d}^{t}$, is about the new x-axis through the angle d such that the new z-axis again coincides with the reference **Z**-axis, and hence contains the vertex A of the triangle, as shown in Figure 10.8(a). As was the case with the first rotation, this rotation direction (in the rh-rule sense) is negative. Hence, the transpose and the minus-d. Figure 10.8(b) shows where this rotation occurs in the sequence.

The sixth and final rotation is about the z-axis through an angle $\gamma = \alpha + \pi$. The reader should carefully check Figure 10.8(c) & (d)

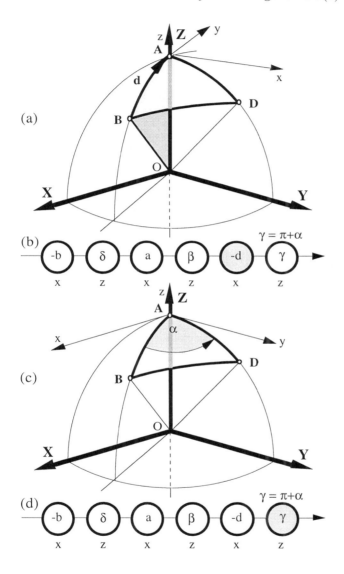

Figure 10.8: Fifth & Sixth Rotations — **d** & α

merely to confirm that after this last rotation in the sequence, the rotated xyz frame once again coincides with the original **XYZ** reference frame. The final yz plane is therefore again coincident with the **YZ** plane and contains both the vertex A and the side b of the triangle. Note also that the composite transformation represented by this sequence of rotations is equivalent to the identity. Therefore, as again illustrated in Figure 10.9, we 'close-the-loop' on this entire rotation sequence.

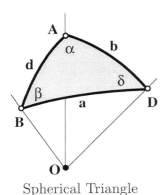

Spherical Triangle

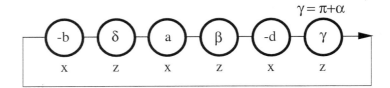

Figure 10.9: Closed-Loop Rotation Sequence

We use this fact, that *the product of these six rotations is the identity*, to find relationships between the various angles which characterize a spherical triangle. We first analyze this sequence of rotations algebraically using matrices, and follow that with an analysis using the quaternion operator.

10.4 Rotation Matrix Analysis

Matrix Format

The matrices used in this section were derived in Chapter 3, as Equations 3.1 and 3.3.

For each of the rotations in the *closed* sequence of Figure 10.9 we have a corresponding rotation matrix. The properly ordered product of these matrices must also, of course, be the identity matrix. So we may write

$$R_\gamma R_d^t R_\beta R_a R_\delta R_b^t = I \qquad (10.1)$$

We now analyze this matrix Equation 10.1 in order to determine some relationships between these six angles in a spherical triangle. By suitable pre- and post-multiplications, we may write

$$R_b R_\delta^t R_a^t = R_\gamma R_d^t R_\beta \qquad (10.2)$$

We define the matrix

$$M = R_b R_\delta^t R_a^t \qquad (10.3)$$

which, more specifically, is the product

$$\begin{bmatrix} 1 & 0 & 0 \\ 0 & \cos b & \sin b \\ 0 & -\sin b & \cos b \end{bmatrix} \begin{bmatrix} \cos \delta & -\sin \delta & 0 \\ \sin \delta & \cos \delta & 0 \\ 0 & 0 & 1 \end{bmatrix} \begin{bmatrix} 1 & 0 & 0 \\ 0 & \cos a & -\sin a \\ 0 & \sin a & \cos a \end{bmatrix}$$

In like manner, we define the matrix

$$N = R_\gamma R_d^t R_\beta \qquad (10.4)$$

which more specifically is the product

$$\begin{bmatrix} \cos\gamma & \sin\gamma & 0 \\ -\sin\gamma & \cos\gamma & 0 \\ 0 & 0 & 1 \end{bmatrix} \begin{bmatrix} 1 & 0 & 0 \\ 0 & \cos d & -\sin d \\ 0 & \sin d & \cos d \end{bmatrix} \begin{bmatrix} \cos\beta & \sin\beta & 0 \\ -\sin\beta & \cos\beta & 0 \\ 0 & 0 & 1 \end{bmatrix}$$

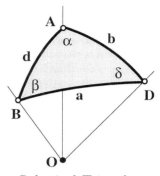

Spherical Triangle

We now compute the matrix products for M and N in order to get the detailed expressions for their respective elements. And, since matrices M and N are equal, we equate their corresponding elements m_{ij} and n_{ij}. This results in nine equations which together define the relationships between the six angles which comprise every spherical triangle:

$$
\begin{aligned}
m_{11} &= \cos\delta &&= \sin\alpha\cos d\sin\beta - \cos\alpha\cos\beta &&= n_{11} \\
m_{12} &= \sin\delta\cos a &&= \cos\alpha\sin\beta + \sin\alpha\cos d\cos\beta &&= n_{12} \\
m_{13} &= \sin a\sin\delta &&= \qquad\qquad\qquad \sin\alpha\sin d &&= n_{13} \\
m_{21} &= \cos b\sin\delta &&= \sin\alpha\cos\beta - \cos\alpha\cos d\sin\beta &&= n_{21} \\
m_{22} &= \cos b\cos\delta\cos a + \sin b\sin a &&= \\
& &&= \sin\alpha\sin\beta - \cos\alpha\cos d\cos\beta &&= n_{22} \\
m_{23} &= \sin b\cos a - \cos b\cos\delta\sin a &&= \cos\alpha\sin d &&= n_{23} \\
m_{31} &= \sin b\sin\delta &&= \qquad\qquad\qquad \sin d\sin\beta &&= n_{31} \\
m_{32} &= \cos b\sin a - \sin b\cos\delta\cos a &&= \sin d\sin\beta &&= n_{32} \\
m_{33} &= \cos d &&= \sin a\cos\delta\sin b + \cos a\cos b &&= n_{33}
\end{aligned}
$$

$\Longleftarrow \quad M = N$

We may now use these equations to find some well-known identities in spherical trigonometry. For example, from the fact that $m_{13} = n_{13}$ and $m_{31} = n_{31}$, we get the familiar expressions

$$\frac{\sin a}{\sin\alpha} = \frac{\sin b}{\sin\beta} = \frac{\sin d}{\sin\delta} \tag{10.5}$$

Further, the equalities $m_{33} = n_{33}$ and $m_{11} = n_{11}$ give us the two companion cosine laws of spherical trigonometry:

$$\cos d = \sin a\cos\delta\sin b + \cos a\cos b \tag{10.6}$$

$$\cos\delta = \sin\alpha\cos d\sin\beta - \cos\alpha\cos\beta \tag{10.7}$$

10.4.1 Right Spherical Triangle

For a *Right Spherical Triangle* $(\delta = \pi/2)$ these identities become

$$\sin d = \frac{\sin a}{\sin\alpha} = \frac{\sin b}{\sin\beta} \tag{10.8}$$

$$\tan\alpha\cos d\tan\beta = 1 \tag{10.9}$$

$$\cos d = \cos a\cos b \tag{10.10}$$

Other identities can be found using various combinations of the other elements of M and N.

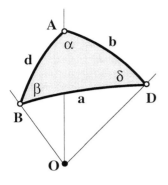

Spherical Triangle

10.4.2　Isoceles Spherical Triangle

For an *Isoceles Spherical Triangle* we take $\alpha = \beta$ and $a = b$. Using the foregoing matrix identities we get

$$\sin^2 a \;=\; \frac{1 - \cos d}{1 - \cos \delta} \quad \text{and} \quad \sin^2 \alpha \;=\; \frac{1 + \cos \delta}{1 + \cos d} \tag{10.11}$$

Other identities may readily be derived in this fashion. However, we turn next to the quaternion approach.

10.5　Quaternion Analysis

We now analyze the same spherical triangle using quaternions. According to Theorem 5.4, the product of the quaternions associated with each of the rotations shown in Figure 10.10 must also be equiv-

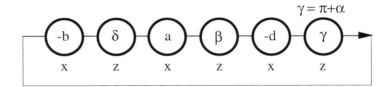

Figure 10.10: Same Sequence for Quaternions

alent to the identity, as before. This means that we must have

$$e = q^*_{x,b}\, q_{z,\delta}\, q_{x,a}\, q_{z,\beta}\, q^*_{x,d}\, q_{z,\gamma} = 1 \tag{10.12}$$

By appropriate pre-multiplications, we may write

$$p = q_{z,\beta}\, q^*_{x,d}\, q_{z,\gamma} = q^*_{x,a}\, q^*_{z,\delta}\, q_{x,b} = q \tag{10.13}$$

Here, for convenience we designate the left-hand side and the right-hand side of Equation 10.13 as p and q, respectively.

The quaternions for each rotation in this sequence are given by

$$q_{x,b} \;=\; \cos\frac{b}{2} + \mathbf{i}\sin\frac{b}{2}$$

$$q_{z,\delta} \;=\; \cos\frac{\delta}{2} + \mathbf{k}\sin\frac{\delta}{2}$$

$$q_{x,a} \;=\; \cos\frac{a}{2} + \mathbf{i}\sin\frac{a}{2}$$

$$q_{z,\beta} \;=\; \cos\frac{\beta}{2} + \mathbf{k}\sin\frac{\beta}{2}$$

$$q_{x,d} \;=\; \cos\frac{d}{2} + \mathbf{i}\sin\frac{d}{2}$$

$$q_{z,\gamma} \;=\; \cos\frac{\gamma}{2} + \mathbf{k}\sin\frac{\gamma}{2}$$

Since the angle $\gamma = \pi + \alpha$ we may write the quaternion expression related to the rotation γ about the z-axis as a function of the desired interior angle rotation α, as follows

$$
\begin{aligned}
q_{z,\gamma} &= \cos\frac{\gamma}{2} + \mathbf{k}\sin\frac{\gamma}{2} \\
&= -\sin\frac{\alpha}{2} + \mathbf{k}\cos\frac{\alpha}{2} \quad\quad (10.14)
\end{aligned}
$$

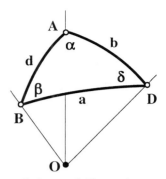

Spherical Triangle

We now compute and equate the two quaternion products, p and q, indicated in Equation 10.13. These products are

$$
\begin{aligned}
p &= p_0 + \mathbf{i}p_1 + \mathbf{j}p_2 + \mathbf{k}p_3 = q_{z,\beta}q^*_{x,d}q_{z,\gamma} \\
&= (\cos\frac{\beta}{2} + \mathbf{k}\sin\frac{\beta}{2})(\cos\frac{d}{2} - \mathbf{i}\sin\frac{d}{2})(-\sin\frac{\alpha}{2} + \mathbf{k}\cos\frac{\alpha}{2}) \\
q &= q_0 + \mathbf{i}q_1 + \mathbf{j}q_2 + \mathbf{k}q_3 = q^*_{x,a}q^*_{z,\delta}q_{x,b} \\
&= (\cos\frac{a}{2} - \mathbf{i}\sin\frac{a}{2})(\cos\frac{\delta}{2} - \mathbf{k}\sin\frac{\delta}{2})(\cos\frac{b}{2} + \mathbf{i}\sin\frac{b}{2})
\end{aligned}
$$

Using our definition of the quaternion product, the elements of the quaternion p are given by

$$
\begin{aligned}
p_0 &= -\cos\frac{\beta}{2}\cos\frac{d}{2}\sin\frac{\alpha}{2} - \sin\frac{\beta}{2}\cos\frac{d}{2}\cos\frac{\alpha}{2} \\
&= -\cos\frac{d}{2}\sin\frac{\alpha+\beta}{2} \\
p_1 &= \cos\frac{\beta}{2}\sin\frac{d}{2}\sin\frac{\alpha}{2} - \sin\frac{\beta}{2}\sin\frac{d}{2}\cos\frac{\alpha}{2} \\
&= \sin\frac{d}{2}\sin\frac{\alpha-\beta}{2} \\
p_2 &= \sin\frac{\beta}{2}\sin\frac{d}{2}\sin\frac{\alpha}{2} + \cos\frac{\beta}{2}\sin\frac{d}{2}\cos\frac{\alpha}{2} \\
&= \sin\frac{d}{2}\cos\frac{\alpha-\beta}{2} \\
p_3 &= \cos\frac{\beta}{2}\cos\frac{d}{2}\cos\frac{\alpha}{2} - \sin\frac{\beta}{2}\cos\frac{d}{2}\cos\frac{\alpha}{2} \\
&= \cos\frac{d}{2}\cos\frac{\alpha+\beta}{2}
\end{aligned}
$$

and the elements of the quaternion q are

$$
\begin{aligned}
q_0 &= \cos\frac{a}{2}\cos\frac{\delta}{2}\cos\frac{b}{2} + \sin\frac{a}{2}\cos\frac{\delta}{2}\sin\frac{b}{2} \\
&= \cos\frac{\delta}{2}\cos\frac{a-b}{2} \\
q_1 &= \cos\frac{a}{2}\cos\frac{\delta}{2}\sin\frac{b}{2} - \sin\frac{a}{2}\cos\frac{\delta}{2}\cos\frac{b}{2}
\end{aligned}
$$

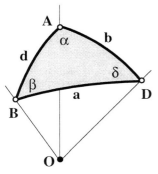

Spherical Triangle

$$= -\cos\frac{\delta}{2}\sin\frac{a-b}{2}$$

$$q_2 = -\cos\frac{a}{2}\sin\frac{\delta}{2}\sin\frac{b}{2} - \sin\frac{a}{2}\sin\frac{\delta}{2}\cos\frac{b}{2}$$

$$= -\sin\frac{\delta}{2}\sin\frac{a+b}{2}$$

$$q_3 = \sin\frac{a}{2}\sin\frac{\delta}{2}\sin\frac{b}{2} - \cos\frac{a}{2}\sin\frac{\delta}{2}\cos\frac{b}{2}$$

$$= -\sin\frac{\delta}{2}\cos\frac{a+b}{2}$$

The indicated triple quaternion products define each element of the composite quaternion p as functions of the rotations α, β and d. Similarly, the elements of the composite quaternion q are functions of a, b and δ. Since the two resulting composite quaternions p and q are equal, equating their corresponding elements yields the following relationships.

Equating q_0 and p_0 gives

$$-\cos\frac{d}{2}\sin\frac{(\alpha+\beta)}{2} = \cos\frac{\delta}{2}\cos\frac{(a-b)}{2} \qquad (10.15)$$

Equating q_1 and p_1 gives

$$\sin\frac{d}{2}\sin\frac{(\alpha-\beta)}{2} = -\cos\frac{\delta}{2}\sin\frac{(a-b)}{2} \qquad (10.16)$$

Equating q_2 and p_2 gives

$$\sin\frac{d}{2}\cos\frac{(\alpha-\beta)}{2} = -\sin\frac{\delta}{2}\sin\frac{(a+b)}{2} \qquad (10.17)$$

Equating q_3 and p_3 gives

$$\cos\frac{d}{2}\cos\frac{(\alpha+\beta)}{2} = -\sin\frac{\delta}{2}\cos\frac{(a+b)}{2} \qquad (10.18)$$

The preceding equations may now be used to derive several well-known identities which hold for spherical trigonometry. The sine law, for example, may be obtained as follows. If we multiply Equations 10.15 and 10.17, and use an appropriate trigonometric identity, we may write

$$\sin d \sin\alpha + \sin d \sin\beta = \sin\delta\sin a + \sin\delta\sin b$$

If we multiply Equations 10.16 and 10.18 we may obtain the expression

$$\sin d \sin\alpha - \sin d \sin\beta = \sin\delta\sin a - \sin\delta\sin b$$

If we now add and, in turn, subtract these expressions we easily obtain the well-known sine law

$$\frac{\sin d}{\sin \delta} = \frac{\sin a}{\sin \alpha} = \frac{\sin b}{\sin \beta}$$

As another example, if we square and add Equations 10.15 and 10.18, and make good use of the half-angle formulas

$$\sin^2 \frac{\theta}{2} = \frac{1 - \cos \theta}{2}$$
$$\cos^2 \frac{\theta}{2} = \frac{1 + \cos \theta}{2}$$

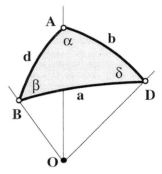

Spherical Triangle

we obtain the cosine law

$$\cos d = \cos a \cos b + \sin a \cos \delta \sin b$$

If we do the same thing with Equations 10.15 and 10.16 we obtain the analogous cosine law for δ

$$\cos \delta = \sin \alpha \cos d \sin \beta - \cos \alpha \cos \beta$$

Other well-known identities may also be obtained. For example, dividing equation 10.15 by equation 10.18 gives

$$\tan \frac{(\alpha + \beta)}{2} \tan \frac{\delta}{2} = \frac{\cos \frac{1}{2}(a - b)}{\cos \frac{1}{2}(a + b)} \tag{10.19}$$

and dividing equation 10.16 by equation 10.17 gives

$$\tan \frac{(\alpha - \beta)}{2} \tan \frac{\delta}{2} = \frac{\sin \frac{1}{2}(a - b)}{\sin \frac{1}{2}(a + b)} \tag{10.20}$$

Dividing equation 10.17 by equation 10.18 gives

$$\frac{\cos \frac{1}{2}(\alpha - \beta)}{\cos \frac{1}{2}(\alpha + \beta)} = \frac{\tan \frac{1}{2}(a + b)}{\tan \frac{1}{2}d} \tag{10.21}$$

and dividing equation 10.16 by equation 10.15 gives

$$\frac{\sin \frac{1}{2}(\alpha - \beta)}{\sin \frac{1}{2}(\alpha + \beta)} = \frac{\tan \frac{1}{2}(a - b)}{\tan \frac{1}{2}d} \tag{10.22}$$

In some mathematical handbooks the two preceding identities are known as *Napier's Analogies*, which are useful in solving for certain unknown quantities in the spherical triangle, given certain other information about the triangle.

Finally, dividing equation 10.20 by equation 10.19 gives

$$\frac{\tan\frac{1}{2}(\alpha - \beta)}{\tan\frac{1}{2}(\alpha + \beta)} = \frac{\tan\frac{1}{2}(a - b)}{\tan\frac{1}{2}(a + b)} \tag{10.23}$$

This identity is known in spherical trigonometry as the *law of tangents*. Note that equation 10.23 is independent of the magnitude of one of the three angles and its associated opposite side.

10.5.1 Right Spherical Triangle

We note finally some special cases for these identities, namely, cases in which $\delta = \pi/2$. For the resulting right spherical triangle the equations which were derived using the quaternion operator become

$$\tan\frac{(\alpha + \beta)}{2} = \frac{\cos\frac{1}{2}(a - b)}{\cos\frac{1}{2}(a + b)} \tag{10.24}$$

$$\tan\frac{(\alpha - \beta)}{2} = \frac{\sin\frac{1}{2}(a - b)}{\sin\frac{1}{2}(a + b)} \tag{10.25}$$

From these equations we now solve for α and β independent of the other. If we let

$$S = \alpha + \beta \qquad \text{and} \qquad D = \alpha - \beta$$

$$\alpha = \frac{S}{2} + \frac{D}{2} \qquad \text{and} \qquad \beta = \frac{S}{2} - \frac{D}{2}$$

$$\text{Then} \quad \tan\alpha = \tan(\frac{1}{2}S + \frac{1}{2}D)$$

$$= \frac{\tan\frac{1}{2}S + \tan\frac{1}{2}D}{1 - \tan\frac{1}{2}S\tan\frac{1}{2}D} = \frac{\tan a}{\sin b}$$

$$\text{and} \quad \tan\beta = \tan(\frac{1}{2}S - \frac{1}{2}D)$$

$$= \frac{\tan\frac{1}{2}S - \tan\frac{1}{2}D}{1 + \tan\frac{1}{2}S\tan\frac{1}{2}D} = \frac{\tan b}{\sin a}$$

10.5.2 Isoceles Spherical Triangle

For an *Isoceles Spherical Triangle* we take $\alpha = \beta$ and $a = b$. Then using the foregoing identities which were derived using the quaternion rotation operator the reader is invited to show that

$$\tan\frac{d}{2} = \sin\alpha\sin a\tan\frac{\delta}{2}$$

Furthermore, for an *isoceles right triangle*

$$\tan\frac{d}{2} = \sin\alpha\sin a$$

Spherical Triangle

10.6   Regular n-gons on the Sphere

An equilateral spherical triangle is a regular n-gon, for $n = 3$. It is represented by the *closed* rotation sequence illustrated in Figure 10.11. Note, this sequence consists of three repeated two-rotation subsequences; we denote each by the matrix

$$T = R_x(\sigma)R_z(a)$$
$$= \begin{bmatrix} \cos a & \sin a & 0 \\ -\sin a \cos \sigma & \cos a \cos \sigma & \sin \sigma \\ \sin a \sin \sigma & -\cos a \sin \sigma & \cos \sigma \end{bmatrix} \quad (10.26)$$

where a =   is the radian length of a triangle side

and $\sigma = \pi - \alpha$ =   supplement of interior angle α.

Then, for an equilateral triangle on the surface of the sphere, we may write

$$T^3 = I = \text{Identity} \quad (10.27)$$

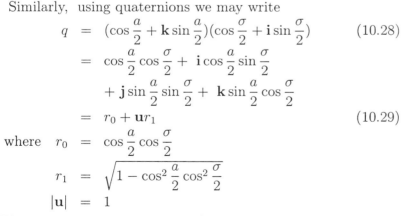

a = radian length of triangle side

$\sigma = \pi{-}\alpha$ where α = triangle interior angle

Figure 10.11:   Equilateral Triangle Rotation Sequence

Similarly, using quaternions we may write

$$q = (\cos\frac{a}{2} + \mathbf{k}\sin\frac{a}{2})(\cos\frac{\sigma}{2} + \mathbf{i}\sin\frac{\sigma}{2}) \quad (10.28)$$
$$= \cos\frac{a}{2}\cos\frac{\sigma}{2} + \mathbf{i}\cos\frac{a}{2}\sin\frac{\sigma}{2}$$
$$+ \mathbf{j}\sin\frac{a}{2}\sin\frac{\sigma}{2} + \mathbf{k}\sin\frac{a}{2}\cos\frac{\sigma}{2}$$
$$= r_0 + \mathbf{u}r_1 \quad (10.29)$$

where $r_0 = \cos\frac{a}{2}\cos\frac{\sigma}{2}$

$$r_1 = \sqrt{1 - \cos^2\frac{a}{2}\cos^2\frac{\sigma}{2}}$$
$$|\mathbf{u}| = 1$$

Then, analogous to Equation 10.27, a rotation sequence for equilateral spherical triangles, using quaternions, says

$$q^3 = \text{the quaternion Identity} \quad (10.30)$$
$$= 1 + \mathbf{i}0 + \mathbf{j}0 + \mathbf{k}0$$

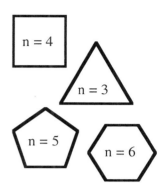

Regular n-gons

Supplemental Angle

It is convenient to use the supplemental angle $\sigma = \pi - \alpha$ in these formulations, where α is an interior angle of the equilateral triangle.

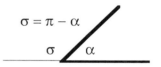
$$\sigma = \pi - \alpha$$

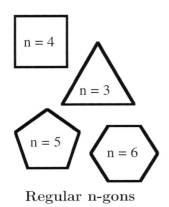

Regular n-gons

Again, these formulations represent two distinct analytical approaches, one using matrices and the other using quaternions.

Spherical Regular n-gon using Matrices

We first relate parameters, σ and a, using Equation 10.27.

$$\text{Since} \quad T^3 \;=\; I \quad \Rightarrow \quad T^2 \;=\; T^t$$

which further implies these respective matrix elements are equal, and we need only compute and equate, say, $[T^2]_{(31)}$ and $[T]_{(13)}$.

We then may write

$$
\begin{aligned}
[T^2]_{(31)} &\;=\; \sin a \sin \sigma (\cos a + \cos a \cos \sigma + \cos \sigma) \\
&\;=\; 0 \;=\; [T]_{(13)}
\end{aligned}
$$

which reduces to

$$\cos \sigma \;=\; \frac{-\cos a}{1 + \cos a} \tag{10.31}$$

See Equation 10.31

This equation may also be written in terms of α, the interior angle of the equilateral triangle. Since the angle σ is the supplement of the interior angle α, it follows that

$$
\begin{aligned}
\cos \sigma &\;=\; \cos(\pi - \alpha) \\
&\;=\; -\cos \alpha
\end{aligned}
$$

Therefore

$$\cos \alpha \;=\; \frac{\cos a}{1 + \cos a}$$

The same method will determine the relationship between σ, the supplement of the interior angle α, and the side, a, for a regular 4-gon on the sphere. The required rotation sequence (similar to that illustrated in 10.11) is now a loop which consists of four repeated subsequences. That is,

$$T^4 = I \quad \Rightarrow \quad T^2 = [T^2]^t$$

This means we may equate the corresponding elements in

$$M = [T^2]_{(ij)} = [T^2]_{(ji)} = M^t$$

$$
\begin{aligned}
m(1,2) &\;=\; \cos a \sin a + \cos a \sin a \cos \sigma \;=\; m(2,1) \\
m(1,3) &\;=\; \sin a \sin \sigma \;=\; m(3,1) \\
m(2,1) &\;=\; \sin a \sin^2 \sigma - \cos a \sin a \cos \sigma - \\
&\qquad \sin a \cos a \cos^2 \sigma \;=\; m(1,2) \\
m(2,3) &\;=\; \cos a \cos \sigma \sin \sigma + \cos \sigma \sin \sigma \;=\; m(3,2) \\
m(3,1) &\;=\; \cos a \sin a \sin \sigma + \cos a \sin a \cos \sigma \sin \sigma + \\
&\qquad \sin a \cos \sigma \sin \sigma \;=\; m(1,3) \\
m(3,2) &\;=\; \sin^2 a \sin \sigma - \cos^2 a \cos \sigma \sin \sigma - \\
&\qquad \cos a \cos \sigma \sin \sigma \;=\; m(2,3)
\end{aligned}
$$

For example, since $m(1,3) = m(3,1)$ we may write

$$
\begin{aligned}
\sin a \sin \sigma \;=\; & \cos a \sin a \sin \sigma + \\
& \cos a \sin a \cos \sigma \sin \sigma + \quad (10.32) \\
& \sin a \cos \sigma \sin \sigma
\end{aligned}
$$

Reducing 10.32, we get

$$
\cos \sigma \;=\; \frac{(1 - \cos a)}{(1 + \cos a)} = -\cos \alpha \qquad (10.33)
$$

Equation 10.33 defines the relationship between the radian length of the side, a, and interior angle, $\alpha = \pi - \sigma$, for any regular 4-gon on the sphere — a *square on the sphere* as $a \to 0$.

The relationship between the radian length of the side and the corresponding interior angle-supplement for any regular n-gon on the sphere may in like manner be determined from

$$
T^n = I \qquad \Rightarrow \qquad [T]^{n-k} = [T^t]^k
$$

where the matrix T is that given in Equation 10.26.

Regular Spherical n-gons using Quaternions

We now use the quaternion expressions in Equation 10.28 and Equation 10.30 and again determine the *side-angle* parameter relationship for an equilateral spherical triangle and 4-gon.

$$
\begin{aligned}
\text{We write} \qquad q \;=\;& r_0 + \mathbf{u} r_1 \\
\text{Then} \qquad q^2 \;=\;& (r_0^2 - r_1^2) + \mathbf{u}(2 r_0 r_1) \\
q^3 \;=\;& r_0(r_0^2 - r_1^2) - (2 r_0 r_1^2) + \\
& \mathbf{u}[r_1(r_0^2 - r_1^2) + 2 r_0^2 r_1] \quad (10.34) \\
\text{and} \qquad q^4 \;=\;& [(r_0^2 - r_1^2)^2 - 4 r_0^2 r_1^2] + \\
& \mathbf{u}[4 r_0 r_1(r_0^2 - r_1^2)] \quad (10.35)
\end{aligned}
$$

We shall now solve for and obtain algorithms which define the relationship between the side, a, and supplemental angle, σ, for both the 3-gon and the 4-gon.

Since we have $q^3 = 1$ for the equilateral spherical triangle (or regular 3-gon), the vector part of Equation 10.34 must be equal to zero.

$$
\begin{aligned}
\text{That is,} \qquad r_1(r_0^2 - r_1^2) + 2 r_0^2 r_1 \;=\;& 0 \\
\text{or} \qquad (r_0^2 - r_1^2) + 2 r_0^2 \;=\;& 0
\end{aligned}
$$

Equation 10.33

In this equation it is worth a moments reflection on the values one gets for various values of a and σ. For example,

$$
\begin{aligned}
\text{for} \quad a \to 0 \quad &\Rightarrow \quad \sigma \to \frac{\pi}{2} \\
\text{for} \quad a = \frac{\pi}{2} \quad &\Rightarrow \quad \sigma = 0 \\
\text{for} \quad a = \frac{\pi}{3} \quad &\Rightarrow \quad \cos \sigma = \frac{1}{3} \\
\text{for} \quad 0 \le a \le \frac{\pi}{2} \quad &\Rightarrow \quad \frac{\pi}{2} \ge \sigma \ge 0
\end{aligned}
$$

In Equation 10.36 it is worth noting how a and σ relate for easily computed values. For example,

$$\text{as} \quad a \to 0 \quad \Rightarrow \quad \sigma \to \frac{2\pi}{3}$$

$$\text{when} \quad a = \frac{\pi}{2} \quad \Rightarrow \quad \sigma = \frac{\pi}{2}$$

$$\text{if} \quad a = \frac{2\pi}{3} \quad \text{then} \quad \cos\sigma = 0$$

$$0 \le a \le \frac{2\pi}{3} \quad \Rightarrow \quad \frac{2\pi}{3} \ge \sigma \ge 0$$

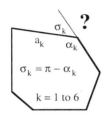

$$\sigma_k = \pi - \alpha_k$$

$$k = 1 \text{ to } 6$$

Regular Spherical 4-gons

In Equation 10.37 we note how the parameters a and σ relate for certain easily determined values in the 4-gon. For example,

$$\text{as} \quad a \to 0 \quad \text{then} \quad \sigma \to \frac{\pi}{2}$$

$$\text{for} \quad a = \frac{\pi}{2} \quad \Rightarrow \quad \sigma = 0$$

$$\text{for} \quad a = \frac{\pi}{3} \quad \Rightarrow \quad \cos\sigma = \frac{\sqrt{2}}{\sqrt{3}}$$

$$0 \le a \le \frac{\pi}{2} \quad \Rightarrow \quad \frac{\pi}{2} \ge \sigma \ge 0$$

and since
$$\cos\frac{a}{2}\cos\frac{\sigma}{2} = r_0$$

and
$$\sqrt{1 - \cos^2\frac{a}{2}\cos^2\frac{\sigma}{2}} = r_1$$

upon substitution and simplification we get

$$\cos\frac{a}{2}\cos\frac{\sigma}{2} = \frac{1}{2} \qquad (10.36)$$

which relates side and angle for any *regular* spherical 3-gon.

Similarly, for the regular spherical 4-gon we have $q^4 = 1$, and this again says that the vector part of Equation 10.35 must be equal to zero.

That is,
$$4r_0 r_1 (r_0^2 - r_1^2) = 0$$

or
$$r_0^2 - r_1^2 = 0$$

Then with the same substitutions as used above, we get

$$\cos\frac{a}{2}\cos\frac{\sigma}{2} = \frac{\sqrt{2}}{2} \qquad (10.37)$$

which relates side and angle for *regular* spherical 4-gons.

No doubt much more could be done along these lines; we might find answers to some interesting questions suggested by the n-gon pictured in the margin. However, we simply remark that in this chapter we have amply demontrated that both matrix and quaternion rotation operators are indeed useful in solving problems relating to surface geometries on the sphere.

10.7   Area and Volume

Before we conclude this chapter on spherical trigonometry we derive expressions for the *area* of a spherical triangle and the *volume* of a volume segment of a sphere which is bounded by a spherical triangle on the surface of the sphere and the radial lines to the vertices of the triangle. Although these derivations do not directly involve quaternions they do answer interesting questions relating to spherical triangles.

First, consider the great circle which contains vertices of the spherical triangle, say A and B. A rotation of this great circle about the diameter which contains point A, through an angle α generates a *double lune* whose area is equal to 4α. This result is easy to verify:

*A **double lune** area is linear with angle and it clearly is equal to 4π (the area of a unit sphere) if the angle of rotation is π.* QED

We now rotate great-circles on the unit sphere about diameters defined by OA, OB, and OD, through angles α, β, and δ, respectively. The rotation α generates a double-lune whose area is equal to 4α. This generated area includes twice the area A of the spherical triangle (see Figure 10.12).

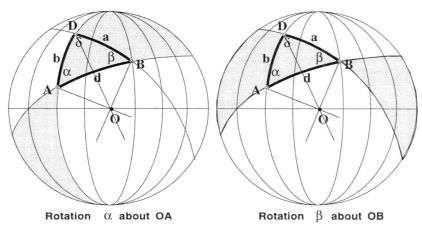

Rotation α about OA   **Rotation β about OB**

Figure 10.12: Great-Circle Double-Lune Rotations

A similar great-circle rotation through the angle β about the diameter which contains the vertex B generates a double-lune area equal to 4β. This double-lune area also contains twice the area A. Similarly, the third great-circle rotation through the angle δ about a diameter which contains the vertex D, produces a double-lune which again includes twice the area A of the spherical triangle.

These last two rotations through the angles β and δ, both include double the area of the spherical triangle — a total *excess* equal to $4 \times$ (the area A of the spherical triangle).

In summary, the *sum* of the total area covered by each of these three great-circle double-lune rotations is

$$4\pi + 4A = 4(\alpha + \beta + \delta)$$
$$\text{therefore,} \quad A = (\alpha + \beta + \delta - \pi)$$
$$= \text{area of a triangle on a unit sphere, or}$$
$$A = R^2(\alpha + \beta + \delta - \pi)$$
$$= \text{the triangle area on a sphere of radius R.}$$

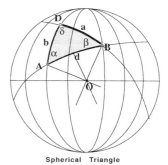

Spherical Triangle

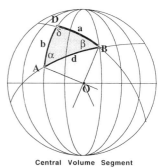

Central Volume Segment

Area and Central Volume
Regions of the Sphere

Using very similar constructions and arguments it is not difficult to show that the volume bounded by a spherical triangle and the three radii from the origin to the three vertices is

$$V = \frac{1}{3}(\alpha + \beta + \delta - \pi)R^3$$

Exercises for Chapter 10

1. Verify Equation 10.11.

2. An alternative rotation sequence, illustrated below, uses *supplemental angles* (see note in the margin) instead of interior

Supplemental Angle Notation

The supplemental angles at the three vertices of the triangle ABD, we define respectively, to be

$$\rho = \pi - \alpha$$
$$\sigma = \pi - \beta$$
$$\tau = \pi - \delta$$

Then

$$\sin \rho = \sin \alpha$$
$$\cos \rho = -\cos \alpha$$
$$\text{etc.}$$

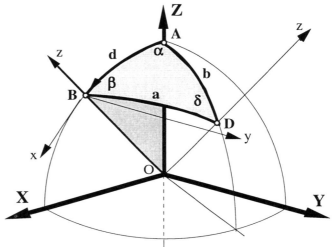

ρ, σ, τ are supplements of α, β, δ, respectively

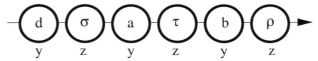

New Rotation Sequence for Spherical Triangle

angles (see Figures 10.1 and 10.2). Using this sequence, derive new expressions for and then verify Equations 10.1, 10.2, 10.3, and 10.4.

3. Find the parameter relationship between an interior angle and the radian length of a side for an equilateral spherical triangle using matrix element T(2,1) instead of T(3,1).

4. Express the interior angle of a spherical regular 4-gon as a function of the radian length of a side, using the T^2 matrix elements $m(2,3)$ and $m(3,2)$.

5. Express the interior angle of a spherical regular 4-gon as a function of the radian length of a side, using quaternion methods.

6. Investigate the spherical regular hexagon (the pentagon is more tedious) using quaternions. Compare with the matrix method.

7. Find the area of a regular 3-gon on the sphere; $r = 1$.

8. Find the area of a regular 4-gon on the sphere; $r = 1$.

9. **A challenging exercise:** Find the relationship between the parameters, a and σ, for a regular 5-gon on the sphere:
(a) using matrices   (b) using quaternions.

10. **A more challenging exercise:** Find a generalized equation which expresses the relationship between the parameters, a and σ, as a function of n for any regular n-gon on the sphere.

Chapter 11

Quaternion Calculus for Kinematics and Dynamics

11.1   Introduction

In this chapter we derive some algorithms which are useful in the study and development of mathematical models encountered in many dynamic and kinematic applications. Two of the most important features of a properly designed model for an application is its capacity to represent the system with integrity, and to provide a basis for the detection and study of significant errors. Toward this end, in this chapter we consider and develop the following Calculus-based ideas, along with relevant applications:

1. Derivative of a Direction Cosine Matrix

2. Derivative of a Quaternion Rotation Operator

3. Perturbations in an Euler Angle Rotation Sequence

4. Perturbations in a Quaternion Rotation Sequence

The application of rotation sequences in dynamic and/or kinematic models often demands answers to questions such as:

> *How do angular errors in one element of a rotation sequence affect the other elements of the sequence?*

> *How do the angular rates about the axes of a specified (body) frame in a rotation sequence influence the angular rates of other elements in the sequence?*

Time-derivatives of the rotation matrix and also of the quaternion rotation operator are derived. This will help provide answers to these questions. Some useful techniques are then developed, using examples, in the study of dynamic/kinematic models.

11.2   Derivative of the Direction Cosine Matrix

$$
\begin{aligned}
\text{With} \quad \mathbf{v} &= \text{some meaningful vector} \\
\mathbf{v}_r &= \mathbf{v} \text{ defined in the reference (fixed) frame} \\
\mathbf{v}_b &= \mathbf{v} \text{ defined in the body (moving) frame} \\
T &= \text{rotation matrix which takes } \mathbf{v}_r \text{ into } \mathbf{v}_b \\
\mathbf{w}_b &= \text{angular rate of the body frame in fixed frame}
\end{aligned}
$$

In what follows we shall keep in mind that each element in the rotation matrix, T, may have been defined by an Euler angle-axis sequence which is meaningful to a particular application. We begin with

$$
\mathbf{v}_b \;=\; T\mathbf{v}_r \tag{11.1}
$$

Since $\overset{\circ}{\mathbf{v}}_r = 0$, differentiating both sides of this equation gives

$$
\overset{\circ}{\mathbf{v}}_b \;=\; \overset{\circ}{T}\,\mathbf{v}_r + T\,\overset{\circ}{\mathbf{v}}_r = \overset{\circ}{T}\,\mathbf{v}_r
$$

We know $\qquad \mathbf{v}_r \;=\; T^{-1}\mathbf{v}_b \tag{11.2}$

Remember!

The vectors $\mathbf{v}_r$ and $\mathbf{v}_b$ are the same vector viewed from different frames, one fixed and one rotating.

therefore $\qquad \overset{\circ}{\mathbf{v}}_b \;=\; \overset{\circ}{T}\,T^{-1}\mathbf{v}_b \tag{11.3}$

It is well-known, however, that the rate-of-change of a vector, that is, its derivative, may be represented by the equation

$$
\overset{\circ}{\mathbf{v}}_r \;=\; \overset{\circ}{\mathbf{v}}_b + \mathbf{w}_b \times \mathbf{v}_b \tag{11.4}
$$

Since $\overset{\circ}{\mathbf{v}}_r = 0$ it follows that $\quad \overset{\circ}{\mathbf{v}}_b = -\mathbf{w}_b \times \mathbf{v}_b \quad$ or,

$$
\begin{bmatrix} \overset{\circ}{v}_x \\ \overset{\circ}{v}_y \\ \overset{\circ}{v}_z \end{bmatrix} =
\begin{bmatrix} 0 & \omega_z & -\omega_y \\ -\omega_z & 0 & \omega_x \\ \omega_y & -\omega_x & 0 \end{bmatrix}
\begin{bmatrix} v_x \\ v_y \\ v_z \end{bmatrix} \tag{11.5}
$$

that is $\qquad \overset{\circ}{\mathbf{v}}_b \;=\; \Omega\, \mathbf{v}_b$

From this, and using Equation 11.3 for $\overset{\circ}{\mathbf{v}}_b$, we write

$$\overset{\circ}{\mathbf{v}}_b = \overset{\circ}{T}\, T^{-1}\mathbf{v}_b = \Omega\, \mathbf{v}_b$$

to give $\qquad \overset{\circ}{T}\, T^{-1} = \Omega$

or $\qquad \overset{\circ}{T} = \Omega\, T \qquad\qquad$ (11.6)

The rotation matrix, T, in Equation 11.6, may be defined by any one of the twelve (12) possible Euler angle/axis sequences.

Angular Rates

See Section 4.4 for other possible Euler angle-axes sequences which define T. The non-zero elements of the matrix Ω are angular rates about the indicated bady axis.

11.3 Body-Axes $\xrightarrow{?}$ Euler Angle Rates

We have just shown that $\overset{\circ}{T} = \Omega\, T$. Using this result and the Aerospace Euler angle representation for the matrix, T, namely, $T = T_\phi T_\theta T_\psi$, we now relate the body-axis angular rates to the associated Euler angle angular rates for this

sequence. We write

Aerospace Sequence

See Section 4.4 where we considered the Aerospace sequence in considerable detail.

$$\Omega = \overset{\circ}{T}\, T^{-1} = \overset{\circ}{T}\, T^T$$
$$= \frac{dT_\phi}{d\phi}T_\phi^T\frac{d\phi}{dt} + T_\phi\frac{dT_\theta}{d\theta}T_\theta^T T_\phi^T\frac{d\theta}{dt}$$
$$+ T_\phi T_\theta\frac{dT_\psi}{d\psi}T_\psi^T T_\theta^T T_\phi^T\frac{d\psi}{dt} \qquad (11.7)$$

$\Longleftarrow$ Some algebraic steps have been omitted in obtaining Equation 11.7.

where

$$\left[\frac{dT}{d\phi}\right]^T = \begin{bmatrix} 0 & \begin{pmatrix} \cos\psi\sin\theta\cos\phi \\ +\sin\psi\sin\phi \end{pmatrix} & \begin{pmatrix} -\cos\psi\sin\theta\sin\phi \\ +\sin\psi\cos\phi \end{pmatrix} \\ 0 & \begin{pmatrix} \sin\psi\sin\theta\cos\phi \\ -\cos\psi\sin\phi \end{pmatrix} & \begin{pmatrix} -\sin\psi\sin\theta\sin\phi \\ -\cos\psi\cos\phi \end{pmatrix} \\ 0 & \cos\theta\cos\phi & -\cos\psi\cos\phi \end{bmatrix}$$

$$\frac{dT}{d\theta} = \begin{bmatrix} -\cos\psi\sin\theta & -\sin\psi\sin\theta & -\cos\theta \\ \cos\psi\cos\theta\sin\phi & \sin\psi\cos\theta\sin\phi & -\sin\theta\sin\phi \\ \cos\psi\cos\theta\cos\phi & \sin\psi\cos\theta\cos\phi & -\sin\theta\cos\phi \end{bmatrix}$$

$$\frac{dT}{d\psi} = \begin{bmatrix} -\sin\psi\cos\theta & \cos\psi\cos\theta & 0 \\ \begin{pmatrix} -\sin\psi\sin\theta\sin\phi \\ -\cos\psi\cos\phi \end{pmatrix} & \begin{pmatrix} \cos\psi\sin\theta\sin\phi \\ -\sin\psi\cos\phi \end{pmatrix} & 0 \\ \begin{pmatrix} -\sin\psi\sin\theta\cos\phi \\ +\cos\psi\sin\phi \end{pmatrix} & \begin{pmatrix} \cos\psi\sin\theta\cos\phi \\ \sin\psi\sin\phi \end{pmatrix} & 0 \end{bmatrix}$$

To get from these to the desired expression for the time-derivative for each of the Euler angles is algebraically tedious, but eventually the result can be put in the following form.

$$\begin{bmatrix} \overset{\circ}{\phi} \\ \overset{\circ}{\theta} \\ \overset{\circ}{\psi} \end{bmatrix} = \begin{bmatrix} 1 & \sin\phi\tan\theta & \cos\phi\tan\theta \\ 0 & \cos\phi & -\sin\phi \\ 0 & \sin\phi\sec\theta & \cos\phi\sec\theta \end{bmatrix} \begin{bmatrix} \omega_x \\ \omega_y \\ \omega_z \end{bmatrix} \qquad (11.8)$$

In the next section we derive this result in another way, which will demonstrate another important analytical technique — that of using perturbations. Equation 11.8, which relates the Euler-angle angular rates to the Body-axis angular rates, does suggest an interesting and related question, namely,

> *How do finite body axes perturbations propagate into finite changes in the indicated heading and attitude orientation Euler angles?*

In order to answer this question, which relates small body-axis angle increments to the resulting small Euler angle increments, we use an angular perturbation approach which conceptually may be more intuitive.

1st Question

How do angular errors in one element of a closed rotation sequence affect the other elements?

11.4   Perturbations in a Rotation Sequence

We here consider the effect of finite perturbations which relate to answering the first question posed at the beginning of this chapter. For this purpose we again choose (of 12 possible) the familiar Aerospace rotation sequence which defines, relative to the Earth's

Heading & Attitude

We define the rotation angles:

ψ = Aircraft **Heading** Angle always about the z-axis

ϕ, θ = the **Attitude** Angles

where:

ϕ = Aircraft Bank Angle always about the x-axis

θ = Aircraft Elevation Angle always about the y-axis

These angles should not to be confused with or referred to as Yaw, Pitch, and Roll. In Aerospace Engineering parlance Yaw, Pitch, and Roll are terms reserved for perturbations or incremental rotations about the z, y, x, aircraft body axes, respectively.

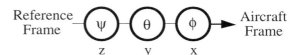

Figure 11.1: Heading & Attitude Rotation Sequence

local tangent-plane and North, the *Heading* and *Attitude* of an aircraft. Its *rotation sequence* is defined in Figure 11.1 and its corresponding *body frame orientation* is illustrated in Figure 11.2.

We first compute the rotation matrix T in terms of its constituent Euler angle rotations, heading, ψ, and attitude angles, θ and ϕ. Therefore

$$T = \begin{bmatrix} a_{11} & a_{12} & a_{13} \\ a_{21} & a_{22} & a_{23} \\ a_{31} & a_{32} & a_{33} \end{bmatrix} = T_\phi^x T_\theta^y T_\psi^z$$

$$= T_\phi \begin{bmatrix} \cos\theta & 0 & -\sin\theta \\ 0 & 1 & 0 \\ \sin\theta & 0 & \cos\theta \end{bmatrix} \begin{bmatrix} \cos\psi & \sin\psi & 0 \\ -\sin\psi & \cos\psi & 0 \\ 0 & 0 & 1 \end{bmatrix}$$

$$= \begin{bmatrix} 1 & 0 & 0 \\ 0 & \cos\phi & \sin\phi \\ 0 & -\sin\phi & \cos\phi \end{bmatrix} \begin{bmatrix} \cos\theta\cos\psi & \cos\theta\sin\psi & -\sin\theta \\ -\sin\psi & \cos\psi & 0 \\ \sin\theta\cos\psi & \sin\theta\sin\psi & \cos\theta \end{bmatrix}$$

$$= \begin{bmatrix} \cos\psi\cos\theta & \sin\psi\cos\theta & -\sin\theta \\ \begin{pmatrix} \cos\psi\sin\theta\sin\phi \\ -\sin\psi\cos\phi \end{pmatrix} & \begin{pmatrix} \sin\psi\sin\theta\sin\phi \\ +\cos\psi\cos\phi \end{pmatrix} & \cos\theta\sin\phi \\ \begin{pmatrix} \cos\psi\sin\theta\cos\phi \\ +\sin\psi\sin\phi \end{pmatrix} & \begin{pmatrix} \sin\psi\sin\theta\cos\phi \\ -\cos\psi\sin\phi \end{pmatrix} & \cos\theta\cos\phi \end{bmatrix}$$

We have already encountered this matrix in Chapter 4. So at this point we should be familiar with the elementary notion

$$T^{-1}\,T \;=\; T^T\,T \;=\; \text{Identity}$$

which may be illustrated in a closed rotation sequence where the first three rotations are followed by their inverses — *negative* rota-

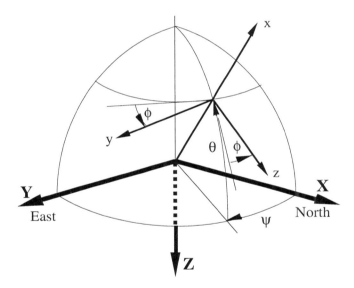

Figure 11.2: Aerospace Rotation Geometry

tions of the same angles in reverse order. But now we complicate the matter. We introduce some perturbations — incremental rotations about each of the *body-axes* as represented in Figure 11.3. We will determine how these body-axis perturbations result in Heading and Attitude perturbations, $\Delta\psi$, $\Delta\theta$, and $\Delta\phi$.

In our analysis, the matrix T defines the body frame orientation relative to the reference frame. We let R represent the body-frame perturbation — the composite incremental rotations, involving all

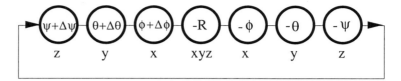

Figure 11.3: Perturbed Heading & Attitude Sequence

Incremental Rotation Notation

Since for a very small angle $\Delta\theta$, $\sin\Delta\theta = \Delta\theta$ and $\cos\Delta\theta = 1$, we may write

$$S_\phi = T_{\phi+\Delta\phi} = T_\phi T_{\Delta\phi}$$
$$S_\theta = T_{\theta+\Delta\theta} = T_\theta T_{\Delta\theta}$$
$$S_\psi = T_{\psi+\Delta\psi} = T_\psi T_{\Delta\psi}$$

$$T_{\Delta\phi} = \begin{bmatrix} 1 & 0 & 0 \\ 0 & 1 & \Delta\phi \\ 0 & -\Delta\phi & 1 \end{bmatrix}$$
$$= I + \Phi$$
$$= \text{incremental Bank}$$

$$\text{with } \Phi = \begin{bmatrix} 0 & 0 & 0 \\ 0 & 0 & \Delta\phi \\ 0 & -\Delta\phi & 0 \end{bmatrix}$$

$$T_{\Delta\theta} = \begin{bmatrix} 1 & 0 & -\Delta\theta \\ 0 & 1 & 0 \\ \Delta\theta & 0 & 1 \end{bmatrix}$$
$$= I + \Theta$$
$$= \text{incremental Elevation}$$

$$\text{with } \Theta = \begin{bmatrix} 0 & 0 & -\Delta\theta \\ 0 & 0 & 0 \\ \Delta\theta & 0 & 0 \end{bmatrix}$$

$$T_{\Delta\psi} = \begin{bmatrix} 1 & \Delta\psi & 0 \\ -\Delta\psi & 1 & 0 \\ 0 & 0 & 1 \end{bmatrix}$$
$$= I + \Psi$$
$$= \text{incremental Heading}$$

$$\text{with } \Psi = \begin{bmatrix} 0 & \Delta\psi & 0 \\ -\Delta\psi & 0 & 0 \\ 0 & 0 & 0 \end{bmatrix}$$

$$\Phi\Theta = \Theta\Psi = \Psi\Phi = 0$$

Here, we let

$$\Omega = \text{Body increments}$$
$$= \begin{bmatrix} 0 & \Delta w & -\Delta v \\ -\Delta w & 0 & \Delta u \\ \Delta v & -\Delta u & 0 \end{bmatrix}$$

$$\Delta u = \text{roll increment } \Delta_x$$
$$\Delta v = \text{pitch increment } \Delta_y$$
$$\Delta w = \text{yaw increment } \Delta_z$$

3 axes. We will let $S = RT =$ the new composite Heading and Attitude rotation, which includes the incremental Euler angle changes due to the body being perturbed by R. Then, we have

$$S^T R T = \text{Identity} = I = T^T R^T S \quad \text{as in Figure 11.3}$$

$$\text{or} \quad R T_\phi T_\theta T_\psi = S_\psi S_\theta S_\phi$$
$$= T_{\phi+\Delta\phi} T_{\theta+\Delta\theta} T_{\psi+\Delta\psi}$$
$$R T_\phi T_\theta = T_\phi T_{\Delta\phi} T_\theta T_{\Delta\theta} T_{\Delta\psi}$$
$$= T_{\Delta\phi} T_\phi T_\theta T_{\Delta\theta} T_{\Delta\psi}$$

To simplify, we let $A = T_\phi T_\theta = \text{Body Attitude}$. Then

$$R = I + \Omega = [I+\Phi]A[I+\Theta][I+\Psi]A^T$$
$$= [I+\Phi]A[I+\Theta+\Psi]A^T$$
$$I + \Omega = [I+\Phi][I + A\Theta A^T + A\Psi A^T]$$
$$\Omega = \Phi + A[\Theta+\Psi]A^T \quad \text{or}$$

$$\begin{bmatrix} 0 & \Delta w & -\Delta v \\ -\Delta w & 0 & \Delta u \\ \Delta v & -\Delta u & 0 \end{bmatrix} = \Phi + A[\Theta+\Psi]A^T$$

From this we get
$$\Delta u = \Delta\phi - \Delta\psi \, \sin\theta$$
$$\Delta v = \Delta\theta \cos\phi + \Delta\psi \cos\theta \sin\phi$$
$$\Delta w = \Delta\psi \cos\theta \cos\phi - \Delta\theta \sin\phi$$

These last three equations relate the angular perturbations about the body-axes to the resulting Euler angle increments. If we now divide both sides of each of these last three equations by Δt, and take the limit of both sides of each equation as $\Delta t \to 0$, we have

$$\lim_{\Delta t \to 0}\frac{\Delta u}{\Delta t} = \lim_{\Delta t \to 0}\frac{\Delta\phi}{\Delta t} - \lim_{\Delta t \to 0}\frac{\Delta\psi}{\Delta t} \, \sin\theta$$

$$\lim_{\Delta t \to 0}\frac{\Delta v}{\Delta t} = \lim_{\Delta t \to 0}\frac{\Delta\theta}{\Delta t} \cos\phi + \lim_{\Delta t \to 0}\frac{\Delta\psi}{\Delta t} \cos\theta \sin\phi$$

$$\lim_{\Delta t \to 0}\frac{\Delta w}{\Delta t} = \lim_{\Delta t \to 0}\frac{\Delta\psi}{\Delta t} \cos\theta \cos\phi - \lim_{\Delta t \to 0}\frac{\Delta\theta}{\Delta t} \sin\phi$$

The result gives us three equations which relate the body-axes angular rates to the resulting Euler angle angular rates,

$$\omega_x = \overset{\circ}{\phi} - \overset{\circ}{\psi} \, \sin\theta$$

$$\omega_y = \overset{\circ}{\theta} \cos\phi + \overset{\circ}{\psi} \cos\theta \sin\phi$$

$$\omega_z = \overset{\circ}{\psi} \cos\theta \cos\phi - \overset{\circ}{\theta} \sin\phi$$

or in matrix form

$$\begin{bmatrix} \omega_x \\ \omega_y \\ \omega_z \end{bmatrix} = \begin{bmatrix} 1 & 0 & -\sin\theta \\ 0 & \cos\phi & \cos\theta\sin\phi \\ 0 & -\sin\phi & \cos\theta\cos\phi \end{bmatrix} \begin{bmatrix} \overset{\circ}{\phi} \\ \overset{\circ}{\theta} \\ \overset{\circ}{\psi} \end{bmatrix} \qquad (11.9)$$

If we invert Equation 11.9, we get

$$\begin{bmatrix} \overset{\circ}{\phi} \\ \overset{\circ}{\theta} \\ \overset{\circ}{\psi} \end{bmatrix} = \begin{bmatrix} 1 & \tan\theta\sin\phi & \tan\theta\cos\phi \\ 0 & \cos\phi & -\sin\phi \\ 0 & \sec\theta\sin\phi & \sec\theta\cos\phi \end{bmatrix} \begin{bmatrix} \omega_x \\ \omega_y \\ \omega_z \end{bmatrix} \qquad (11.10)$$

We emphasize again that although the rotations denoted S and T in the foregoing have identical angle-axis sequences, they differ, in general, because of the *body-axis rotational perturbations*, R. The following definitions may be helpful in this context:

$$T = T_{old} = T_\phi T_\theta T_\psi$$

$$S = T_{new} = T_{\phi+\Delta\phi} T_{\theta+\Delta\theta} T_{\psi+\Delta\psi} = R\, T_{old}$$

Here the composite body-axis *angular* perturbations may be represented by

$$R = \begin{bmatrix} 1 & \Delta_z & -\Delta_y \\ -\Delta_z & 1 & \Delta_x \\ \Delta_y & -\Delta_x & 1 \end{bmatrix}$$

where the subscripts indicate the axis of the angular perturbation.

We now have a general solution to the problem posed by our 1st question. We obtain better insight, however, by introducing only, say, a pitch angle, Δ_y, about the body y-axis, that is, $\Delta_x = \Delta_z = 0$. This single body-axis perturbation produces the following changes in the three Euler angles:

$$\Delta\phi = \Delta_y \tan\theta \sin\phi$$

$$\Delta\theta = \Delta_y \cos\phi \qquad (11.11)$$

$$\Delta\psi = \Delta_y \sec\theta \sin\phi$$

We now consider these matters again using the quaternion.

Heading & Attitude Geometry

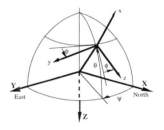

11.5 Derivative of the Quaternion

Recall that any two unitary quaternions may be related by some transition quaternion. Therefore, we may relate $q(t)$ and $q(t + \Delta t)$

as follows

$$q(t + \Delta t) = q(t)\Delta r(t)$$
$$\text{where} \quad \Delta r(t) = \cos(\Delta\alpha) + \mathbf{v}(t)\sin(\Delta\alpha)$$

Transition Quaternion

In this discussion we assume the elements of the quaternion are functions of time.

For a discussion of transition quaternions, see Section 8.3. on pages 179f.

What about

$$\frac{d}{dt}\mathbf{v}(t) \quad ?$$

is the incremental transition quaternion. Its rotation angle is $2\Delta\alpha$, about an axis defined by the unit vector $\mathbf{v}(t)$. For our purposes, we assume that Δt, and therefore

$$\Delta r(t) = \cos(\Delta\alpha) + \mathbf{v}(t)\sin(\Delta\alpha)$$
$$= 1 + \mathbf{v}(t)(\Delta\alpha)$$
$$\text{Then} \quad q(t + \Delta t) = q(t)[1 + \mathbf{v}(t)(\Delta\alpha)]$$
$$\text{and} \quad q(t + \Delta t) - q(t) = q(t)\mathbf{v}(t)(\Delta\alpha)$$

Dividing both sides by Δt and invoking the limit gives

$$\frac{dq}{dt} = \lim_{\Delta t \to 0} \frac{q(t + \Delta t) - q(t)}{\Delta t}$$
$$= \lim_{\Delta t \to 0} \frac{q(t)\mathbf{v}(t)(\Delta\alpha)}{\Delta t}$$
$$= q(t)\mathbf{v}(t)\omega(t) = q(t)\overline{\omega}(t)$$

Derivative

Note that we here define the derivative in the usual way, that is, as the *limit* of a difference quotient. Moreover, it can be shown that the ordinary derivative rule for products also holds for quaternion products. See Exercises.

where

$$\omega(t) = \overset{\circ}{\alpha} = \text{scalar angular rate about direction vector } \mathbf{v}(t)$$
$$\overline{\omega}(t) = \overset{\circ}{\alpha}\mathbf{v}(t) = \text{angular rate vector of quaternion } \Delta r$$

11.6   Derivative of the Conjugate

To find the derivative of the conjugate, q^*, we first write

$$\frac{d}{dt}[q^*q] = 0$$

Using the product rule and the foregoing results, it follows that

$$(\frac{d}{dt}q^*)q + q^*\frac{dq}{dt} = 0 \Rightarrow \frac{d}{dt}(q^*) = -q^*\frac{dq}{dt}q^* = -q^*q\overline{\omega}q^*$$

$$\text{Thus} \quad \frac{d}{dt}[q(t)] = q(t)\overline{\omega}(t)$$
$$\frac{d}{dt}[q^*(t)] = -\overline{\omega}(t)q^*(t)$$

Note that the derivative of a quaternion is itself a quaternion. In matrix form the derivative of a quaternion may be written:

$$\frac{dq}{dt} = q(t)\overline{\omega}(t) = \begin{bmatrix} 0 & -\omega_1 & -\omega_2 & -\omega_3 \\ \omega_1 & 0 & \omega_3 & -\omega_2 \\ \omega_2 & -\omega_3 & 0 & \omega_1 \\ \omega_3 & \omega_2 & -\omega_1 & 0 \end{bmatrix} \begin{bmatrix} q_0 \\ q_1 \\ q_2 \\ q_3 \end{bmatrix}$$

$$
= \begin{bmatrix} q_0 & -q_1 & -q_2 & -q_3 \\ q_1 & q_0 & -q_3 & q_2 \\ q_2 & q_3 & q_0 & -q_1 \\ q_3 & -q_2 & q_1 & q_0 \end{bmatrix} \begin{bmatrix} 0 \\ \omega_1 \\ \omega_2 \\ \omega_3 \end{bmatrix} \quad (11.12)
$$

Remember!
These are
Quaternion Products

The two derivative expressions $q(t)\overline{\omega}(t)$ and $-\overline{\omega}(t)q(t)$ involve quaternion products and therefore are quite distinctly different. You recall, in general, quaternion products do not commute.

Similarly, for the derivative of the conjugate, we have

$$
\frac{dq^*}{dt} = -\overline{\omega}(t)q^*(t) = \begin{bmatrix} 0 & \omega_1 & \omega_2 & \omega_3 \\ -\omega_1 & 0 & \omega_3 & -\omega_2 \\ -\omega_2 & -\omega_3 & 0 & \omega_1 \\ -\omega_3 & \omega_2 & -\omega_1 & 0 \end{bmatrix} \begin{bmatrix} q_0 \\ -q_1 \\ -q_2 \\ -q_3 \end{bmatrix}
$$

$$
= \begin{bmatrix} q_0 & -q_1 & -q_2 & -q_3 \\ -q_1 & -q_0 & q_3 & -q_2 \\ -q_2 & -q_3 & -q_0 & q_1 \\ -q_3 & q_2 & -q_1 & -q_0 \end{bmatrix} \begin{bmatrix} 0 \\ \omega_1 \\ \omega_2 \\ \omega_3 \end{bmatrix} (11.13)
$$

11.7 Derivative of Quaternion Operators

Now that we have expressions for the derivative of the quaternion and of its conjugate, we may write the time derivative for a *quaternion rotation operator*. We begin with a simple coordinate transformation, $T = q^* I q$, which we have come to know quite well.

Recall from the earlier chapters, when we had quaternions operating on a matrix, it was understood that the matrix merely represented a set of column vectors. For the operator, $T = q^* I q$, the matrix is the identity matrix, denoted I, whose columns represent the standard basis vectors **i**, **j**, and **k** of the reference frame in R^3. In this context, these three column vectors are *pure* quaternions, that is, they are quaternions whose real part is zero.

We will perform the indicated quaternion operations, in turn, on each of these pure quaternions, **i**, **j**, and **k**. We get pure quaternions, $\mathbf{e}_1$, $\mathbf{e}_2$, and $\mathbf{e}_3$, respectively. These represent the basis vectors of the new coordinate frame. We write

$$
T = \begin{bmatrix} | & | & | \\ \mathbf{e}_1 & \mathbf{e}_2 & \mathbf{e}_3 \\ | & | & | \end{bmatrix} = \begin{bmatrix} | & | & | \\ q^*\mathbf{i}q & q^*\mathbf{j}q & q^*\mathbf{k}q \\ | & | & | \end{bmatrix} \quad (11.14)
$$

From earlier results obtained in Section 5.14, we may write

$$
T = \begin{bmatrix} 2q_0^2 - 1 & 2q_1q_2 & 2q_1q_3 \\ +2q_1^2 & +2q_0q_3 & -2q_0q_2 \\ \\ 2q_1q_2 & 2q_0^2 - 1 & 2q_2q_3 \\ -2q_0q_3 & +2q_2^2 & +2q_0q_1 \\ \\ 2q_1q_3 & 2q_2q_3 & 2q_0^2 - 1 \\ +2q_0q_2 & -2q_0q_1 & +2q_3^2 \end{bmatrix}
$$

If we define $\overset{\circ}{T}$ to be the ordinary limit of a difference quotient,

it follows $\qquad \overset{\circ}{T} = \begin{bmatrix} | & | & | \\ \overset{\circ}{\mathbf{e}}_1 & \overset{\circ}{\mathbf{e}}_2 & \overset{\circ}{\mathbf{e}}_3 \\ | & | & | \end{bmatrix}$

So, to find $\overset{\circ}{T}$, we find the derivative for each $\mathbf{e}_k$ separately.

$$
\begin{aligned}
\text{First,} \quad \frac{d}{dt}\mathbf{e}_1 &= \frac{d}{dt}[q^*\mathbf{i}q] \\
&= \frac{d}{dt}[q^*]\mathbf{i}q + q^*\mathbf{i}\frac{d}{dt}[q] \\
&= -\overline{\omega}[q^*\mathbf{i}q] + [q^*\mathbf{i}q]\overline{\omega}
\end{aligned}
$$

$$
\begin{aligned}
\text{So,} \quad \overset{\circ}{\mathbf{e}}_1 &= -\overline{\omega}\mathbf{e}_1 + \mathbf{e}_1\overline{\omega} & (11.15) \\
&= 2\mathbf{e}_1 \times \overline{\omega} & (11.16)
\end{aligned}
$$

In exactly the same way we may write

$$
\begin{aligned}
\overset{\circ}{\mathbf{e}}_2 &= -\overline{\omega}\mathbf{e}_2 + \mathbf{e}_2\overline{\omega} & (11.17) \\
&= 2\mathbf{e}_2 \times \overline{\omega} & (11.18) \\
\text{and} \quad \overset{\circ}{\mathbf{e}}_3 &= -\overline{\omega}\mathbf{e}_3 + \mathbf{e}_3\overline{\omega} & (11.19) \\
&= 2\mathbf{e}_3 \times \overline{\omega} & (11.20)
\end{aligned}
$$

It follows that $\qquad \overset{\circ}{T} = \dfrac{d}{dt}\left(q^* \begin{bmatrix} | & | & | \\ \mathbf{i} & \mathbf{j} & \mathbf{k} \\ | & | & | \end{bmatrix} q \right)$

$$
\begin{aligned}
&= -\overline{\omega}(t)q^*(t) \begin{bmatrix} | & | & | \\ \mathbf{i} & \mathbf{j} & \mathbf{k} \\ | & | & | \end{bmatrix} q(t) \\
&\quad + q^*(t) \begin{bmatrix} | & | & | \\ \mathbf{i} & \mathbf{j} & \mathbf{k} \\ | & | & | \end{bmatrix} q(t)\overline{\omega}(t) \\
&= -\overline{\omega}(t)T + T\overline{\omega}(t) & (11.21)
\end{aligned}
$$

If we rewrite these equations in terms of quaternion elements,

$$
\begin{aligned}
\text{we get} \quad \overset{\circ}{\mathbf{e}}_1 &= 2\mathbf{i}(e_{21}\omega_3 - e_{31}\omega_2) \\
&\quad + 2\mathbf{j}(e_{31}\omega_1 - e_{11}\omega_3)
\end{aligned}
$$

$$+ 2\mathbf{k}(e_{11}\omega_2 - e_{21}\omega_1)$$

Similarly $\qquad \overset{\circ}{\mathbf{e}}_2 = 2\mathbf{e}_2 \times \overline{\omega}$ $\qquad\qquad$ (11.22)

$$= 2\mathbf{i}(e_{22}\omega_3 - e_{32}\omega_2)$$
$$+ 2\mathbf{j}(e_{32}\omega_1 - e_{12}\omega_3)$$
$$+ 2\mathbf{k}(e_{12}\omega_2 - e_{22}\omega_1)$$

and $\qquad \overset{\circ}{\mathbf{e}}_3 = 2\mathbf{e}_3 \times \overline{\omega}$ $\qquad\qquad$ (11.23)

$$= 2\mathbf{i}(e_{23}\omega_3 - e_{33}\omega_2)$$
$$+ 2\mathbf{j}(e_{33}\omega_1 - e_{13}\omega_3)$$
$$+ 2\mathbf{k}(e_{13}\omega_2 - e_{23}\omega_1)$$

Angular Rate Components

Here ω_1, ω_2, and ω_3 are the components of the angular rate vector, $\overline{\omega}(t)$.

We can now write our result more concisely in matrix form as

$$\overset{\circ}{T} = \begin{bmatrix} | & | & | \\ \overset{\circ}{\mathbf{e}}_1 & \overset{\circ}{\mathbf{e}}_2 & \overset{\circ}{\mathbf{e}}_3 \\ | & | & | \end{bmatrix} = -\overline{\omega}(t)T + T\overline{\omega}(t) \qquad (11.24)$$

$$= \begin{bmatrix} 0 & \omega_3 & -\omega_2 \\ -\omega_3 & 0 & \omega_1 \\ \omega_2 & -\omega_1 & 0 \end{bmatrix} \begin{bmatrix} e_{11} & e_{12} & e_{13} \\ e_{21} & e_{22} & e_{23} \\ e_{31} & e_{32} & e_{33} \end{bmatrix}$$

Quaternion Rotation Operator

For application of the quaternion rotation operator to sets of vectors, see Section 7.10.2.

It is instructive if we compare this result with the result we obtained earlier in Equation 11.6, namely,

$$\overset{\circ}{T} = \Omega\, T$$

where the matrix T is expressed in terms of an Euler angle sequence. Comparison of these two equations should be confirming and afford some insight into the relationship between these two alternative analytical approaches — the quaternion rotation operator approach and the direction cosine matrix approach. That is, we compare

$$\overset{\circ}{T} = -\overline{\omega}(t)T + T\overline{\omega}(t) \qquad \text{and} \qquad \overset{\circ}{T} = \Omega\, T$$

One interesting application of Equation 11.21 is that it suggests how one might specify the control law for stabilizing the attitude of an orbiting spacecraft. Gyroscopes of one sort or another, in general, provide a measure of the spacecraft orientation and/or its angular rate, both with respect to an inertial frame.

For example, an important related reference frame might be the orbit frame — a frame established by the plane of the orbit and, say, the local geocentric line. Then one possible control law would seek to mainain the spacecraft in a specified attitude with respect to the earth local tangent plane, say for observation purposes.

11.8   Perturbations in Quaternions

Many applications of dynamics and kinematics involve closed sequences of rotation operators. In general, the related mathematical models are non-linear. In this section we consider linearizing these models using perturbations on the appropriate parameters in these operators. We illustrate the technique using the oversimplified example described in the following question.

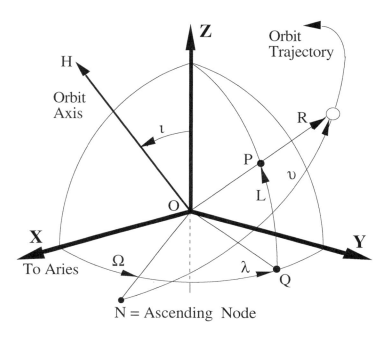

Figure 11.4: Orbit and Ephemeris Geometry

Orbit Anomolies

In our example we might expect that the tracking station would see the satellite again in the same position after one orbital period. Remember, the planet, Lars, is not rotating, and is in free space; no other planets or external forces and so the orbit is fixed — if the gravitational field is uniformly Lars-centric.

For our purposes, however, we will assume such is not the case; Lars is not only non-spherical but is also non-homogeneous. This, in general, will result in some precession of the Nels orbital plane.

An Interesting Note

Several years ago, a near-polar orbit of a research weather satellite was specifically designed to precess about the oblate spheroid called Earth to help keep the Earth-Sun line in the plane of the orbit.

A low-altitude satellite, Nels, is tracked by several tracking stations at known locations on a non-rotating planet, Lars. The data from these stations on Lars is processed to give values for the ephemeris: Latitude and Longitude. From this the orbit parameters are estimated. We ask: How do we relate small changes or errors, $\Delta\lambda$, ΔL, and $\Delta\alpha$, in the Orbit Ephemeris to the required corrections, $\Delta\Omega$, $\Delta\iota$, and $\Delta\nu$, in the Orbit parameters?

In this example we let p and q be *unit* quaternions which denote the Orbit parameter rotation sequence and the Ephemeris parameter rotation sequence, respectively. The geometric relationship between the parameters in these two quaternions is illustrated in Figure 11.4.

The definition of and the relationship between the quaternions p and q have already been discussed in Section 9.3.2. The required sequence of rotations is illustrated in Figure 11.5. Recall that the angle α is an auxilliary angle necessary to relate the Ephemeris to the Orbit parameters. Note, in particular, that closure of the loop implies $p = q$, so that we may equate their quaternion elements. That is, we have

$$qp^*Ipq^* = I \Rightarrow pq^* = p^*q = 1 \Rightarrow p = q$$
$$p = q \Rightarrow p_k = q_k \quad \text{for} \quad k = 0, 1, 2, 3$$

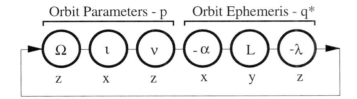

Figure 11.5: Orbit and Ephemeris Sequence

In the earlier section we obtained equations which related the orbit parameters to functions of the ephemeris parameters. These equations are clearly non-linear, and generally not suitable for our present purposes. Note also that the quaternions p and q represent the same composite rotation, but are composed of distinct rotation sequences. Then we write

$$q^*p = qp^* = 1 \quad \Rightarrow \quad q^*pIp^*q = I \quad \text{or} \quad qp^*Ipq^* = I$$

Whenever the computed orbit parameters correspond precisely to the ephemeris data, fine. However, if the tracking stations detect errors ΔL and $\Delta \lambda$ in the computed measure of the Latitude and Longitude, respectively, of the orbiting satellite, then the corresponding computed values for the orbit parameters must also be in error.

We now answer our question, stated above, by determining the corrections required in the Orbit parameters. We begin by recalling, as in Section 9.3, that

$$p = (\cos\Omega + \mathbf{k}\sin\Omega)(\cos\iota + \mathbf{i}\sin\iota)(\cos\nu + \mathbf{k}\sin\nu)$$
$$= p_0 + \mathbf{i}\,p_1 + \mathbf{j}\,p_2 + \mathbf{k}\,p_3 \qquad (11.25)$$
$$\text{where} \quad p_0 = \cos\Omega\cos\iota\cos\nu - \sin\Omega\cos\iota\sin\nu$$
$$p_1 = \cos\Omega\sin\iota\cos\nu + \sin\Omega\sin\iota\sin\nu$$

Extending the Model

The models discussed in this chapter are a bit simplistic — circular orbits and strictly kinematic. A somewhat more realistic model could take into account orbit eccentricity, so we briefly review:

Kepler's Laws

1. A planet orbit is an ellipse with the sun at a focus *Trajectories are conics.*

2. The sun-to-planet vector sweeps out equal areas in equal times. That is, $\frac{dA}{dt} = \text{constant}.$

3. $\frac{a^3}{T^2}$ is a constant, where T is an orbit's period, $2a$ is an orbit's major-axis.

From these laws an orbit trajectory may easily be derived for an arbitrary eccentricity, using only the Law of cosines. And then with $\theta = \nu - \nu_0$ we may write

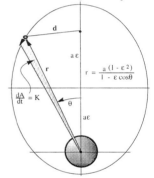

$$r = \frac{a(1-\epsilon^2)}{1-\epsilon\cos(\nu-\nu_0)}$$
$$\overset{\circ}{\nu} = K[1 + \epsilon\cos(\nu-\nu_0)]^2$$

Constants: semi-major axis a, ϵ, and ν_0, are best estimates.

These equations may be fitted with appropriate parameters so as to represent most any situation. In fact, these ideas can be extended to provide a nominal kinematic account to non-spherical bodies.

Remember!

In all of these angle functions the angles are all *half-angles*:

$$\Omega \quad \text{represents} \quad \frac{\Omega}{2}$$

$$\lambda \quad \text{represents} \quad \frac{\lambda}{2}$$

$$\text{etc}$$

Half-Angles

It is understood that the sequence of angles shown in all of the triple-products for p and q are each actually half-angles, that is, each should be shown divided by 2. For example, the quaternion representing the first factor in the composite Orbit rotation sequence for p should properly be written

$$p_\Omega = \cos\frac{\Omega}{2} + \mathbf{k}\sin\frac{\Omega}{2}$$

We omit the divisor 2 on each angle simply to make the notation less congested.

Further, in Section 4.5.2 the rotation angles used in the Orbit and Ephemeris sequences are defined.

**Perturbations
or
Errors?**

Small changes in the behaviour of dynamical systems of this sort are variously referred to as errors or perturbations, depending upon the application and whether the changes are intentional or unintentional.

Incremental Angles

In these computations once again we have used the small-angle approximation for very small angles, that is,

$$\sin\theta = \theta$$
$$\cos\theta = 1$$

$$p_2 = \sin\Omega\sin\iota\cos\nu - \cos\Omega\sin\iota\sin\nu$$
$$p_3 = \sin\Omega\cos\iota\cos\nu + \cos\Omega\cos\iota\sin\nu$$

$$\text{and} \quad q = (\cos\lambda + \mathbf{k}\sin\lambda)(\cos L - \mathbf{j}\sin L)(\cos\alpha + \mathbf{i}\sin\alpha)$$
$$= q_0 + \mathbf{i}\,q_1 + \mathbf{j}\,q_2 + \mathbf{k}\,q_3 \qquad (11.26)$$
$$\text{where} \quad q_0 = \cos\lambda\cos L\cos\alpha - \sin\lambda\sin L\sin\alpha$$
$$q_1 = \sin\lambda\sin L\cos\alpha + \cos\lambda\cos L\sin\alpha$$
$$q_2 = \sin\lambda\cos L\sin\alpha - \cos\lambda\sin L\cos\alpha$$
$$q_3 = \sin\lambda\cos L\cos\alpha + \cos\lambda\sin L\sin\alpha$$

These equations define the elements p_k and q_k for $k = 0, 1, 2, 3$ for the quaternions p and q, in terms of the angle parameters represented in their respective rotation sequences.

If we now introduce a change, say r, in either p or q, we may write $\qquad p_{new} = p_{old}\,r = q_{old}\,r = q_{new}$

Here $r = r_0 + \mathbf{r}$ may be a small or a large change. In our example, however, we are concerned with a small change, say Δr. Gravitational anomalies in the planet, Lars, perturb the orbital parameters in the elements of the quaternion p. These perturbations are observed by the tracking stations as ephemeris parameter errors in q. Since Δr is a very *small error* quaternion, we may write

$$\Delta r = r_0 + \mathbf{r} = \cos\beta + \mathbf{u}\sin\beta$$
$$\cong 1 + \Delta\mathbf{r} \cong 1 + \mathbf{u}\beta$$

Here β is small, since it represents the composite of the small errors or perturbations in the respective parameters which define p and q. The parameter changes in our example are very small, so we may relate the change in a quaternion element to the changes in its parameters, using the total differential. The partial derivatives in this total differential are sometimes called *sensitivity coefficients* in the resulting linearized error equations. These equations linearly express the small error in each of the quaternion-elements as a linear combination of the small parameter errors in its rotation sequence.

Since $\qquad \Delta p = \Delta r = 1 + \Delta\mathbf{r}$

we may write, $\quad \Delta p = 1 + \Delta\mathbf{p} = 1 + \Delta\mathbf{r} = 1 + \mathbf{u}\beta$

$\qquad\qquad \text{and} \quad \Delta\mathbf{p} = \mathbf{i}\Delta p_1 + \mathbf{j}\Delta p_2 + \mathbf{k}\Delta p_3$

$\qquad \text{So, for} \quad k = 1, 2, 3$

$$\Delta p_k = \frac{\partial p_k}{\partial\Omega}\Delta\Omega + \frac{\partial p_k}{\partial\iota}\Delta\iota + \frac{\partial p_k}{\partial\nu}\Delta\nu \qquad (11.27)$$

Similarly, $\quad \Delta q = 1 + \mathbf{i}\Delta q_1 + \mathbf{j}\Delta q_2 + \mathbf{k}\Delta q_3 = 1 + \Delta\mathbf{q}$

where, again for $\quad k = 1, 2, 3 \quad$ we write

$$\Delta\mathbf{q}_n = \frac{\partial q_k}{\partial\lambda}\Delta\lambda + \frac{\partial q_k}{\partial L}\Delta L + \frac{\partial q_k}{\partial\alpha}\Delta\alpha \qquad (11.28)$$

In order to equate the corresponding quaternion error elements: $\Delta p_k = \Delta q_k$ for $k = 1, 2, 3$, we first must determine the sensitivity of each element of these quaternions to small perturbations or errors in their constituent angle parameters. So for the quaternion

Remember!

In all of these angle functions

$\Omega \quad$ represents $\quad \dfrac{\Omega}{2}$

$\lambda \quad$ represents $\quad \dfrac{\lambda}{2}$

etc

$$p = p_0 + \mathbf{i}\, p_1 + \mathbf{j}\, p_2 + \mathbf{k}\, p_3$$

we compute $\quad \Delta p_1 = \dfrac{\partial p_1}{\partial\Omega}\Delta\Omega + \dfrac{\partial p_1}{\partial\iota}\Delta\iota + \dfrac{\partial p_1}{\partial\nu}\Delta\nu$

$$\Delta p_2 = \frac{\partial p_2}{\partial\Omega}\Delta\Omega + \frac{\partial p_2}{\partial\iota}\Delta\iota + \frac{\partial p_2}{\partial\nu}\Delta\nu$$

$$\Delta p_3 = \frac{\partial p_3}{\partial\Omega}\Delta\Omega + \frac{\partial p_3}{\partial\iota}\Delta\iota + \frac{\partial p_3}{\partial\nu}\Delta\nu$$

Similarly, for $\quad q = q_0 + \mathbf{i}\, q_1 + \mathbf{j}\, q_2 + \mathbf{k}\, q_3$

we compute $\quad \Delta q_1 = \dfrac{\partial q_1}{\partial\lambda}\Delta\lambda + \dfrac{\partial q_1}{\partial L}\Delta L + \dfrac{\partial q_1}{\partial\alpha}\Delta\alpha$

$$\Delta q_2 = \frac{\partial q_2}{\partial\lambda}\Delta\lambda + \frac{\partial q_2}{\partial L}\Delta L + \frac{\partial q_2}{\partial\alpha}\Delta\alpha$$

$$\Delta q_3 = \frac{\partial q_3}{\partial\lambda}\Delta\lambda + \frac{\partial q_3}{\partial L}\Delta L + \frac{\partial q_3}{\partial\alpha}\Delta\alpha$$

These equations may conveniently be expressed in matrix form

$$\begin{bmatrix} \Delta p_1 \\ \Delta p_2 \\ \Delta p_3 \end{bmatrix} = \begin{bmatrix} \dfrac{\partial p_1}{\partial\Omega} & \dfrac{\partial p_1}{\partial\iota} & \dfrac{\partial p_1}{\partial\nu} \\ \dfrac{\partial p_2}{\partial\Omega} & \dfrac{\partial p_2}{\partial\iota} & \dfrac{\partial p_2}{\partial\nu} \\ \dfrac{\partial p_3}{\partial\Omega} & \dfrac{\partial p_3}{\partial\iota} & \dfrac{\partial p_3}{\partial\nu} \end{bmatrix} \begin{bmatrix} \Delta\Omega \\ \Delta\iota \\ \Delta\nu \end{bmatrix}$$

and

$$\begin{bmatrix} \Delta q_1 \\ \Delta q_2 \\ \Delta q_3 \end{bmatrix} = \begin{bmatrix} \dfrac{\partial q_1}{\partial\lambda} & \dfrac{\partial q_1}{\partial L} & \dfrac{\partial q_1}{\partial\alpha} \\ \dfrac{\partial q_2}{\partial\lambda} & \dfrac{\partial q_2}{\partial L} & \dfrac{\partial q_2}{\partial\alpha} \\ \dfrac{\partial q_3}{\partial\lambda} & \dfrac{\partial q_3}{\partial L} & \dfrac{\partial q_3}{\partial\alpha} \end{bmatrix} \begin{bmatrix} \Delta\lambda \\ \Delta L \\ \Delta\alpha \end{bmatrix}$$

But, as stated earlier, $\Delta p_k = \Delta q_k$ for $k = 1, 2, 3$ so we may write

$$\begin{bmatrix} \dfrac{\partial p_1}{\partial\Omega} & \dfrac{\partial p_1}{\partial\iota} & \dfrac{\partial p_1}{\partial\nu} \\ \dfrac{\partial p_2}{\partial\Omega} & \dfrac{\partial p_2}{\partial\iota} & \dfrac{\partial p_2}{\partial\nu} \\ \dfrac{\partial p_3}{\partial\Omega} & \dfrac{\partial p_3}{\partial\iota} & \dfrac{\partial p_3}{\partial\nu} \end{bmatrix} \begin{bmatrix} \Delta\Omega \\ \Delta\iota \\ \Delta\nu \end{bmatrix} = \begin{bmatrix} \dfrac{\partial q_1}{\partial\lambda} & \dfrac{\partial q_1}{\partial L} & \dfrac{\partial q_1}{\partial\alpha} \\ \dfrac{\partial q_2}{\partial\lambda} & \dfrac{\partial q_2}{\partial L} & \dfrac{\partial q_2}{\partial\alpha} \\ \dfrac{\partial q_3}{\partial\lambda} & \dfrac{\partial q_3}{\partial L} & \dfrac{\partial q_3}{\partial\alpha} \end{bmatrix} \begin{bmatrix} \Delta\lambda \\ \Delta L \\ \Delta\alpha \end{bmatrix} \qquad (11.29)$$

For these equations, we must now compute the indicated required partial derivatives or sensitivity coefficients. These are

$$\frac{\partial p_1}{\partial \Omega} = -\sin \Omega \sin \iota \cos \nu + \cos \Omega \sin \iota \sin \nu$$

$$\frac{\partial p_1}{\partial \iota} = +\cos \Omega \cos \iota \cos \nu + \sin \Omega \cos \iota \sin \nu$$

$$\frac{\partial p_1}{\partial \nu} = -\cos \Omega \sin \iota \sin \nu + \sin \Omega \sin \iota \cos \nu$$

$$\frac{\partial p_2}{\partial \Omega} = +\cos \Omega \sin \iota \cos \nu + \sin \Omega \sin \iota \sin \nu$$

$$\frac{\partial p_2}{\partial \iota} = +\sin \Omega \cos \iota \cos \nu - \cos \Omega \cos \iota \sin \nu$$

$$\frac{\partial p_2}{\partial \nu} = -\sin \Omega \sin \iota \sin \nu - \cos \Omega \sin \iota \cos \nu$$

$$\frac{\partial p_3}{\partial \Omega} = +\cos \Omega \cos \iota \cos \nu - \sin \Omega \cos \iota \sin \nu$$

$$\frac{\partial p_3}{\partial \iota} = -\sin \Omega \sin \iota \cos \nu - \cos \Omega \sin \iota \sin \nu$$

$$\frac{\partial p_3}{\partial \nu} = -\sin \Omega \cos \iota \sin \nu + \cos \Omega \cos \iota \cos \nu$$

and

$$\frac{\partial q_1}{\partial \lambda} = +\cos \lambda \sin L \cos \alpha - \sin \lambda \cos L \sin \alpha$$

$$\frac{\partial q_1}{\partial L} = +\sin \lambda \cos L \cos \alpha - \cos \lambda \sin L \sin \alpha$$

$$\frac{\partial q_1}{\partial \alpha} = -\sin \lambda \sin L \sin \alpha + \cos \lambda \cos L \cos \alpha$$

$$\frac{\partial q_2}{\partial \lambda} = +\cos \lambda \cos L \sin \alpha + \sin \lambda \sin L \cos \alpha$$

$$\frac{\partial q_2}{\partial L} = -\sin \lambda \sin L \sin \alpha - \cos \lambda \cos L \cos \alpha$$

$$\frac{\partial q_2}{\partial \alpha} = +\sin \lambda \cos L \cos \alpha + \cos \lambda \sin L \sin \alpha$$

$$\frac{\partial q_3}{\partial \lambda} = +\cos \lambda \cos L \cos \alpha - \sin \lambda \sin L \sin \alpha$$

$$\frac{\partial q_3}{\partial L} = -\sin \lambda \sin L \cos \alpha + \cos \lambda \cos L \sin \alpha$$

$$\frac{\partial q_3}{\partial \alpha} = -\sin \lambda \cos L \sin \alpha + \cos \lambda \sin L \cos \alpha$$

Half-Angles

In what follows, it will be understood that the sequence of angles shown in Figure 11.5 when used in its quaternion requires half-angles. For example the quaternion which represents the rotation of the 1st Orbit parameter is

$$q_{\Omega}^{\vec{z}} = \cos \frac{\Omega}{2} + \mathbf{k} \sin \frac{\Omega}{2}$$

In the quaternion algebra and formulations employed in this section we omit the divisor 2 on each angle simply to make the notation less congested.

Further, in Section 4.5.2 the rotation angles used in the Orbit and Ephemeris sequences are defined.

It seems clear that to find a general solution for Equation 11.29, in closed-form would, at best, be exceedingly tedious, or at worst, impossible! However, in any practical situation we would evaluate these sensitivity coefficients at known points on the orbit. In that case, these coefficient matrices would have numerical entries, and the resulting equations would be solved in the usual way.

We illustrate this procedure with a simplified case. We suppose, in our example, that an error, $\Delta\lambda$, is detected in the ephemeris data. The ascending node was expected at λ_0, but it actually occured at $\lambda = \lambda_0 + \Delta\lambda$. It is clear from Figure 11.4 that, at the ascending node, $L = 0$, $\nu = 0$, and $\alpha = \iota$. With $\Delta L = \Delta\alpha = 0$, Equation 11.29 reduces to

$$
\begin{bmatrix}
\dfrac{\partial p_1}{\partial\Omega} & \dfrac{\partial p_1}{\partial\iota} & \dfrac{\partial p_1}{\partial\nu} \\[2mm]
\dfrac{\partial p_2}{\partial\Omega} & \dfrac{\partial p_2}{\partial\iota} & \dfrac{\partial p_2}{\partial\nu} \\[2mm]
\dfrac{\partial p_3}{\partial\Omega} & \dfrac{\partial p_3}{\partial\iota} & \dfrac{\partial p_3}{\partial\nu}
\end{bmatrix}
\begin{bmatrix}
\Delta\Omega \\[2mm] \Delta\iota \\[2mm] \Delta\nu
\end{bmatrix}
=
\begin{bmatrix}
\dfrac{\partial q_1}{\partial\lambda} \\[2mm]
\dfrac{\partial q_2}{\partial\lambda} \\[2mm]
\dfrac{\partial q_3}{\partial\lambda}
\end{bmatrix}
\begin{bmatrix} \Delta\lambda \end{bmatrix}
$$

If at the ascending node $\lambda = \lambda_0$, $\iota = \iota_0$, and $\Omega = \Omega_0$, the sensitivity coefficients may be written

$$\frac{\partial p_1}{\partial\Omega} = -\sin\Omega\sin\iota = -\sin\Omega_0\sin\iota_0$$

$$\frac{\partial p_1}{\partial\iota} = +\cos\Omega\cos\iota = +\cos\Omega_0\cos\iota_0$$

$$\frac{\partial p_1}{\partial\nu} = +\sin\Omega\sin\iota = +\sin\Omega_0\sin\iota_0$$

$$\frac{\partial p_2}{\partial\Omega} = +\cos\Omega\sin\iota = +\cos\Omega_0\sin\iota_0$$

$$\frac{\partial p_2}{\partial\iota} = +\sin\Omega\cos\iota = +\sin\Omega_0\cos\iota_0$$

$$\frac{\partial p_2}{\partial\nu} = -\cos\Omega\sin\iota = -\cos\Omega_0\sin\iota_0$$

$$\frac{\partial p_3}{\partial\Omega} = +\cos\Omega\cos\iota = +\cos\Omega_0\cos\iota_0$$

$$\frac{\partial p_3}{\partial\iota} = -\sin\Omega\sin\iota = -\sin\Omega_0\sin\iota_0$$

$$\frac{\partial p_3}{\partial\nu} = +\cos\Omega\cos\iota = +\cos\Omega_0\cos\iota_0$$

and

$$\frac{\partial q_1}{\partial\lambda} = -\sin\lambda\sin\alpha = -\sin\lambda_0\sin\iota_0$$

$$\frac{\partial q_2}{\partial\lambda} = +\cos\lambda\sin\alpha = +\cos\lambda_0\sin\iota_0$$

$$\frac{\partial q_3}{\partial\lambda} = +\cos\lambda\cos\alpha = +\cos\lambda_0\cos\iota_0$$

Thus, in this simplified case we must solve the following matrix equation

$$
\begin{bmatrix}
-\sin\Omega_0\sin\iota_0 & \cos\Omega_0\cos\iota_0 & \sin\Omega_0\sin\iota_0 \\
\cos\Omega_0\sin\iota_0 & \sin\Omega_0\cos\iota_0 & -\cos\Omega_0\sin\iota_0 \\
\cos\Omega_0\cos\iota_0 & -\sin\Omega_0\sin\iota_0 & \cos\Omega_0\cos\iota_0
\end{bmatrix}
\begin{bmatrix}
\Delta\Omega \\
\Delta\iota \\
\Delta\nu
\end{bmatrix} =
$$

$$
\begin{bmatrix}
-\sin\lambda_0\sin\iota_0 \\
+\cos\lambda_0\sin\iota_0 \\
+\cos\lambda_0\cos\iota_0
\end{bmatrix}
\begin{bmatrix} \Delta\lambda \end{bmatrix}
$$

Even solving this matrix equation for this simplified case would require considerable effort if it were not for programs such as Maple© or Mathematica©, which do the tedious manipulations. Using these aids we get

$$
\Delta\Omega = \frac{1}{2}\left[\frac{\tan\Omega_0\sin(\Omega_0-\lambda_0)}{2\cos^2\iota_0} + 2\cos(\Omega_0-\lambda_0)\right]\Delta\lambda
$$

$$
\Delta\iota = [\tan\iota_0\sin(\Omega_0-\lambda_0)]\Delta\lambda
$$

$$
\Delta\nu = \frac{\tan\Omega_0}{\cos^2\iota_0}[\sin(\Omega_0-\lambda_0)]\Delta\lambda
$$

Although this is an oversimplified example, it does demonstrate the computational mechanics involved in the perturbation approach. In whatever more realistic applications one might encounter, computations similar to those in our example must be made.

Usually, realistic applications of the perturbation approach involve some type of iterative process. If the foregoing computations represent merely one step of an on-going dynamic cyclic iteration process, then the *most recent* parameter values (determined in the last iteration) are used to define the required parameters for the next iteration.

Recall, at the outset our satellite Nels orbits the planet Lars, which in fact may be an oblate spheroid. Ordinarily, external gravitational influences in such a situation will cause the orbit to precess. It is this precession that introduced the errors we discussed in our example. If the data shows that this precession rate is constant, our mathematical model could be enhanced to predict the orbit and ephemeris parameters accurately. The errors then would converge to zero. Unfortunately, however, the natural world we live in is never quite so simple.

Exercises for Chapter 11

1. Using matrix Equation 11.9 verify that matrix in Equation 11.10 gives the inverse relationship, that is, it takes *body axis* angular rates into the *Euler angle* angular rates.

2. Using the perturbation method employed in Section 11.4 derive a matrix equation similar to Equation 11.9 for the Orbit Parameter angle/axis sequence, zxz. First, construct the rotation sequence diagram similar to that in Figure 11.3.

3. Determine the changes that occur in the Euler angles for a perturbation in yaw. What changes occur for a body axis roll perturbation? See Equation 11.11.

4. Explain the zero main-diagonal of the first equality in Equation 11.12.

5. Prove that the ordinary product rule for derivatives holds for quaternion products.

6. Verify that Equation 11.15 may be expressed as it is written in Equation 11.16.

7. A $+\Delta\lambda$ and a $-\Delta\alpha$ error are detected at the orbit *descending* node. Compute the appropriate orbit parameter changes which are required? What options if only the $-\Delta\alpha$ error is detected?

8. The ephemeris point, $(\lambda, L, \alpha) = (45°W, 30°N, -30°)$ was expected by the ground tracking station but the following parameter errors $\Delta L = -1°$ and $\Delta\alpha = 0.5°$ were detected. What corrections are required in the nominal orbit parameters? *Hint:* First find the nominal orbit parameters for the expected ephemeris point.

Chapter 12

Rotations in Phase Space

12.1 Introduction

The scalar or dot product representation of Ordinary Differential Equations (ODE's), $\mathbf{a}^T\mathbf{x} = \mathbf{a} \bullet \mathbf{x} = f$, offers some geometric insight into the nature of its solution space (see margin). For 2nd order ODE's, solutions may be generated kinematically by two synchronously rotating planes defined in the R^3 solution space.

For 2nd order ODE's, the familiar phase-plane solution trajectories are simply projections, onto the xy-plane, of the more general trajectories generated in its R^3 solution space. Two planes, called *Kyper-planes*, which rotate synchronously, define the instantaneous directions of the state vector and its velocity vector by their respective intersections with the solution plane.

This kinematic solution-generating scheme also offers some geometric insight into the nature of the solutions, even for the more general non-autonomous, forced, and/or non-linear ODE's. In this chapter we explore these matters in some detail.

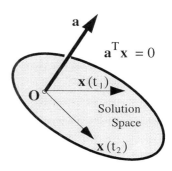

$\mathbf{a}^T\mathbf{x} = 0$
Solution Space

12.2 Constituents in the ODE Set

The universal set of ODE's may be partitioned into two disjoint subsets: **Non-Linear ODE's** (NLDE), and the relatively miniscule but important subset called **Linear ODE's** (LDE).

Further, LDE's may be partitioned into two disjoint subsets: **Autonomous LDE's**, which do not involve terms or coefficients which are functions of the independent variable explicitly, and

ODE's

Perhaps the most challenging, if not disturbing, statement which can be made about Ordinary Differential Equations is —

of all the Ordinary Differential Equations one can write, almost none of them can be solved.

Non-autonomous LDE's which may involve *time-varying* coefficients and/or are driven by some *forcing function*.

LDE's are important because their solutions are completely known and the autonomous LDE's, especially, offer some insight into the behavior of *local solutions* for certain NLDE's.

The general n^{th} order NLDE may be written

$$F[\overset{(n)}{x}(t), \overset{(n-1)}{x}(t), \cdots, \overset{(1)}{x}(t), x(t), f(t)] = 0 \qquad (12.1)$$

where $x(t)$ is the dependent variable, $\overset{(k)}{x}(t)$ is its kth derivative, and t, in this instance, is the independent variable.

The nth order LDE is often written in the scalar form

$$\sum_{k=0}^{n} a_k(t) \overset{(k)}{x}(t) = f(t) \qquad (12.2)$$

The nth order LDE may also be written, quite elegantly, as a first order linear vector-matrix differential equation

$$\frac{d}{dt}\mathbf{z}(t) = A(t)\mathbf{z}(t) + \mathbf{F}(t) \qquad (12.3)$$

For our present purposes, however, these LDE's and many NLDE's may also be expressed in the form of the scalar or dot product

$$\mathbf{a}(t)^T\mathbf{x}(t) = \mathbf{a} \bullet \mathbf{x} = f(t) \qquad (12.4)$$

Expressing it in this form affords some interesting insight into the geometry of the solution space, as we shall see.

In the foregoing equations, superscript T means transpose and

$$
\begin{aligned}
\mathbf{a} &= \text{col}[a_0(t), a_1(t), \cdots, a_n(t)] \\
\mathbf{x} &= \text{solution vector in } R^{n+1} \\
&= \text{col}[x_0(t), x_1(t), \cdots, x_n(t)] \\
x_k &= k^{th} \text{ derivative of scalar function } x(t) = \overset{(k)}{x} \\
f(t) &= \text{scalar forcing function} \\
\mathbf{z}(t) &= \text{col}[x_0, x_1, \cdots, x_{n-1}] \\
A(t) &= n \times n \text{ coefficient matrix} \\
\mathbf{F}(t) &= \text{n-vector forcing function}
\end{aligned}
$$

Our primary purpose is to consider 2nd order ODE's written in the form represented by Equation (12.4); however, we first give a brief review of standard techniques for finding the general solutions for n^{th}-Order LDE's. For LDE's written in matrix form, as in Equation 1.3, we review the eigenvalue and eigenvector ideas which are important in these solution techniques. We then identify the canonical solution types which provide a reference for characterizing local solutions of NLDE's. We begin our review by rearranging terms in Equation 12.3

$$\overset{\circ}{\mathbf{z}}(t) = A(t)\mathbf{z}(t) + \mathbf{F} \qquad (12.5)$$

to give $\qquad \overset{\circ}{\mathbf{z}}(t) - A(t)\mathbf{z}(t) = \mathbf{F}(t) \qquad \mathbf{z}(0) = \mathbf{c}$

Notation

$\mathbf{z}(0) = \mathbf{c}$ is an initial condition.
$Y(t)$ is an integrating factor.

After staring at this last equation for a while, it might occur to us that an *integrating factor*, $Y(t)$, would be helpful. If we let

$$-A(t) = Y^{-1}(t)\,\overset{\circ}{Y}(t) \qquad (12.6)$$

then we may write

$$\overset{\circ}{\mathbf{z}}(t) + Y^{-1}(t)\,\overset{\circ}{Y}(t)\mathbf{z}(t) = \mathbf{F}$$

Premultiplying both sides by $Y(t)$ gives

$$Y(t)\overset{\circ}{\mathbf{z}}(t) + \overset{\circ}{Y}(t)\mathbf{z}(t) = Y(t)\mathbf{F} \qquad (12.7)$$

Integrating both sides over the interval, say, $[0, t]$, we get

Adjoint Equation

$$Y(t)\mathbf{z}(t) - Y(0)\mathbf{z}(0) = \int_0^t Y(\sigma)\mathbf{F}(\sigma)d\sigma \qquad (12.8)$$

Equation 12.9 is called the adjoint equation of Equation 3.5 — it provides the integrating factors for Equation 12.5

This is the solution we seek. However, to get the required integrating factor, $Y(t)$, we must solve the linear homogeneous 1st-order matrix differential equation

$$\overset{\circ}{Y}(t) = -Y(t)A(t) \qquad (12.9)$$

where we may let $\quad Y(0) = I = \text{Identity} \qquad (12.10)$

So, we write the solution of the Nonhomogeneous LDE as

$$\mathbf{z}(t) = Y(t)^{-1}[\mathbf{c} + \int_0^t Y(\sigma)\mathbf{F}(\sigma)d\sigma] \qquad (12.11)$$

Having solved the non-autonomous LDE we now consider a more general class of ODE's — in particular, Non-Linear ODE's, and later even those which are non-autonomous. We begin, in the next section, by reviewing some of the conventions and underlying ideas of the traditional and familiar *phase plane* approach.

12.3   The Phase Plane

The term *phase plane*, coordinatized (x,y), where $y = \overset{\circ}{x}$, suggests the more general notion of *phase space*. Both refer to

> $\cdots$ *a multidimensional space, each point of which determines completely the phase of the system, that is, the values of the generalized coordinates.*

For the necessary background on what follows, we review the ordinary phase *plane* trajectories of 2nd order ODE's.

$$\text{For the NLDE} \qquad f(\overset{\circ\circ}{x}, \overset{\circ}{x}, x) \;=\; 0$$

$$\text{where, if we let} \qquad \overset{\circ}{x} \;=\; y \tag{12.12}$$

$$\text{we may usually write} \qquad \overset{\circ}{y} \;=\; g(x,y) \tag{12.13}$$

Dividing Equation 12.13 by Equation 12.12 gives

$$\frac{dy}{dt} \Big/ \frac{dx}{dt} \;=\; \frac{dy}{dx} \;=\; \frac{g(x,y)}{y} \;=\; F(x,y) \tag{12.14}$$

Direction Fields

The direction of the solution trajectory is defined by $\frac{dy}{dx}$ at every point (x,y) in the domain of Equation 12.14 and Equation 12.18. That is, associated with each point is a direction defined by $\frac{dy}{dx}$. These directions define a *direction field*.

Equation 12.14 defines the *slope* of the solution trajectory at any point (x, y) in its domain. The phase plane view of the solution trajectories for a 2nd order LDE (with constant coefficients) provides a useful reference for identifying the qualitative nature of *local* solutions of some NLDE's. So, we review the solution trajectories of those homogeneous autonomous 2nd order LDE's of the form

$$\overset{\circ\circ}{x} + b\,\overset{\circ}{x} + cx \;=\; 0 \tag{12.15}$$

$$\text{Here, if we let} \qquad \overset{\circ}{x} \;=\; y \tag{12.16}$$

$$\text{then} \qquad \overset{\circ}{y} \;=\; -cx - by \tag{12.17}$$

Dividing Equation 12.17 by Equation 12.16 gives

$$\frac{dy}{dt} \Big/ \frac{dx}{dt} \;=\; \frac{dy}{dx} \;=\; \frac{-cx - by}{y} \tag{12.18}$$

Isoclines

If we consider those points (x,y) which lie on the line $y = mx$, then Equation 12.18 shows that every solution trajectory crosses this line in the *same* direction; hence, the name *iso*-cline. With these ideas the *direction field* may be easily sketched

Now we can easily find those directions, called *eigen-directions*, in which the direction $\frac{dy}{dx}$ of the *velocity* vector coincides with the direction of its *state* vector (x, y), in this homogeneous case. To find these common directions we let $y = mx$ and $\frac{dy}{dx} = m$ in Equation 12.18, which gives

$$m^2 + bm + c \;=\; 0 \tag{12.19}$$

This quadratic equation, of course, has two roots, which are either real or complex. These roots are given by

$$m = -\frac{b}{2} \, [1 \pm \sqrt{1 - \frac{4c}{b^2}} \,]$$

If $b^2 > 4c$ and $c > 0$, we get two real negative roots. If $b^2 > 4c$ and $c < 0$, we get one real positive root and one real negative root. Finally, if $b^2 < 4c$ then we get a pair of complex conjugate roots. The importance of these distinctions is illustrated in what follows. Real eigen-directions, however, are especially significant in that geometrically all solution trajectories in phase space approach these directions asymptotically, as we shall see.

Reference Trajectories

Based upon the nature of the roots just described, we classify linearized 2nd order ODE solution trajectories into three types: Nodes, Saddles, and Foci. We illustrate each type, defined below, using a specific LDE.

> (1) **Stable Nodes**, where we have
> $b^2 > 4c$ with $c > 0$, illustrated
> using equation $\overset{\circ\circ}{x} + 3\,\overset{\circ}{x} + 2x = 0$

> (2) **Saddles**, where we have
> $b^2 > 4c$ with $c < 0$, illustrated
> using equation $2\,\overset{\circ\circ}{x} + \overset{\circ}{x} - x = 0$

> (3) **Stable Foci**, where we have
> $b^2 < 4c$ with $b > 0$, illustrated
> using equation $\overset{\circ\circ}{x} + \overset{\circ}{x} + 2x = 0$

We should also note, however, that conventional phase plane reference trajectories, when listed, often include the Unstable Node, the Unstable Focus, and the Center. These types are closely related to those we have listed, so we do not illustrate them here. Note that the saddle, as a type, is always unstable.

From each figure illustrating the stable case, the reader can visualize the related unstable trajectories by merely changing the direction of the arrows and then flipping the figure about its x-axis.

12.3.1   Phase Plane Stable Node

As an example of a Stable Node we use the equation

$$\overset{\circ\circ}{x} + 3\,\overset{\circ}{x} + 2x = 0$$

So in Equation 12.15 we must take $b = 3$ and $c = 2$, which gives two real negative eigenvalues characterizing a Stable Node. With $y = mx$, Equation 12.18 becomes

$$\frac{dy}{dx} = \frac{-2x - 3y}{y} = -\frac{2 + 3m}{m}$$

If we assign various values to the parameter m, we can plot the direction field illustrated in Figure 12.1.   Based upon this direction

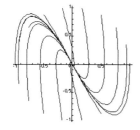

Phase Plane
Solution Trajectories
for
Stable Node
$$\overset{\circ\circ}{x} + 5\,\overset{\circ}{x} + 6x = 0$$

which may be written
$$\overset{\circ}{x} = y$$
$$\overset{\circ}{y} = -6x - 5y$$
with

Eigendirections
$$y = -2x$$
$$y = -3x$$

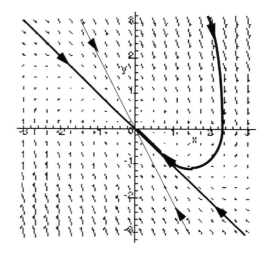

Figure 12.1: Stable Node $\overset{\circ\circ}{x} + 3\,\overset{\circ}{x} + 2x = 0$

field, one may sketch the solution trajectory as illustrated.

From the family of solution trajectories shown in the margin, one should note that all trajectories, for a large *state vector*, have velocity directions asymptotic to the eigendirection $y = -2x$. Similarly, as each solution trajectory approaches the origin, the slope of the trajectory approaches the eigendirection $y = -x$.

12.3.2   Phase Plane Saddle

If we take $b = 1$ and $c = -2$ in Equation 12.15, we get one positive and one negatve eigenvalue, namely, $+1$ and -2. This results in the phase plane direction field which characterizes a Saddle, shown in Figure 12.2, which also shows two solution trajectories.

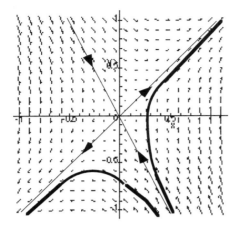

Figure 12.2: Saddle $\overset{\circ\circ}{x} + \overset{\circ}{x} - 2x = 0$

Saddle
with
Eigendirections
$$y = x$$
$$y = -2x$$

12.3.3 Phase Plane Stable Focus

Finally, we let $b = 2$ and $c = 10$. The resulting direction field and

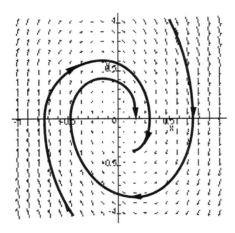

Figure 12.3: Stable Focus $\overset{\circ\circ}{x} + 2\overset{\circ}{x} + 10x = 0$

Phase Plane Solution
for
$$\overset{\circ\circ}{x} + 2\overset{\circ}{x} + 10x = 0$$
is a
Stable Focus
with
Eigendirections
$$y = (-1 \pm \mathbf{i}3)x$$

solution trajectories, represent the Stable *Focus* in Figure 12.3.

The three examples we have considered here are, of course, trivial and are easily solved using any of the standard methods. These examples are merely intended to illustrate specific canonical representations for direction fields and their implicit solution trajectories for homogeneous *2nd* order LDE with constant coefficients. This is useful because similar trajectories often characterize the behaviour of non-linear differential equations (NLDE's) in sufficiently small *linearized* neighborhoods of their singular points. This review will provide the necessary background and perspective for our more comprehensive *unifying* R^3 phase space approach which we begin in the next section. The utility of the conventional phase plane method is that it sometimes provides insight into the nature of the solution of a NLDE's when it seems they can not be solved any other way.

The usefulness of the conventional phase plane analysis vanishes, however, or at least becomes very fuzzy indeed, if the ODE under consideration is either non-homogeneous or, worse, non-autonomous. In other words, the conventional phase plane or phase space offers virtually no insight into the nature of the solution of an ODE that is forced and/or has time-varying coefficients. On the other hand, the scalar product representation for the ODE, namely,

$$\mathbf{a}^T\mathbf{x} \;=\; \mathbf{a} \circ \mathbf{x} \;=\; f \qquad (12.20)$$

not only provides a new perspective, but also offers some geometric insight into the nature of the solution space for a larger set of ODE's, as we shall see. The general solution of specific 2nd order LDE's are more easily viewed as a subset of the set of all possible general solutions of 2nd order differential equations, including those which are non-autonomous and non-linear.

Figure for
Equation 12.22
$\mathbf{a}^T\mathbf{x} = 0$
Solution Space

R$^{(n+1)}$ Trajectories

The conventional phase plane trajectories are simply a projection of the trajectories which are generated in the $R^{(n+1)}$ solution space.

12.4   Some Preliminaries

To simplify the notation in what follows, the order of the derivatives, which are the elements of the *solution* vector $\mathbf{x}(t)$, is denoted by subscripts rather than by *dots* or parenthesized superscripts. That is, $\mathbf{x} = (x, \overset{\circ}{x}, \overset{\circ\circ}{x}) = (x_0, x_1, x_2)$ where

$$x_0 \;=\; \overset{(0)}{x} \;=\; x \;=\; \text{the dependent variable}$$
$$x_1 \;=\; \overset{(1)}{x} \;=\; \overset{\circ}{x} \;=\; \text{its first derivative}$$
$$x_2 \;=\; \overset{(2)}{x} \;=\; \overset{\circ\circ}{x} \;=\; \text{its second derivative}$$

Vector Elements

As usual these subscripts denote the order of the vector-elements. In the case of the solution vector $\mathbf{x}(t)$, however, the subscript also denotes the order of the derivative.

The discussion is confined, for the most part, to second order ODE's, for geometrically obvious reasons. The extension to n^{th} order ODE's should be a natural one — planes become hyperplanes, etc. — admittedly, a shameful oversimplification.

We now define S to be a vector space over the reals, isomorphic to the set of points, say, in R^n. Then, in our abbreviated notation for the homogeneous ODE, namely,

$$\mathbf{a} \circ \mathbf{x} = \mathbf{a}^T \mathbf{x} = 0 \qquad (12.21)$$

we will understand that $\mathbf{a} = \mathbf{a}(x,t)$ and $\mathbf{x} = \mathbf{x}(t)$ are real-valued vectors in S for each time $t \ \epsilon \ T$. In order to establish some of the relevant geometry in R^3, we adopt the *standard* basis, so that the points (or vectors)

$$\mathbf{x} = col[x_0, x_1, x_n]$$
$$\mathbf{a} = col[a_0, a_1, a_s]$$

which characterize Equation 12.21, become more meaningful in S geometrically.

As stated above, the subscripts on the elements of $\mathbf{x}(t)$ signify the order of the derivative of the scalar function $x(t)$. The vector $\mathbf{x}(t)$ is called the *solution vector* and the vector $\mathbf{a}(x,t)$ is called the *coefficient vector*. Clearly, the solution of Equation 12.21, for each time $t \ \epsilon \ T$, is an element of a *solution plane*, a subspace of S which, at this particular time t, is orthogonal to the coefficient vector, $\mathbf{a}(x,t)$. We state this important notion as a theorem.

Theorem 1   If $\mathbf{a}^T\mathbf{x} = 0$ is an nth order ODE where
the coefficients $\mathbf{a} = \mathbf{a}(x,t)$ and
the solution $\mathbf{x} = \mathbf{x}(t)$ are real-valued
vectors in R^{n+1} for each time t ϵ T
then for each t, the solution $\mathbf{x}(t)$, if it exists,
is contained in a subspace orthogonal to $\mathbf{a}$(x,t).

12.5   Linear Differential Equations

As in any introductory exposition of ODE's, that small subset of LDE's with constant coefficients is often discussed first, simply because its solutions are completely known and understood. So, here we consider the *homogeneous constant coefficient* case

$$\mathbf{a}^T\mathbf{x} = 0 \qquad (12.22)$$

where $\mathbf{a}$ is a *constant coefficient vector*, and $\mathbf{x}$ is the *solution vector*. Equation 12.22, when expanded, reads

$$\mathbf{a}^T\mathbf{x} = \mathbf{a} \circ \mathbf{x} = a_0 x_0 + a_1 x_1 + a_2 x_2 = 0$$

Solution Planes in R^3

What are they?

Well, if the *coefficient vector* $\mathbf{a}$ is a constant vector, then the solution space for

$$\mathbf{a}^T\mathbf{x} = 0$$

is a *plane* in R^3. This plane (an R^2 subspace of R^3) is orthogonal to the *fixed* coefficient vector $\mathbf{a}$. All the solutions to the given equation are necessarily contained in this plane. In fact, each point in this plane is a potential initial condition, so that the vector associated with this point is a solution.

On the other hand, if the *coefficient vector* $\mathbf{a}$ is *not* a constant, then $\mathbf{a}(x,t)$ will have a time-varying direction. Then at each $t \epsilon T$, the solution plane must assume a corresponding time-varying orientation orthogonal to $\mathbf{a}(x,t)$ while the solution trajectory is being generated in R^3.

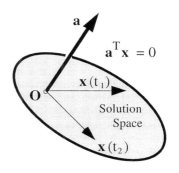

Figure for
$\mathbf{a}^T\mathbf{x} = 0$
Solution Space

NonHomogeneous ODE

If in Eqn 12.20 $\mathbf{a}(t)^T\mathbf{x}(t) = f(t)$ is a linear non-homogeneous 2nd-order ODE it may, of course, be written as a 1st-order matrix equation

$$\overset{\circ}{\mathbf{x}} = A\mathbf{x} + \mathbf{f}$$
$$A = \begin{bmatrix} 0 & 1 \\ -a_0 & -a_1 \end{bmatrix}$$

and $\quad \mathbf{x} = \text{col}[x_0, x_1]$

and $\quad \mathbf{f} = \text{col}[0, f(t)]$

However, for our purposes we choose to write it in R^3 as

$$\overset{\circ}{\mathbf{x}} = B\mathbf{x} + \overset{\circ}{\mathbf{f}}$$

where

$$B = \begin{bmatrix} 0 & 1 & 0 \\ 0 & 0 & 1 \\ b_{31} & b_{32} & b_{33} \end{bmatrix}$$

with $\quad b_{31} = -\overset{\circ}{a_0}/a_2$

$\qquad b_{32} = -(a_0 + \overset{\circ}{a_1})/a_2$

$\qquad b_{33} = -(a_1 + \overset{\circ}{a_2})/a_2$

$\qquad \mathbf{x} = \text{col}[x_0, x_1, x_2]$

$\qquad \overset{\circ}{\mathbf{x}} = \text{col}[\overset{\circ}{x_0}, \overset{\circ}{x_1}, \overset{\circ}{x_2}]$

and $\quad \overset{\circ}{\mathbf{f}} = \text{col}[0, 0, \overset{\circ}{f(t)}]$

If the 3×3 matrix B is a constant, then in this case it is also singular, of course, because the solution is contained in an R^2 subspace. These representations, however, provides us with useful insight and information into the nature of this *planar* solution subspace in R^3, as we shall see.

Equation 12.24

An argument can be made that the two Equations 12.23 and 12.24 are equivalent. Of course, by definition they are. However, it is important here to view Equation 12.24 as relating the *incremental changes*, Δx_0 and Δx_1. These relative changes occur along the x_0 and x_1 axes, and are specified by the plane, CEB, in Figure 12.4. These *incremental changes* instantaneously define the solution velocity vector which also must lie in the solution plane.

Again, by definition $\qquad \overset{\circ}{x_0} = x_1$

$$\overset{\circ}{x_1} = x_2$$

Here the dot notation means the derivative with respect to the independent variable, which is time, t. We now divide the first of

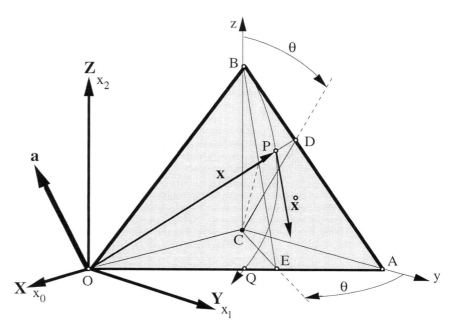

Figure 12.4: $\mathbf{a}^T\mathbf{x} = 0$ Solution Plane Geometry

these equations by the second, and call the result p. For each time $t \in T$ there exists a $|p| < \infty$ such that

$$\frac{\overset{\circ}{x_0}}{\overset{\circ}{x_1}} = \frac{x_1}{x_2} = p, \qquad \overset{\circ}{x_1} = x_2 \neq 0$$

Then we may write $\qquad x_1 = px_2 \qquad\qquad (12.23)$

$$\overset{\circ}{x_0} = p\overset{\circ}{x_1} \qquad\qquad (12.24)$$

or $\qquad \Delta x_0 \simeq p\Delta x_1$

Equation 12.23 defines a plane which contains the x_0-axis, and rotates about the x_0-axis as p varies in time. Similarly, Equation 12.24 may be viewed as a plane which contains the x_2-axis, and rotates about the x_2-axis, governed by the same time-varying p. It is then clear that, *however* this time-varying function p describes the rotation of these two planes, they must rotate synchronously.

The intersections of each of these two rotating *kyper*planes with the solution plane define the directions of the state vector **x** and

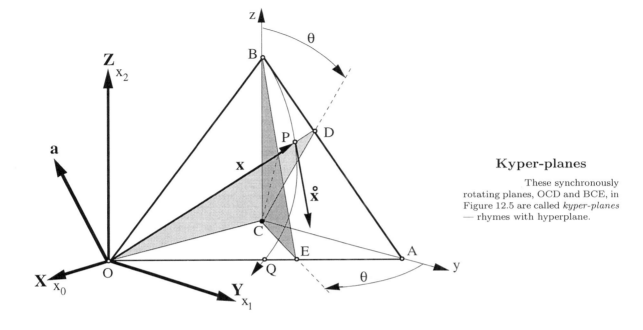

Figure 12.5: Rotating *Kyper*plane Geometry

its velocity vector $\overset{\circ}{\mathbf{x}}$ in S, respectively. Figure 12.4 illustrates the solution space geometry for a homogeneous 2nd order LDE with a constant coefficient vector, **a**. The solution space S is the plane BOA. The solution vector **x**, and also its velocity vector $\overset{\circ}{\mathbf{x}}$, are contained in the planar solution space, S, that is, in this homogeneous case these vectors remain orthogonal to the coefficient vector **a** for all t. The planes OCD and BCE are the rotating *kyper-planes* of Equation 12.23 and Equation 12.24 shown in Figure 12.5.

If the coefficient vector $\mathbf{a} = \mathbf{a}(\mathbf{x}, t)$ varies with time and/or **x**, the solution space S is, in general, not a plane. However, any solution vector $\mathbf{x}(t)$, *at each instant*, still lies in a plane which, *at that particular instant*, is orthogonal to **a**.

Since, in this case, the coefficient vector **a** varies with time, so does the orientation of this plane. Thus, in this time-varying case, given an *initial point* $\mathbf{x}(0)$, the solution trajectory $\mathbf{x}(t)$ is generated point-wise in the solution space by the three time-varying planes.

Kyper-planes

These synchronously rotating planes, OCD and BCE, in Figure 12.5 are called *kyper-planes* — rhymes with hyperplane.

Equation 12.22
$$\mathbf{a}^T \mathbf{x} = 0$$
Solution plane BOA

Equation 12.23
$$x_1 = p x_2$$
plane COD

Equation 12.24
$$\overset{\circ}{x}_0 = p \overset{\circ}{x}_1$$
plane CEB

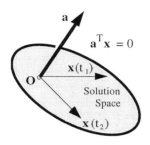

$$\mathbf{a}^T\mathbf{x} = 0$$
Solution Space

In summary, we note in Figure 1.3 that:

- Equation 12.23 and Equation 12.24 represent the *kyper*planes COD and CEB, respectively, in R^3.

- The *kyper-plane*, COD, in Equation 12.23 contains the x_0-axis *and rotates about* this axis as the rotation angle θ varies. (See the note in the margin on Rotating Planes.)

- The intersection of the plane (Equation 12.23) with the Solution plane (Equation 12.22) *contains* the solution vector, $\mathbf{x}(t)$.

- The second plane, CEB (Equation 12.24), contains *and rotates about* the x_2-axis as θ varies.

- The intersection of the *kyper*plane, CEB, with the Solution plane, BOA, specifies the *direction* of the solution velocity vector, $\overset{\circ}{\mathbf{x}}(t)$. See Equation 12.24 comment in margin.

- The two planes, COD and CEB, rotate *synchronously* with a time-varying p or θ, to generate a solution trajectory from any initial point in the solution plane, BOA.

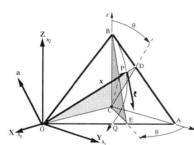

Rotating Planes

The geometric orientation of the two planes represented in Equations 12.23 and 12.24 is time-varying. The rotation of both are governed by the same function, $p(t)$, which makes their rotation synchronous. From this point on, however, it is convenient to let

$$p(t) = \tan\theta(t)$$

for the geometric insight and control it affords.

It is also important to note, however, any instantaneous changes in the position of point P due to the time-varying nature of $\mathbf{a}(\mathbf{x}, t)$, occur first along the z-axis. Corresponding changes in θ also affect the $\mathbf{x}$ and $\overset{\circ}{\mathbf{x}}$ directions. We will develop these matters further — but first in the context of LDE's with constant coefficients.

The two synchronously rotating *kyper-planes*, Equation 12.23 and 12.24, are conveniently represented using their respective time-varying normal vectors. These are

$$\mathbf{b}_0 = col[0, \ -\cos\theta, \ \sin\theta]$$
$$\mathbf{b}_1 = col[-\cos\theta, \ \sin\theta, \ 0]$$

The solution plane and the rotating *kyper*-plane defined by

$$\mathbf{b}_0^T\mathbf{x} = 0 \quad \Rightarrow \quad \mathbf{r}_{dir} \qquad (12.25)$$

Kyper-planes

The rotating *kyper*-plane machine generates not only a direction field; if an initial point is specified, then a continuous solution trajectory is generated by the machine.

intersect to determine the state vector direction, $\mathbf{r}_{dir}$, in S; and the solution plane and the synchronously rotating *kyper*-plane

$$\mathbf{b}_1^T\mathbf{x} = 0 \quad \Rightarrow \quad \overset{\circ}{\mathbf{r}}_{dir} \qquad (12.26)$$

intersect to determine the velocity vector direction, $\overset{\circ}{\mathbf{r}}_{dir}$, in S.

The *direction* of the solution vector is determined by

$$\mathbf{x}_{dir} = \mathbf{b}_0 \times \mathbf{a} = det \begin{vmatrix} \mathbf{i} & \mathbf{j} & \mathbf{k} \\ 0 & -\cos\theta & \sin\theta \\ a_0 & a_1 & a_2 \end{vmatrix}$$

$$= -\mathbf{i}(a_2\cos\theta + a_1\sin\theta)$$
$$+ \mathbf{j}(a_0\sin\theta) + \mathbf{k}(a_0\cos\theta) \qquad (12.27)$$

The *direction* of the velocity vector is determined by

$$\overset{\circ}{\mathbf{x}}_{dir} = \mathbf{b}_1 \times \mathbf{a} = det \begin{vmatrix} \mathbf{i} & \mathbf{j} & \mathbf{k} \\ -\cos\theta & \sin\theta & 0 \\ a_0 & a_1 & a_2 \end{vmatrix}$$

$$= +\mathbf{i}(a_2\sin\theta) + \mathbf{j}(a_2\cos\theta)$$
$$- \mathbf{k}(a_1\cos\theta + a_0\sin\theta) \qquad (12.28)$$

These Directions

We emphasize, both of these *directions* are functions of the *kyper*plane rotation angle θ, where θ must, of course, implicitly be some function of the original independent variable, usually time.

12.6   Initial Conditions

To this point we have defined only the *directions* of the solution vector and its velocity vector, $\mathbf{x}_{dir}$ and $\overset{\circ}{\mathbf{x}}_{dir}$, respectively, in the solution plane $\mathbf{a}^T\mathbf{x} = 0$. To get a particular solution we must specify two *scalar* initial conditions, namely, the *initial* values $x_0(0) = c_0$ and $\overset{\circ}{x}_0(0) = x_1(0) = c_1$. Using these scalar values, we may write the initial state and velocity vectors in our R^3 context as

$$\mathbf{x}(0) = \mathbf{c} = col[c_0,\ c_1,\ c_2]$$
$$\text{where} \quad c_2 = -\frac{(a_0c_0 + a_1c_1)}{a_2}$$
$$\text{and} \quad \overset{\circ}{\mathbf{x}}(0) = col[c_1,\ c_2,\ c_3]$$
$$\text{where} \quad c_3 = -\frac{(a_0c_1 + a_1c_2)}{a_2}$$

These same initial conditions may also be expressed in terms of an initial value, θ_0, for the angle θ in the synchronously rotating *kyper-plane* system. To make this initial angle θ_0 consistent with the specified initial conditions, we start by letting R be an appropriate scalar, and writing the *initial* state vector as

$$\mathbf{x}(\theta_0) = -R\mathbf{x}(\theta_0)_{dir}$$
$$= -\mathbf{i}R(a_2\cos\theta_0 + a_1\sin\theta_0)$$
$$+ Ra_0(\mathbf{j}\sin\theta_0 + \mathbf{k}\cos\theta_0)$$
$$= \mathbf{i}c_0 + \mathbf{j}c_1 + \mathbf{k}c_2$$

Initial Conditions

A particular solution to any 2nd order ODE requires two initial conditions. However, since every solution we seek is constrained to the solution plane, $a^T x = 0$, an appropriate 3-tuple may be identified as the *initial point* in this plane.

The initial conditions for a scalar 2nd order ODE are specified at some point in time (usually) and written as

$$x(t_0) = c_0$$
$$\overset{\circ}{x}(t_0) = c_1$$

Since $\mathbf{a}^T\mathbf{c} = 0$, it follows that

$$c_2 = -\frac{(a_0c_0 + a_1c_1)}{a_2}$$

We then solve for the initial angle θ_0 and the scalar R by writing

$$Ra_0 \sin \theta_0 = c_1$$

$$\text{and} \quad Ra_0 \cos \theta_0 = c_2$$

$$\text{so that} \quad Ra_0 = \sqrt{c_1^2 + c_2^2}$$

from which the scalar R may be determined.

We have $\quad \tan \theta_0 = \dfrac{c_1}{c_2} = -\dfrac{a_2 c_1}{(a_0 c_0 + a_1 c_1)}$

from which the initial angle θ_0 may be determined.

We confirm these results by writing the *initial velocity vector*

Velocity Vector

Using the initial velocity vector has the advantage of introducing $\overset{\circ}{\theta}$, which is the rate of rotation of the rotating *kyper*plane system. Further, whether $\overset{\circ}{\mathbf{x}}$ means $\frac{d\mathbf{x}}{d\theta}$ or $\frac{d\mathbf{x}}{dt}$ is clear from the context.

$$\text{as} \quad \overset{\circ}{\mathbf{x}}(\theta_0) = R \overset{\circ}{\theta} \overset{\circ}{\mathbf{x}}(\theta_0)_{dir}$$

$$= R \overset{\circ}{\theta} a_2 (\mathbf{i} \sin \theta_0 + \mathbf{j} \cos \theta_0)$$

$$- \mathbf{k} R \overset{\circ}{\theta} (a_1 \cos \theta_0 + a_0 \sin \theta_0)$$

$$= \mathbf{i} c_1 + \mathbf{j} c_2 + \mathbf{k} c_3$$

$$\text{Again,} \quad \tan \theta_0 = \frac{c_1}{c_2} \text{ and } R \overset{\circ}{\theta} a_2 = \sqrt{c_1^2 + c_2^2} = Ra_0$$

Finally, after some tedious algebra, we may summarize the general expressions for the state vector and its velocity vector as:

$$\mathbf{x}(\theta) = -\mathbf{i}R[a_2 \sin(\theta + \theta_0) + a_1 \cos(\theta + \theta_0)]$$

$$+ Ra_0[\mathbf{j} \cos(\theta + \theta_0) + \mathbf{k} \sin(\theta + \theta_0)] \qquad (12.29)$$

$$\overset{\circ}{\mathbf{x}}(\theta) = + Ra_2[\mathbf{i} \cos(\theta + \theta_0) + \mathbf{j} \sin(\theta + \theta_0)]\frac{d\theta}{dt}$$

$$- \mathbf{k}R[a_1 \sin(\theta + \theta_0) + a_0 \cos(\theta + \theta_0)]\frac{d\theta}{dt} \quad (12.30)$$

where $\tan \theta_0$ and R are as determined above

12.7   Partitions in R^3 Phase Space

In Section 12.3 we introduced the notion of eigenvalues and eigendirections associated with the solutions of the general 2nd order HLDE. There, three categories (nodes, foci, saddles) were noted, based upon the eigenvalues (positive or negative, real or complex). In subsequent sections we develop a similar criterion which, in R^3 phase space, geometrically partitions all possible real coefficient vectors into these same categories.

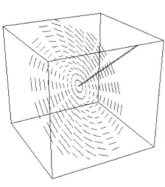

**Direction Field
for a
Focus**

First, in this section, we find a criterion which geometrically partitions R^3 into those coefficient vectors which give complex eigendirections and those which have real eigendirections.

Recall the traditional phase-plane methods for 2nd order LDE's, which were written as a 1st order matrix equation, $\overset{\circ}{\mathbf{x}} = A\mathbf{x}$. We asked whether there are points in the solution space where the velocity vector is a scalar multiple of the state vector, that is, $\overset{\circ}{\mathbf{x}} = \lambda\mathbf{x}$. To find the eigenvalues, λ, and the associated eigenvectors, $\mathbf{x}$, we solve

$$\lambda\mathbf{x} = A\mathbf{x} \qquad \Rightarrow \qquad [A - \lambda I]\mathbf{x} = 0$$

In our R^3 *phase space* context, we ask the same question, but approach it in a slightly different manner. We compute the cross-product of the vectors in Equations 12.27 and 12.28, which, of course, must be a scalar multiple of the coefficient vector $\mathbf{a}$. We get

$$\mathbf{x}_{dir} \times \overset{\circ}{\mathbf{x}}_{dir} = C\mathbf{a}$$
$$\text{where} \qquad C = -(a_0 \sin^2 \theta$$
$$+ a_1 \cos \theta \sin \theta$$
$$+ a_2 \cos^2 \theta) = (\mathbf{b}_0 \times \mathbf{b}_1)^T \mathbf{a}$$

In this context, eigendirections occur when $C = 0$, so we write

Figure 12.6

If the coefficient vector is *inside* the conical locus shown in Figure 12.6, then there are *no real* eigenvectors.

If the coefficient vector is *on* the surface of the conical locus, then there is one *repeated* real eigenvector.

If the coefficient vector is *exterior to* the conical locus, then there are two distinct real eigenvectors.

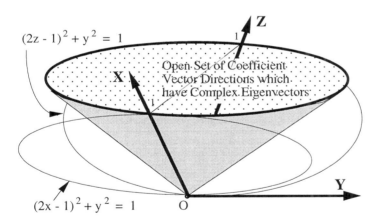

Figure 12.6: Coefficient Vector Directions

$$a_0 \sin^2 \theta + a_1 \sin \theta \cos \theta + a_2 \cos^2 \theta = 0$$
$$\text{or} \qquad a_0 \tan^2 \theta + a_1 \tan \theta + a_2 = 0$$

If we let $|u| = |\tan \theta| < \infty$, we get the polynomial

$$a_0 u^2 + a_1 u + a_2 = 0$$

This equation has real-valued solutions *iff*

$$a_1^2 - 4a_0 a_2 \geq 0 \qquad (12.31)$$

The locus of those coefficient vectors, $\mathbf{a}$, for which $a_1^2 = 4a_0a_2$ is the cone $y^2 = 4xz$ in Figure 12.6. Since this cone is symmetric with respect to the zx-plane, we invoke an auxilliary constraint, $x + z = 1$, so that in Figure 12.6 we may more easily see the boundary for the set of coefficient vector directions which give complex eigendirections, namely, the interior of the cone.

We now write the coefficient vector, $\mathbf{a}$, in terms of its direction angles, α and β. We get

$$\mathbf{a} = \mathbf{i}\cos\alpha\cos\beta + \mathbf{j}\sin\alpha\cos\beta + \mathbf{k}\sin\beta \qquad |\mathbf{a}| = 1$$

Here $\mathbf{a}$ is a point on the unit sphere for each α and β. Using the criteria listed in the margin, we can now partition the set of all possible coefficient vectors into the now familiar canonical behaviour

Solution Space

The coordinate frame, $\mathbf{XYZ}$, illustrated in Figure 3.4, is defined over the reals with a standard basis, $\{\mathbf{i},\mathbf{j},\mathbf{k}\}$. At some $t \,\epsilon\, T$, a direction is defined in this space by the real-valued coefficient vector $\mathbf{a}[\mathbf{x}(t), t]$. The three coordinate axes, $\mathbf{XYZ}$, are assigned to represent, at this same $t \,\epsilon\, T$, the scalar values of the solution $x(t)$ and its 1st and 2nd time derivatives. These are the respective components of the solution vector, $\mathbf{x}(t)$.

Coefficient Vector Directions define System Behaviour

R^3 is partitioned in Figure 12.7 to show how the direction of the coefficient vector $\mathbf{a} = col[a_0, a_0, a_0]$ relates to regions on the unit ball which define familiar canonical system behaviour (all regions are symmetric through the origin):

Stable Node
$a_0a_2 > 0$ and $a_0 > 0$
$a_1^2 \geq 4a_0a_2$

Unstable Node
$a_0a_2 > 0$ and $a_0 < 0$
$a_1^2 \geq 4a_0a_2$

Stable Foci
$a_0a_2 > 0$ and $a_1 > 0$
$a_1^2 < 4a_0a_2$

Unstable Foci
$a_0, a_2 > 0$ and $a_1 < 0$
$a_1^2 < 4a_0a_2$

Saddle
$a_0a_2 < 0$

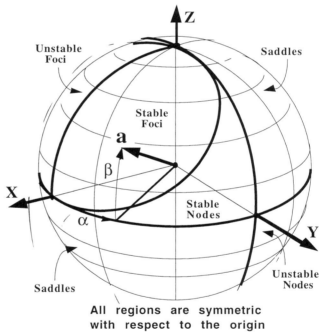

All regions are symmetric
with respect to the origin
Coefficient Vector & Canonical Behavior
Figure 12.7

categories, which were discussed earlier in connection with the phase plane. The results of this R^3 Phase Space partitioning are illustrated in Figure 12.7. Observe that each region on this unit sphere is symmetric with repect to the origin. By selecting the appropriate α and β angles for the direction of the coefficient vector $\mathbf{a}$, the category indicated in Figure 12.7 is determined. Note especially that the region indicated on the sphere at any point in time by $\mathbf{a}(\mathbf{x}, t)$ determines the system behaviour at that instant.

12.8   Space-filling Direction Field in R^3

By now it is clear that $\mathbf{a}(x,t)^T\mathbf{x}(t) = 0$ is the equation of a *time-varying* plane which contains the origin, and has the *time-varying* coefficient vector $\mathbf{a} = \mathbf{a}(\mathbf{x},t)$ as its normal vector. In our present context, the direction of a solution trajectory in R^3 for a 2nd Order ODE is generated in this plane for each $t \; \epsilon \; T$, by the synchronously rotating *kyper*planes. However, for most of us ordinary motals these time-varying directions for s solution trajectory are quite difficult to visualize.

To gain insight we reconsider the constant coefficient case. The intersection directions which the two *kyper*planes make in every *fixed* solution-plane define a direction field continuum in that plane. See Figure 12.8. However, these rotating *kyper*planes define a direction

Equation 12.22
$$\mathbf{a}^T\mathbf{x} = 0$$
Solution plane BOA

Equation 12.23
$$x_1 = px_2$$
plane COD
$$\mathbf{b}_0^T\mathbf{x} = 0 \text{ where}$$
$$\mathbf{b}_0 = col[0, \; -\cos\theta, \; \sin\theta]$$

Equation 12.24
$$\overset{\circ}{x}_0 = p\,\overset{\circ}{x}_1$$
plane CEB
which at the origin is
$$\mathbf{b}_1^T\mathbf{x} = 0 \text{ where}$$
$$\mathbf{b}_1 = col[-\cos\theta, \; \sin\theta, \; 0]$$

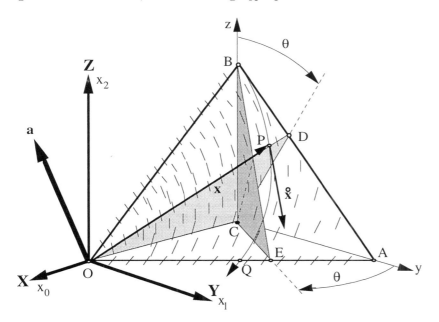

Figure 12.8: Solution Plane Direction Field

field for *every possible* solution plane, $\mathbf{a}^T\mathbf{x} = 0$. Thus, the direction field is defined *everywhere* in R^3. That is, the direction field *fills* R^3. Given the equation $\mathbf{a}(x,t)^T\mathbf{x}(t) = 0$, these rotating *kyper*planes become an important part of

A Kinematic Solution-generating Machine

In summary, at least in the constant coefficient case, a solution trajectory from any arbitrary initial point is generated in the plane $\mathbf{a}^T\mathbf{x} = 0$ by the rotating *kyper*planes:

$$\mathbf{b}_0^T\mathbf{x} = 0$$
$$\mathbf{b}_1^T\overset{\circ}{\mathbf{x}} = 0$$

$$\text{where} \quad \mathbf{a} = col[a_0, a_1, a_2] = \text{ODE coefficient vector}$$
$$\mathbf{b}_0 = col[0, -\cos\theta, \sin\theta]$$
$$\mathbf{b}_1 = col[-\cos\theta, \sin\theta, 0]$$
$$\text{and} \quad \mathbf{x} = col[x_0, x_1, x_2] = \text{ODE solution state vector}$$
$$\text{and} \quad \overset{\circ}{\mathbf{x}} = col[\overset{\circ}{x}_0, \overset{\circ}{x}_1, \overset{\circ}{x}_2] = \text{ODE velocity vector}$$

Figure 12.9 illustrates the direction field for a focus viewed from

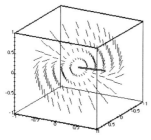

$\mathbf{a} = col[.6, 1, .3]$
Stable Node
(a)

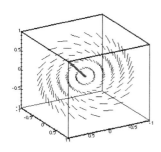

$\mathbf{a} = col[1, 1, 1]$
Stable Focus
(b)

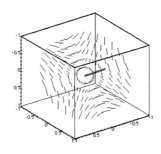

$\mathbf{a} = col[1, .2, -.7]$
Saddle
(c)

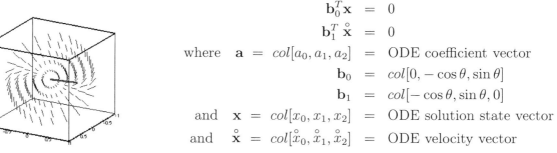

Figure 12.9: Solution Plane Orientation

different perspectives, whereas the figures in the margin illustrate three distinct solution categories. The figures in the margin also suggest that the R^3 direction-field is a continuum in our R^3 phase space. That is, if the orientation of the coefficient vector is time-varying, then the direction field in the time-varying solution plane may smoothly change from being a node, to a focus, to a saddle, etc.

Clearly there are several lines of further investigation in all of this. For example, in the time-varying case, an analysis of the derivatives of the equations

$$\mathbf{a}(x, t)^T\mathbf{x}(t) = 0$$
$$\mathbf{a}(x, t) \times \mathbf{b_0}(t) = \mathbf{x}(t)_{dir}$$
$$\mathbf{a}(x, t) \times \mathbf{b_1}(t) = \overset{\circ}{\mathbf{x}}(t)_{dir}$$

provide some deeper and important insight into the transition directions from one solution plane trajectory to the next. These matters, however provoking and interesting, are beyond the intended purpose and scope of this book.

12.9   Locus of all Real Eigenvectors

The two synchronously rotating *kyper*planes in Figure 12.5 intersect in a line whose direction is

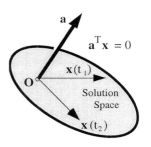

$$\mathbf{b}_1 \times \mathbf{b}_0 = det \begin{vmatrix} \mathbf{i} & \mathbf{j} & \mathbf{k} \\ -\cos\theta & \sin\theta & 0 \\ 0 & -\cos\theta & \sin\theta \end{vmatrix}$$

$$= \mathbf{i}\sin^2\theta + \mathbf{j}\sin\theta\cos\theta + \mathbf{k}\cos^2\theta \qquad (12.32)$$

$\mathbf{a}^T\mathbf{x} = 0$
Solution Space

This direction vector, which is common to both synchronously rotating planes, is an eigenvector *iff* it is also contained in the solution plane. That is, every real solution of the scalar triple product

$$\mathbf{a} \circ \mathbf{b}_1 \times \mathbf{b}_0 = 0 \qquad (12.33)$$

defines a real eigendirection in the solution plane $\mathbf{a} \circ \mathbf{x} = 0$. Moreover, if we let

$$\mathbf{i}\sin^2\theta + \mathbf{j}\sin\theta\cos\theta + \mathbf{k}\cos^2\theta = \mathbf{i}x + \mathbf{j}y + \mathbf{k}z$$

then Equation 12.32 says the eigendirections must lie on the cone

$$y = \sqrt{xz}$$

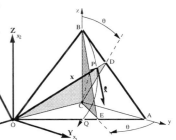

Rotating Planes

which we call the *Eigenvector Cone*. Every pair of two real eigenvectors or eigendirections is defined by the intersection of the solution plane with this cone.

One can easily visualize a solution plane which contains the two distinct eigenvectors illustrated in Figure 12.10. Here again, if we write $x + z = 1$, the projections of the point-set $f(x,y,z) = 0$ in each coordinate plane are more easily seen to be

$$\begin{array}{rcll} x^2 + y^2 - x & = & 0 & \text{a circle in the xy-plane} \\ y^2 + z^2 - z & = & 0 & \text{a circle in the yz-plane} \quad (12.34) \\ \text{for} \quad x, z \geq 0 \quad z + x & = & 1 & \text{a line segment in the zx-plane} \end{array}$$

The set of points defined by these equations represents the locus of all *real* eigendirections. This locus is illustrated in Figure 12.10. Any two distinct eigendirections (or eigenvectors) such as those illustrated in this figure also identify a particular *solution plane* and therefore also the coefficient vector which is, of course, normal to this solution plane.

We now briefly introduce non-autonomous ODE's in this R^3 phase space context for completeness, although it too spans material which is beyond the intended scope of this book.

Eigendirections

The intersection of the rotating *Kyper*-planes generates the conical locus shown in Figure 12.10. The solution plane $\mathbf{a}^T\mathbf{x} = 0$ must intersect this cone to have real eigendirections. If the two vectors shown on the conical surface are eigendirections then the solution plane intersects the cone along those two directions.

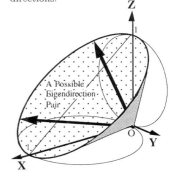

Figure 12.10
Eigenvector Locus

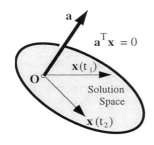

$$\mathbf{a}^T \mathbf{x} = 0$$
Solution Space

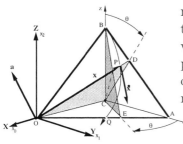

Rotating Planes

$\Longrightarrow$

The solution plane must remain normal to the coeficient vector a(t) such that

$$\mathbf{a}(t)^T \mathbf{x}(t) = 0 \ \forall \ t > t_2$$

12.10   Non Autonomous Systems

As an interesting question at this point we ask: *How is the solution trajectory generated in R^3 at each $t \ \epsilon \ T$ for a system which:*

- has a time-varying coefficient vector, or

- is non-homogeneous, that is, forced as in $\mathbf{a}^T \mathbf{x} = f(t)$ or

- is non-linear as in, for example, VanderPol's equation.

Again, comprehensive answers to these questions are beyond the intended scope of this book as mentioned earlier. However, we can make some helpful remarks and observations. First, we emphasize that as the coefficient vector *waggles about* in R^3, in conformance with $\mathbf{a}(x,t)$, the orientation of the solution generating plane, expressed in its most comprehensive form as $\mathbf{a}(\mathbf{x},t)^T \mathbf{x}(t) = 0$, obviously must remain orthogonal to $\mathbf{a}(\mathbf{x},t)$ for all $t \ \epsilon \ T$. Of course, this means that the eigendirections then are also time-varying.

This raises another interesting question. What constraints must govern the dynamical and/or kinematic motion of a point in this R^3 solution space, especially under these non-linear and/or time-varying conditions. That such constraints do exist is clear from the following simple example: Suppose a unit step-function occurs in the second element of an an otherwise constant coefficient vector $\mathbf{a} = col[a_0, a_1, a_2]$. That is, $a_1(t) = u(t - t_1) = $ a unit-step which occurs at $t = t_1$. This example does entail a constraint on how the solution point may move in this moving solution plane. Remember, the solution plane must remain orthogonal to the coefficient vector, $\mathbf{a}(t)$. However, the solution point in the plane at $t = t_1$ can jump only along the x_2-axis. Otherwise δ-type functions would be generated along the higher derivative axes. In this context we challenge the reader with Vander Pols' equation $\overset{\circ\circ}{x} + \mu(x^2 - 1) \ \overset{\circ}{x} + x = 0$.

For further insight we consider a constant-coefficient LDE which is forced; we write $\mathbf{a}^T \mathbf{x} = f(t)$ We let $f(t) = u(t - t_2)$ which is a unit-step at $t = t_2$. After a moments reflection we conclude that equation again represents a plane which contains the origin for all $t < t_2$. At $t = t_2$ this solution plane jumps to a new position, all the while remaining orthogonal to $\mathbf{a}$. Its new position is dictated by the requirement that for all $\mathbf{x}$ and $t > t_2$ we have $\mathbf{a}^T \mathbf{x} = 1$.

In the next section we end our discussion of these matters by devising rotation sequences to relate significant ODE parameters.

12.11 Phase Space Rotation Sequences

We now consider phase space rotation sequences in the solution space for $\mathbf{a}^T\mathbf{x} = 0$. Theses sequences provide some useful angle relationships, thereby providing further insight into the kinematics of the various parameters which generate the solution trajectory.

12.11.1 Rotation Sequence for State Vector

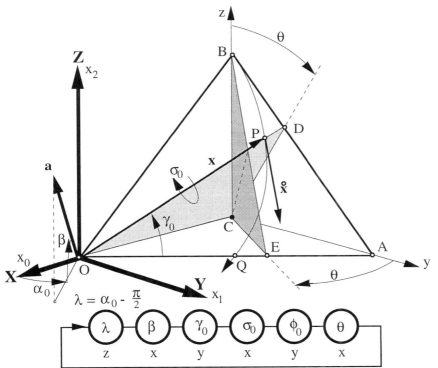

Figure 12.11: Rotation Sequence locates State Vector

A Rotation Sequence which governs the State Vector

This rotation sequence relates the various significant *directions* in the R^3 solution space. The directions specified by the rotations in this particular angle/axes sequence relate various kinematic motions all, say, dependent upon the rotating *Kyper*planes as the input.

1. Rotation α_0 about $\mathbf{Z}$ makes the new x-axis contain OA.

2. Rotation β about x (OA) makes the vector $\mathbf{a}$ the new y-axis.

3. Rotation γ_0 about $\mathbf{a}$ (y-axis) takes x into solution vector $\mathbf{x}$.

4. Rotation σ_0 about $\mathbf{x}$ takes $\mathbf{a}$ into $\mathbf{b_0}$.

5. Rotation ϕ_0 about $\mathbf{b_0}$ takes the x-axis into the reference $\mathbf{X}$.

6. Rotation θ about $\mathbf{X}$ takes the $\mathbf{X}$yz-frame into $\mathbf{XYZ}$. This *closes* the rotation sequence as shown in Figure 12.11.

We now demonstrate the utility of this closed sequence by asking the question

How is the rotation θ related to the rotation γ_0 ?

A careful inspection of the closed-loop in Figure 12.11 suggests that the rotation angle ϕ_0 is not of interest. So at that particular rotation the loop is *opened* and we input the vector $\mathbf{v} = col[0, 1, 0]$ or $y_0 = 1$ at the rotation θ thereby eliminating ϕ_0 from the sequence as shown in Figure 12.12.

**Angle
Definitions**

The angle θ determines the orientation of the *kyper*planes and therefore it determines the directions of the state and velocity vectors in the solution plane.

The angle γ_0 defines the orientation of the *state* vector in the solution plane, measured in solution plane coordinates.

The angle, $\lambda = \alpha_0 - \frac{\pi}{2}$, properly makes the coefficient vector to be the y-axis at that point in the sequence. This is convenient since in the sequence the *kyper*plane normal, $\mathbf{b}_0$, is appropriately also the y-axis.

See Figure 12.11.

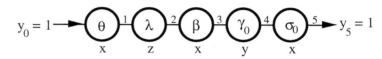

Figure 12.12: Simplified State Vector Rotation Sequence

Operating on the input vector $col[0, 1, 0]$ with the successive rotations indicated in Figure 12.12 we get

$$
\begin{aligned}
y_0 &= 1 \\
y_1 &= \cos\theta \\
z_1 &= -\sin\theta \\
x_2 &= \cos\theta \sin\lambda = -\cos\theta \cos\alpha_0 \\
y_2 &= \cos\theta \cos\lambda = \cos\theta \sin\alpha_0 \\
y_3 &= \cos\theta \sin\alpha_0 \cos\beta - \sin\theta \sin\beta \\
z_3 &= -\sin\theta \cos\beta - \cos\theta \sin\alpha_0 \sin\beta \\
z_4 &= z_3 \cos\gamma_0 + x_2 \sin\gamma_0 \\
x_4 &= x_2 \cos\gamma_0 - z_3 \sin\gamma_0 = 0 \Rightarrow \tan\gamma_0 = \frac{x_2}{z_3} \\
y_5 &= y_3 \cos\sigma_0 + z_4 \sin\sigma_0 = 1 \Rightarrow y_3 = \cos\sigma_0 \\
z_5 &= z_4 \cos\sigma_0 - y_3 \sin\sigma_0 = 0 \Rightarrow z_4 = \sin\sigma_0
\end{aligned}
$$

**See
Note
in the
margin**

$$
\begin{aligned}
\tan\gamma_0 &= \frac{\cos\theta \cos\alpha_0}{\sin\theta \cos\beta + \cos\theta \sin\alpha_0 \sin\beta} \\
\text{or} \qquad \tan\theta &= \frac{\cos\alpha_0 \cos\gamma_0 - \sin\gamma_0 \sin\beta \sin\alpha_0}{\cos\beta \sin\gamma_0}
\end{aligned}
$$

A

These last two equations relate the angle and/or angular motion of the state vector in the solution plane to the angle and/or angular motion of the rotating *kyper*plane, $\mathbf{b}_0(\theta)^T \mathbf{x} = 0$.

12.11.2   Rotation Sequence for Velocity Vector

This book is about rotations and rotation sequences. Here, as in the last section, we show how they may be used to analyze certain kinematic and/or dynamic models. We consider a rotation sequence which relates the parameters governing the orientation of the *velocity vector* in the solution space.

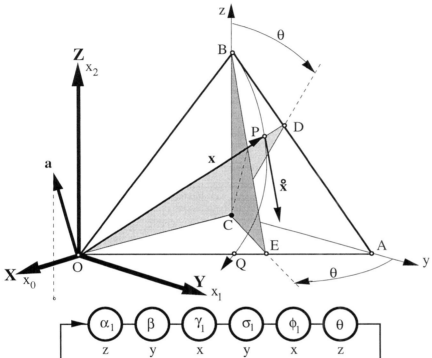

Figure 12.13:   Rotation Sequence locates Velocity Vector

A Rotation Sequence which governs the Velocity Vector

This rotation sequences relates the various significant *directions* in the R^3 solution space. The directions specified by the angles in this particular sequence relate the various kinematic motions which are all dependent upon, say, the rotating *Kyper*planes as the input.

1. Rotation α_1 about the **Z**-axis takes the **Y**-axis into OA.

2. Rotation β about OA (y-axis) makes **a** the new x-axis.

3. Rotation γ_1 about **a** (x-axis) makes $\overset{\circ}{\mathbf{x}}$ the new y-axis.

4. Rotation σ_1 about $\overset{\circ}{\mathbf{x}}$ (y-axis) takes **a** (x-axis) into $\mathbf{b_1}$.

5. Rotation ϕ_1 about $\mathbf{b_1}$ (x-axis) takes the z-axis into **Z**.

6. Finally, rotation θ about the (**Z**-axis) takes the xy-axes into the **XY**-axes, closing the sequence shown in Figure 12.13.

Note in the rotation sequence of Figure 12.13 that the last rotation and the first rotation occur about the same axis, namely the z-axis. This means that for computational purposes the sequence can be simplified if we combine these two rotations. This we do. Now in order to compute γ_1 in terms of the other parameters, we open the sequence at the rotation ϕ_1. At that point we let the input be $col[1,0,0]$ which then eliminates ϕ_1 from the sequence thereby simplifying whatever computations remain. We now solve for the angles γ_1 and σ_1 as in the last section.

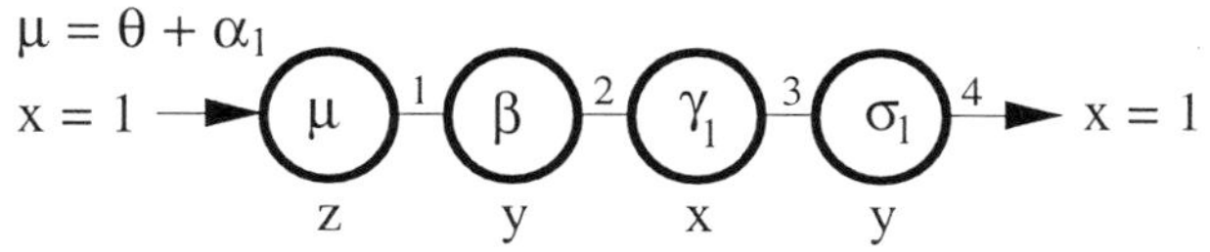

$$\mu = \theta + \alpha_1$$

Figure 12.14: Simplified Velocity Vector Rotation Sequence

$$
\begin{aligned}
x_0 &= 1 \\
x_1 &= \cos\mu \\
y_1 &= -\sin\mu \\
z_2 &= \cos\mu\sin\beta \\
x_2 &= \cos\mu\cos\beta \\
y_3 &= -\sin\mu\cos\gamma_1 + \cos\mu\sin\beta\sin\gamma_1 = 0 \\
z_3 &= \cos\mu\sin\beta\cos\gamma_1 + \sin\mu\sin\gamma_1 \\
z_4 &= z_3\cos\sigma_1 + x_2\sin\sigma_1 = 0 \;\Rightarrow\; z_3 = -\sin\sigma_1 \\
x_4 &= x_2\cos\sigma_1 - z_3\sin\sigma_1 = 1 \;\Rightarrow\; x_2 = \cos\sigma_1 \\
\tan\gamma_1 &= \frac{\sin(\theta+\alpha_1)}{\sin\beta\cos(\theta+\alpha_1)} \\
&= \frac{\sin\theta\cos\alpha_1 + \cos\theta\sin\alpha_1}{\sin\beta(\cos\theta\cos\alpha_1 - \sin\theta\sin\alpha_1)} \\
\tan\theta &= \frac{\sin\gamma_1\sin\beta\cos\alpha_1 - \cos\gamma_1\sin\alpha_1}{\sin\gamma_1\sin\beta\sin\alpha_1 + \cos\gamma_1\cos\alpha_1} \qquad\qquad \mathbf{B} \\
\cos\sigma_1 &= \cos(\theta+\alpha_1)\cos\beta
\end{aligned}
$$

The two *kyper*planes, of course, are rotating synchronously. Therefore, it must be that Equation $\mathbf{A}$ = Equation $\mathbf{B}$, which gives

$$\frac{\cos\alpha_0\cos\gamma_0 - \sin\gamma_0\sin\beta\sin\alpha_0}{\cos\beta\sin\gamma_0} = \frac{\sin\gamma_1\sin\beta\cos\alpha_1 - \cos\gamma_1\sin\alpha_1}{\sin\gamma_1\sin\beta\sin\alpha_1 + \cos\gamma_1\cos\alpha_1}$$

But in writing this equation in this manner we also invoke the *eigendirection-condition*, namely, that of co-directional state and velocity vectors. This condition means we may let $\gamma_0 = \gamma_1 = \gamma$.

Angle Subscripts

The angle designations have, in general, the same meanings in both of these rotation sequences. The subscript *zero (0)* on some angles mean it relates to the state vector whereas the subscript *one (1)* is for those angles related to the velocity directions.

Another Matter

The composite angle used to properly point the coefficient vector **a** in the state vector sequence is α whereas for the velocity vector case it is $\alpha - \frac{\Pi}{2}$. This necessary in order to make the coefficient vectors represent the y-axis and the x-axis, respectively, in their sequences. All rotations are understood to be taken in the positive (right-handed) sense.

After some tedious algebra, we obtain the quadratic equation

$$au^2 + bu + c = 0$$

where

$$u = \tan\gamma$$
$$a = \sin\beta\cos\beta\cos\alpha + \sin^2\beta\sin^2\alpha$$
$$b = -\cos\beta\sin\alpha$$
$$c = -\cos^2\alpha$$

Then

$$u = \frac{\cos\beta\sin\alpha \pm \sqrt{\cos^2\beta\sin^2\alpha + 4(\sin\beta\cos\beta\cos\alpha + \sin^2\beta\sin^2\alpha)\cos^2\alpha}}{2(\sin\beta\cos\beta\cos\alpha + \sin^2\beta\sin^2\alpha)}$$

These two solutions for $\tan\gamma$ give us expressions for the angle γ in terms of the angles α and β, which we presume are given. This, incidentally, is a clear example of how one may use the rotation sequence approach to determine fundamental relationships between parameters in an application.

If this solution, $u = \tan\gamma$, is to be such that the state and velocity directions coincide, that is, the angle γ locates an eigendirection in the solution plane, then clearly it must be real. That is to say, the discriminant $b^2 - 4ac$ must be non-negative. If we use the above values for a, b, and c, the familiar inequality $b^2 - 4ac \geq 0$ reduces to the quadratic

$$y(w) = (4\cos^2\alpha\sin^2\alpha)w^2 + (4\cos^3\alpha)w + \sin^2\alpha \geq 0$$

or

$$y(w) = (4\sin^2\alpha)w^2 + (4\cos\alpha)w + \tan^2\alpha \geq 0 \quad (12.35)$$

$$\text{where} \quad w = \tan\beta$$

We solve for the equality part of the Inequality 12.35 and we get

$$\tan\beta = \frac{-\cos^2\alpha \pm \sqrt{\cos 2\alpha}}{\sin\alpha\sin 2\alpha} \quad (12.36)$$

The discriminant for this solution of the equality part of Equation 12.35 says that for real roots we must have

$$\cos 2\alpha \geq 0 \quad \text{or} \quad |\alpha| \leq \frac{\pi}{4} \quad (12.37)$$

For those values of α which satisfy Equation 12.37 the corresponding values for β are determined by Equation 12.36.

We now analyze the inequality part of 12.35 hueristically by considering its graph. The equation $y = f(\alpha, w)$ is, for any α, clearly that of a parabola which opens upward in the wy-plane. As

**Equation 12.35
Discriminant**

If the discriminant $\cos 2\alpha = 0$ then the angle, of course, must be

$$|\alpha| = \frac{\pi}{4}$$

which corresponds, in some sense, to a repeated root condition.

in elementary algebra the roots, $w_1 \leq w_2$, of the equation are easily illustrated on its graph by where the w-axis crossings (if any) occur. With the graph in mind there are, clearly, three possibilities: The parabola will haveat most 2 crossings, precisely 1 point of tangency (the two roots coelesce for some α to form one), or no crossing. The *discriminant*, $D = \cos 2\alpha$ also determines these cases by $D > 0$, $D = 0$, and $D < 0$, respectively. We summarize these three cases:

1. $D = \cos 2\alpha > 0$. This means $|\alpha| < \frac{\Pi}{4}$ and that real solutions occur only for $\tan \beta < \tan \beta_1$ and $\tan \beta > \tan \beta_2$ where $\beta_1 < \beta_2$.

2. $D = \cos 2\alpha = 0$ so that $|\alpha| = \frac{\Pi}{4}$. This means the graph is tangent to the w-axis.

3. $D = \cos 2\alpha < 0$ means no real solutions can occur and hence no real eigendirections.

The foregoing supplements and confirms the geometric boundaries illustrated in Figure 12.7.

In this chapter we have explored a geometric approach which gives insight into the structure and nature of the solution space of 2nd-order ODE's. We have, of course, only scratched the surface; much remains to be done. But we terminate our discussion at this point and move on to new application matters in the next chapter.

Exercises for Chapter 12

Some of these exercises are intended for those with some familiarity with either Maple© or Mathematica©

1. Choose a coefficient vector having real eigendirections (a node or a saddle). Determine the angle θ for the rotating *kyper*planes where the rotation stops.

2. Plot the direction of the intersection of the two *kyper*planes over one rotation $0 < \theta \leq 2\pi$.

3. Select any coefficient vector and choose some initial condition vector then plot the resulting solution trajectory.

4. Verify the result in Equation 12.35.

5. Confirm Equation 12.36

6. Plot a portion of the limit-cycle trajectory in R^3 for Vander-Pols equation written in the form $\mathbf{a}^T(x)\mathbf{x} = 0$.

Chapter 13

A Quaternion Process

13.1 Introduction

In this chapter we describe a system which uses electromagnetic dipole fields generated at some point fixed in one body and detected at some point fixed in a remote body. Our purpose is to describe a process which measures the relative position and/or orientation between the two independent bodies. More precisely, we describe a six degrees-of-freedom transducer which completely defines how two independent relatively remote bodies are *situated* with respect to each other.

Three essential components comprise this system: (1) a *Source* (sometimes called a Radiator) which is fixed in one body and which generates electromagnetic fields; (2) a similar compatible *Sensor* which is fixed in some relatively remote body and which detects the fields generated by the Source; and (3) a *Processor* whose function is to correlate the Source and Sensor signals. Given these signals, this system determines the relative *position* and *orientation* between the two independent bodies. We now describe the spatial situation geometrically, define appropriate mathematical models for the System components, and finally derive the required process algorithms.

Typically, the System components are distributed as illustrated in Figure 13.1. All six degrees-of-freedom, i.e., the three *translation* parameters (α, β, R) and the three *rotation* parameters (ψ, θ, ϕ) are determined in the process. These six degrees-of-freedom are illustrated in Figure 13.2. They define how two independent bodies are *situated* relative to each other.

A three-axis electromagnetic dipole Source is fixed in and *repre-*

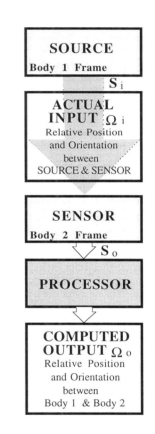

Figure 13.1
System Block Diagram

303

The transducer described here may well be the first integrated six degree-of-freedom transducer in the history of science. This application of **rotation sequences** was conceived, patented, and developed by the author. The original disclosure, dated 17Jun71, was witnessed and signed as being understood by Bill Polhemus, Roland Pittman, and Al DeRuyck. A copy of this original disclosure was cut & pasted and submitted by Polhemus Navigation Sciences, Inc. (PNSI), as a proposal to the USAF. This resulted in a sequence of preliminary breadboard, feasibility, and development contracts.

In 1975, PNSI filed for Chapter 11, Bankruptcy, and shortly thereafter was acquired by Advanced Systems Technology in Roselle, NJ, a division of The Austin Company in Cleveland Ohio. In 1983, the technology was purchased by McDonnell-Douglas who continue to this day to build and develop this technology for a wide variety of applications. It was also subsequently licensed to Kaiser Electronics.

This, my original transducer concept, is today variously known as the SPASYN, Polhemus Tracker, The Bird, or even A Flock of Birds, which is an n-body transducer of the same ilk (This n-body version I originally called TWOWAY). The Bird, a transducer which employs the same technology but with a clever pulsed DC excitation, is being developed and manufactured by Ascension Technology Inc., a company which was started by Ernie Blood and Jack Sculley upon their leaving PNSI.

The technology is also being further developed and applied and manufactured by Ferranti Ltd in Edinborough, Scotland and by Elbit in Israel.

Although this technology was originally funded and developed by and for the military, applications now also range over a wide variety of interesting commercial, educational, and other innovative concepts and interests. Too many to mention here have contributed to this technology but Fred Raab contributed much to compensate environments hostile to EM fields.

sents the frame of BODY 1. This Source generates a time-multiplexed sequence of variously polarized electromagnetic fields which are detected by a similar three-axis dipole Sensor. This Sensor in like manner is fixed in and *represents* the remote BODY 2 frame. The fields (the S_i in Figure 13.1) which are generated by the Source and

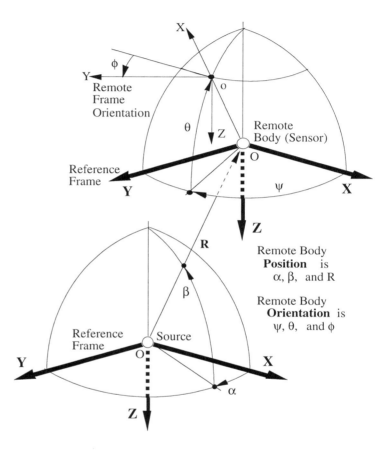

Figure 13.2: Six Degrees-of-Freedom

the corresponding set of signals detected by the Sensor (the S_o in Figure 13.1) are correlated by the Processor. The result is a precise measure of the relative position and orientation *between* the two independent bodies.

Using both matrix and quaternion methods, we will derive algorithms which correlate the sensed data and the *known* source excitation sequence. First, however, we describe the electromagnetic field structure and introduce an important geometric property of the coupling between the Source and the Sensor.

13.2　Dipole Field Structure

A simple coil excited by an electric current generates a *space-filling* electromagnetic field. In free-space this EM field has a polar axis of symmetry which is normal to and centered in the plane of the coil. This dipole magnetic field has the familiar torroidal structure *filled with* concurrent roughly circular flux lines all issuing from

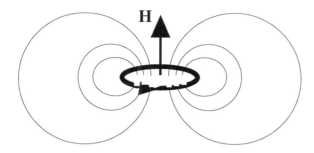

Figure 13.3: Simple Coil EM Field Structure

and returning to the plane circular area enclosed by the coil, as illustrated in Figure 13.3. The intensity and polarization of the field is conveniently represented by a vector **H** whose magnitude is proportional to the excitation current and whose direction is normal to the plane of the coil (in the right-hand sense).

13.3　Electromagnetic Field Coupling

In this section we consider an electromagnetic field phenomenon known as *coupling*. Coupling exists between the signal-emitting device called a *Source* and the signal-receiving device known as a *Sensor*. The peculiar properties of this coupling (in the near-field) are important to the measurement strategy which we are about to develop. Consider for a moment that you are holding in your hand a small AM/FM radio. Let the remote radio station with its antenna (wherever it is) be the Source, and the hand-held radio with its antenna, the Sensor. We all know that the orientation of the radio antenna relative to the electromagnetic field generated by the station antenna directly affects the signal sensed by the radio. That is, if we orient the antenna in a certain way, the signal sensed by the radio is maximal. If, however, we orient the antenna in a different way, perhaps orthogonal to the first orientation, the sensed signal is minimal. No doubt every reader has experienced this phenomenon. This homely example demonstrates that the EM *coupling* between

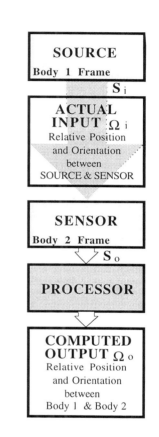

Figure 13.1
System Block Diagram

the Source and the Sensor depends directly upon the relative orientation of the two antennas.

Here we shall represent the EM field **H** generated by the Source and the corresponding signal **S** detected by the Sensor as vectors. Although electromagnetic field theory is beyond the scope of our present discussion, we shall have to make use of some important results from that theory. One of these results is that if we represent these two vectors, **H** and **S**, in the same frame, the components of the sensed signal vector turn out to be certain constant multiples of the components of the source field vector. These constants are called *coupling coefficients*. Our first task will be to determine the nature of these coefficients.

Let both the Source and the Sensor be simple "flat" circular coils whose physical dimensions are very small compared to the distance which separates them. If one coil is the Source and the other the Sensor, then at any fixed separation distance such that *near-field conditions* prevail (terms we shall define in a moment), the magnitude of the signal detected by the Sensor when the two coils

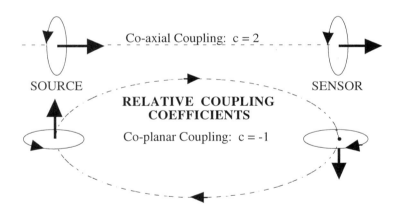

Figure 13.4: Near-Field EM Coupling

are coaxial is twice that detected by the Sensor when the two coils are coplanar. Moreover, if the co-axial coupling is positive, then the co-planar coupling is negative, as the illustration in Figure 13.4 suggests. This fundamental property of electromagnetic field coupling in the near-field is crucial to the measurement strategy and process which we now develop.

13.3.1 Unit Z-axis Source Excitation

We begin by considering a relatively simple case which will be sufficient to suggest what these coupling coefficients are. Consider a unit excitation applied to a Source — a single short dipole element, as illustrated in Figure 13.5. The field generated by the

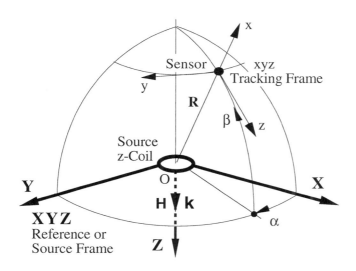

Figure 13.5: Z-axis Dipole Source

Z-Axis Dipole

In Figure 13.5 the Source z-Coil is the Source element, with the magnetic moment vector **H** directed along the reference Z-axis.

unit excitation vector **H**, which in this case is simply the basis vector $\mathbf{k} = col[0, \ 0, \ 1]$, is detected by a remote Sensor whose *location* (R,α,β), *as yet unknown*, is partially defined by the indicated tracking frame. The tracking transformation is defined by the matrix of Equation 13.1 and denoted P, where

Matrix Analysis First

Here we begin by using the more familiar *rotation matrices*; quaternion methods come later. The reader may wish to review the rotation matrix tracking transformation as discussed in Chapter 3. See Equation 3.5.

$$
P = \begin{bmatrix} \cos\beta & 0 & -\sin\beta \\ 0 & 1 & 0 \\ \sin\beta & 0 & \cos\beta \end{bmatrix} \begin{bmatrix} \cos\alpha & \sin\alpha & 0 \\ -\sin\alpha & \cos\alpha & 0 \\ 0 & 0 & 1 \end{bmatrix}
$$
$$
= \begin{bmatrix} \cos\alpha\cos\beta & \sin\alpha\cos\beta & -\sin\beta \\ -\sin\alpha & \cos\alpha & 0 \\ \cos\alpha\sin\beta & \sin\alpha\sin\beta & \cos\beta \end{bmatrix} \tag{13.1}
$$

The matrix operator P takes every signal which is generated by the Source (expressed as a vector in the Source frame) into its equivalent expression in the Tracking frame.

We emphasize at this point that simply because the Source signals are here expressed in the tracking frame (which simply *locates* the Sensor relative to the Source) does not necessarily mean that the *orientation* of the Sensor coincides with that of the tracking frame. It is not only convenient but necessary that we express the coupling relationships between source and sensor in this intermediate frame. The actual source field components are expressed in the tracking frame, and the actual sensor signal components must also be expressed in the tracking frame for correlation purposes.

The electromagnetic field equations for this "short" Z-axis dipole (equivalent dipole length < 0.1 wavelength) illustrated in Figure 13.5 are derived in many standard texts on electromagnetic waves. From the results derived in these references we may state that (for the indicated z-axis excitation) the remotely *sensed* signal components defined in the tracking frame are

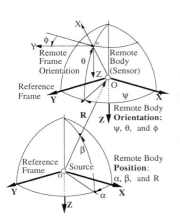

$$
\begin{aligned}
e_x &= -2kp^{-3}(\cos\nu - p\sin\nu)\sin\beta \\
e_y &= 0 \\
e_z &= -kp^{-3}(\cos\nu - p\sin\nu + p^2\cos\nu)\cos\beta
\end{aligned}
\tag{13.2}
$$

where, in these equations along with the position parameters, R, α and β, we have

$$
p = \frac{2\pi R}{\lambda}
$$

where

$$
\begin{aligned}
R &= \text{distance from Source to Sensor} \\
\lambda &= \text{wavelength of Source excitation} \\
\beta &= \text{elevation angle of the radial direction vector,} \\
  &\quad \text{Source-to-Sensor, from the plane normal} \\
  &\quad \text{to the excitation vector (see Figure 13.5)} \\
\nu &= \omega t - p \\
\omega &= \text{radian frequency of Source excitation} \\
k &= \text{attenuation factor}
\end{aligned}
$$

Figure 13.2
Six Degrees-of-Freedom

When we assume that we are operating in the *near-field*, we mean that the distance R is very small relative to the wavelength λ, of the Source excitation. As a consequence the constant p is very small, that is, $p << 1$. And since $|\sin\nu| \le 1$ and $|\cos\nu| \le 1$, Equations 13.2 may be simplified by neglecting the terms $p\sin\nu$ and $p^2\cos\nu$. If we define a parameter $F = p^{-3}\cos\nu$ then Equations 13.2 may be written

$$
\begin{aligned}
e_x &= -2kF\sin\beta \\
e_y &= 0 \\
e_z &= -kF\cos\beta
\end{aligned}
\tag{13.3}
$$

We emphasize that although the components, e_x, e_y, and e_z, are Sensor signals, they are here defined in the tracking frame at that point in the Source space occupied by the Sensor.

It is not difficult to verify that the field components of the unit Z-axis source excitation, $\mathbf{k} = col[0, 0, 1]$, as expressed in the tracking frame, are

$$P\mathbf{k} = col[-\sin\beta, \quad 0, \quad \cos\beta]$$

But now note, these Source components which are defined in the tracking frame couple into the Sensor to produce the respective components indicated in Equations 13.3 (also, defined in the tracking frame). That is,

$$\begin{matrix} \text{Source} \\ \text{Components} \end{matrix} = \begin{bmatrix} -\sin\beta \\ 0 \\ \cos\beta \end{bmatrix} \Rightarrow \begin{bmatrix} -2F\sin\beta \\ 0 \\ -F\cos\beta \end{bmatrix} = \begin{matrix} \text{Sensor} \\ \text{Signals} \end{matrix}$$

From this we conclude that the *relative coupling coefficients* which take the $\mathbf{k}$-vector excitation components into the corresponding sensed components, as defined in the tracking frame, are 2, -1 and -1, respectively. As we would expect, the two *co-planar or transverse* coupling coefficients are the same.

Up to this point, the unit excitation of the source was taken to be the vector $\mathbf{k}$. The analysis of this choice is commonly made in the literature — and this choice makes sense in that vertical source antennas with respect to the *ground plane* are by far the most common orientation in applications. In order to determine the **sensed** field intensity at an arbitrary remote point in space relative to this vertical source antenna, a polar translation, (R, α, β), was employed to define the remote point. The derivation is simplified because the 1st rotation, α, in the polar translation is about the vertical axis which contains the source antenna. This produces a zero as the y-component of the $\mathbf{k}$-axis source excitation expressed in the tracking frame.

13.3.2 Unit X-axis Source Excitation

In the preceding section we began by taking the unit source excitation to be the basis vector $\mathbf{k}$. We did this simply because that is the usual choice in electromagnetic field theory literature. We could just as well have chosen another basis vector, although the choice of $\mathbf{k}$ was particularly convenient. If we take the unit source excitation to be the $\mathbf{i}$-vector (that is, along the x-axis), then the sensed field

Note!

Here we defined $F = p^{-3}\cos\nu$. This suggests that F is a constant. But the reader should note that since ν is time-dependent, so is the parameter F.

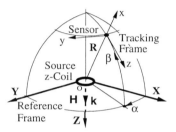

Figure 13.5
Z-axis Dipole Source

Note in this tracking transformation

$$P = YZ$$

that is, first a principal axis rotation about the z-axis, followed by a principal axis rotation about the y-axis. It follows then that $p_{23} = 0$ as indicated in the matrix Equation 13.1.

intensity or signal components defined in the tracking frame would be

$$
\begin{aligned}
e_x &= 2p^{-3}(\cos\nu - p\sin\nu)\cos\alpha\cos\beta \\
e_y &= p^{-3}(\cos\nu - p\sin\nu + p^2\cos\nu)\sin\alpha \\
e_z &= -p^{-3}(\cos\nu - p\sin\nu + p^2\cos\nu)\cos\alpha\sin\beta
\end{aligned}
\tag{13.4}
$$

Equations 13.4

These equations are familiar to those conversant with electromagnetic field theory.

instead of those shown in Equations 13.2. As a result, the simplified equations become

$$
\begin{aligned}
e_x &= 2F\cos\alpha\cos\beta \\
e_y &= F\sin\alpha \\
e_z &= -F\cos\alpha\sin\beta
\end{aligned}
\tag{13.5}
$$

instead of those listed in Equations 13.3. A third choice is, of course,

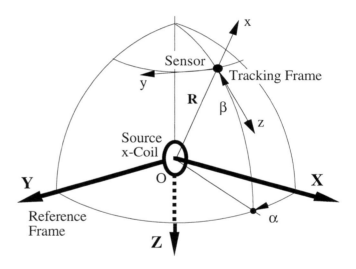

Figure 13.6: X-axis Dipole Source

the basis vector $\mathbf{j}$, and expressions similar to the above could be obtained. In fact, we can summarize all three choices in the following way. We may think of the choices of $\mathbf{i}$, $\mathbf{j}$, and $\mathbf{k}$ as a *time-multiplexed sequence* of unit Source excitations. If we let the columns of the matrix E_i represent the sequence of Source excitations, expressed in the Source frame, then the resulting sensed signals, expressed in the tracking frame, may be represented by the columns of the matrix, E_o, where

$$
E_o = CPE_i = CP\begin{bmatrix} 1 & 0 & 0 \\ 0 & 1 & 0 \\ 0 & 0 & 1 \end{bmatrix}
\tag{13.6}
$$

Here P is the tracking matrix of Equation 13.1 and the *Coupling matrix, C*, is

$$C = \begin{bmatrix} 2 & 0 & 0 \\ 0 & -1 & 0 \\ 0 & 0 & -1 \end{bmatrix} \tag{13.7}$$

Notice that the entries in the matrix C are exactly the coupling coefficients obtained earlier, and that for each excitation component the elements of the matrix C apply appropriately.

13.4 Source-to-Sensor Coupling

In this section we introduce a slight generalization of what we have done so far. The generalized electromagnetic near-field Source-to-

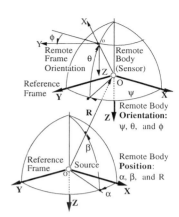

Figure 13.2

Six Degrees-of-Freedom

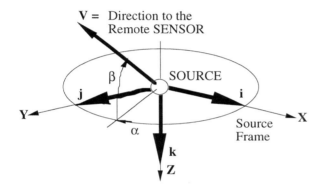

Figure 13.7: Source and Source Frame

Sensor coupling is now derived in the context of the coordinate relationships which define the *relative orientation* between the Source and Sensor frames. In Figure 13.7 the vectors **i**, **j** and **k**, represent not only the orthonormal frame of the body which contains the Source, but also the *sequence* of the three (*time-multiplexed*) orthonormal excitations of the 3-axis Source. This sequence of *unit Source excitations* is conveniently represented by the column vectors of the matrix E_i, which in this case is the identity matrix as in Equation 13.6. However, since the location of the remote Sensor is in the direction of the vector V (defined by the x-axis of the tracking frame) we resolve each of these Source excitations into an axial and two normal components defined in the remote tracking frame. Clearly, these excitation components in the tracking frame are functions of the tracking angles, α and β. The relative coupling coefficients for these resolved components (one axial and two normal)

Coupling Matrix

The columns of the input excitation matrix, E_i, can in principle be any set of three independent time-multiplexed excitation vectors.

in the tracking frame, are 2, -1, and -1, respectively. We emphasize that these values for the coupling coefficients are appropriate when both the Source excitations and the signals sensed by the Sensor are viewed in the same coordinate frame, which in this case is the tracking frame.

We emphasize further that the tracking frame is simply an auxiliary frame within which the Source and Sensor signals are easily related. What we need are the signals detected by the Sensor, *in the Sensor frame*.

First, we recall that the instantaneous position of a remote body is defined by the polar position triple (α, β, R), with respect to the Source frame. Here the parameter R represents the Source-to-Sensor distance. This distance, or Range, is measured using an "inverse-cube-attenuation-with-range" scheme which is discussed in the next section. However, the actual Source-to-Sensor coupling is also dependent upon the *relative orientation* of these two bodies.

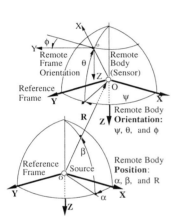

Figure 13.2
Six Degrees-of-Freedom

In order to define the *orientation* of the remote Sensor with respect to the Reference or Source frame, we adopt the familiar Aerospace sequence of three Euler angle rotations, namely, Heading (ψ), Elevation (θ), and Bank Angle (ϕ), about the z, y, and x axes, respectively, as illustrated in Figure 13.2. As we know from the discussion in Section 4.3 and 4.4, this Aerospace sequence may be represented by the rotation matrix

$$A = \begin{bmatrix} a_{11} & a_{12} & a_{13} \\ a_{21} & a_{22} & a_{23} \\ a_{31} & a_{32} & a_{33} \end{bmatrix} = REH \qquad (13.8)$$

$$\text{and} \quad R = \begin{bmatrix} 1 & 0 & 0 \\ 0 & \cos\phi & \sin\phi \\ 0 & -\sin\phi & \cos\phi \end{bmatrix}$$

$$E = \begin{bmatrix} \cos\theta & 0 & -\sin\theta \\ 0 & 1 & 0 \\ \sin\theta & 0 & \cos\theta \end{bmatrix}$$

$$\text{and} \quad H = \begin{bmatrix} \cos\psi & \sin\psi & 0 \\ -\sin\psi & \cos\psi & 0 \\ 0 & 0 & 1 \end{bmatrix}$$

The indicated matrix product for A defines the elements:

$$a_{11} = \cos\psi \cos\theta$$
$$a_{12} = \sin\psi \cos\theta$$

$$a_{13} = -\sin\theta$$
$$a_{21} = \cos\psi\sin\theta\sin\phi - \sin\psi\cos\phi$$
$$a_{22} = \sin\psi\sin\theta\sin\phi + \cos\psi\cos\phi$$
$$a_{23} = \cos\theta\sin\phi \qquad (13.9)$$
$$a_{31} = \cos\psi\sin\theta\cos\phi + \sin\psi\sin\phi$$
$$a_{32} = \sin\psi\sin\theta\cos\phi - \cos\psi\sin\phi$$
$$a_{33} = \cos\theta\cos\phi$$

In summary, the two rotation matrices, P and A, provide the means for relating the Source frame to the Tracking frame and the Source frame to the Sensor frame, respectively.

Our ultimate objective, of course, is to determine the relative position and orientation of the remote Sensor. To do this we can work only with the signals used to excite the Source and those sensed by the Sensor. At this point, we have the sensed signals analytically expressed in the tracking frame by the matrix of Equation 13.6. To get these signals from the tracking frame into the Sensor frame, we merely take them back first to the Source frame, using the inverse of the matrix, P, namely, P^t. We then take them from the Source frame into the Sensor frame using the orientation matrix, A. It then follows that if the time-multiplexed excitations on the Source are given by the columns of the matrix E_i and the corresponding sensed signals are expressed in the Sensor frame by the columns of the matrix E_o, we can write

$$E_o = SE_i \qquad (13.10)$$

where
$$S = kAP^tCP = kQCP \qquad (13.11)$$
and $\quad Q = AP^t \quad$ takes vectors from the Tracking
frame to the Sensor frame

A = the orientation matrix, which relates
the Sensor frame wrt Source frame

P = the tracking matrix, which defines the
direction to the Sensor wrt Source frame

C = the coupling matrix
$\quad = dg[2, -1, -1]$, for near-field

k = the EM field coupling/attenuation factor

Equation 13.10 properly relates any near-field excitation of the Source (given by E_i) to the corresponding signals, E_o, sensed or

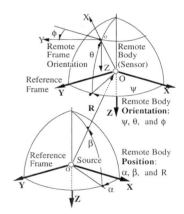

Figure 13.2
Six Degrees-of-Freedom

measured by the Sensor. Since we know E_i àpriori, and for this known input sequence we measure the corresponding output sequence E_o, we can write

$$S \;=\; E_o E_i^{-1} \tag{13.12}$$

It follows that, since the spanning set of excitations E_i is known àpriori, its inverse may be computed àpriori, and since the corresponding E_o is measured, the elements of the matrix, S, are determined.

Recall, however, we earlier developed an analytical expression for this measured signal matrix, S, namely,

$$S = kAP^tCP$$

The only known quantities in this equation are the measured signal matrix, S, and the coupling matrix, C. We now must find algorithms which will determine the attenuation factor, k, the Range, R, and the rotation matrices, P and A. We do this in a rather creative way. This then will completely determine the relative position and orientation of the remote object, which is our goal.

13.5   Source-to-Sensor Distance

The magnitude of the signals detected by the 3-axis Sensor, under near-field conditions, varies inversely with the cube of the distance R, between Source and Sensor, that is, the attenuation factor k is proportional to $1/R^3$.

Algebraic Details

$$
\begin{aligned}
U &= S^tS \\
  &= (kAP^tCP)^t(kAP^tCP) \\
  &= k^2(P^tC^tPA^tAP^tCP) \\
  &= k^2(P^tC^tCP) \\
\text{but}\quad C^t &= C \\
\text{so,}\quad U &= k^2P^tC^2P
\end{aligned}
$$

Trace

A Theorem which applies to Equation 13.13 is that the trace of a matrix, C^2, is invariant under the similarity transformation, $P^t[tr C^2]P$

To determine k, we use the note in the margin. We let

$$U = S^tS \;=\; k^2P^tC^2P \tag{13.13}$$

Recall, the trace of the matrix U, denoted $tr(U)$, is merely the sum of the terms on the main diagonal of the matrix U. So, from Equation 13.13 we get

$$
\begin{aligned}
tr(U) &= tr(S^tS) \\
      &= tr(k^2P^tC^2P) \\
      &= tr(k^2C^2) \\
      &= 6k^2
\end{aligned}
$$

Therefore, since U is computed using the measured signal matrix

S we can write

$$k = [tr(U)/6]^{(1/2)} \qquad (13.14)$$

The measurement of the Range, R, that is, the distance between the Source and the Sensor, uses this computed value for k in the following open-loop range measurement scheme. In this measurement scheme we assume that the excitation of the Source is well regulated and that the electronic attributes of the system are consistent with maintaining a precise open-loop calibration over a wide dynamic range.

Let k_o be the value for k in Equation 13.14 which corresponds to range R_o, the reference range (distance Source-to-Sensor) as determined in the initial calibration of the system. Then, since in the near-field, k varies inversely with R^3 we can express the measure for any range (within some sensible operational domain) to be

$$R = R_o(k_o/k)^{(1/3)} \qquad (13.15)$$

Thus we have determined the *distance R* between the Source and the Sensor. This range R, however, is only one of the six degrees-of-freedom to be determined by this measurement system.

13.6 Angular Degrees-of-Freedom

What remains to be determined are the two tracking angles α and β in the matrix P, and the three orientation angles ψ, θ, and ϕ, in the rotation matrix A.

13.6.1 Preliminary Analysis

We will determine the remaining five degrees of freedom by uncoupling the position and orientation matrices P and A which reside in the measured signal matrix S. We begin by dividing the signal matrix S by the value of k which we have just determined. This gives us a matrix M which is independent of the range R. That is, we write

$$M = AP^tCP \qquad (13.16)$$

Here, at this point in our analysis, it is both somewhat fortuitous and quite important for us to recognize that the coupling coefficient matrix C may be written

$$C = dg[2, -1, -1] = 3E_1 - I \qquad (13.17)$$

Inverse-Cube Relationship

This inverse-cube relationship between range and the detected signal strength means we may write

$$
\begin{aligned}
\alpha &= \text{some constant} \\
&= kR^3 \\
&= k_0 R_0^3 \\
\Rightarrow \quad R^3 &= R_0^3 \frac{k_0}{k}
\end{aligned}
$$

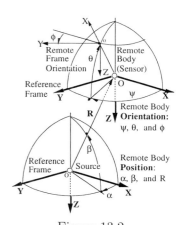

Figure 13.2
Six Degrees-of-Freedom

Here the matrix $I = dg[1, 1, 1]$ is the identity matrix, and the matrix E_1 is the zero matrix except with element $e_{11} = 1$.

Algebraic Details

$$
\begin{aligned}
M^t M &= S^t S \\
&= (AP^t CP)^t (AP^t CP) \\
&= P^t C^t P A^t A P^t CP \\
&= P^t C^t CP \\
\text{but}\quad C^t &= C \quad \text{so we write} \\
M^t M &= P^t C^2 P \\
&= P^t (3E_1 - I)^2 P \\
&= P^t (9E_1^2 - 6E_1 + I)P \\
&= P^t (3E_1 + I)P \\
&= 3 P^t E_1 P + I
\end{aligned}
$$

Remember in all of this, the matrix, M, is the "normalized" matrix of signal measurements.

Using this representation for C, the reader should verify (as shown, in part, in the margin) that

$$M^t M = 3X + I \tag{13.18}$$

$$\text{and} \qquad A^t M = 3X - I \tag{13.19}$$

$$\text{where} \qquad X = P^t E_1 P \tag{13.20}$$

13.6.2   Closed-form Tracking Angle Computation

Note that Equation 13.18 involves only the tracking matrix P, that is, X is independent of the orientation matrix A. If we solve this equation for X we get

$$X = P^t E_1 P = (M^t M - I)/3 \tag{13.21}$$

Since the numerical values in the matrix M are known (measured), so are the numerical values in the matrix X.

However, in our analysis the matrix X was defined as in Equation 13.20 to be

$$X = P^t E_1 P$$

This means that each element of the matrix X must be some function of the tracking angles, α and β.

Elements of X

The reader should analytically verify these expressions for the elements of the matrix X.

$$
\begin{aligned}
\text{In fact,} \qquad x_{11} &= \cos^2 \alpha \cos^2 \beta \\
x_{12} &= \cos \alpha \sin \alpha \cos^2 \beta \\
x_{13} &= -\cos \alpha \sin \beta \cos \beta \\
x_{21} &= \cos \alpha \sin \alpha \cos^2 \beta \\
x_{22} &= \sin^2 \alpha \cos^2 \beta \\
x_{23} &= -\sin \alpha \sin \beta \cos \beta \\
x_{31} &= -\cos \alpha \sin \beta \cos \beta \\
x_{32} &= -\sin \alpha \sin \beta \cos \beta \\
x_{33} &= \sin^2 \beta
\end{aligned}
\tag{13.22}
$$

It is now easy to see that

$$\tan \alpha = \frac{\sin \alpha}{\cos \alpha} = \frac{x_{22}}{x_{12}}$$

and
$$\sin \beta = \pm\sqrt{x_{33}}$$

Since the numerical values for x_{12}, x_{22}, and x_{33}, are known, these equations determine the tracking angles α and β. Any ambiguities in the angles over the operational domain are usually resolved by the physical constraints imposed by the application.

13.6.3 Closed-Form Orientation Angle Computation

To compute the relative *orientation* of the remote Sensor, we shall first write the inverse of the normalized signal matrix M of Equation 13.16 as

$$M^{-1} = [AP^tCP]^{-1} = P^tC^{-1}PA^t$$

To get a useful analytical expression for this inverse we first invent a new expression for the inverse of the coupling matrix C. Since C is a diagonal matrix, it is easily inverted to obtain

$$C^{-1} = dg[\frac{1}{2}, -1, -1] = \frac{1}{2}dg[1, -2, -2]$$

Then using the techniques we employed earlier we may write

$$C^{-1} = \frac{1}{2}[3E_1 - 2I]$$

Inverse Coupling Matrix

This technique was first applied in the expression for the Coupling Matrix in Equation 13.17.

where I is an identity matrix and E_1, as before, is a zero matrix except that $e_{11} = 1$. The inverse of the matrix M then becomes

$$\begin{aligned} M^{-1} &= P^tC^{-1}PA^t \\ &= \frac{1}{2}P^t[3E_1 - 2I]PA^t \\ &= \frac{1}{2}[3P^tE_1P - 2I]A^t \end{aligned}$$

But now, since $P^tE_1P = X$ and $M^tM = 3X + I$, we may write

$$\begin{aligned} M^{-1} &= \frac{1}{2}[3X - 2I]A^t \\ &= \frac{1}{2}[M^tM - 3I]A^t \end{aligned}$$

If we premultiply both sides of this last equation by the signal matrix M, and postmultiply both sides by the orientation rotation matrix A, we finally obtain

$$A = \frac{1}{2}M[M^tM - 3I] \qquad (13.23)$$

Since the numerical values of the elements of the matrix M are known, then so are the numerical values of the rotation matrix A. And since analytical expressions for the elements of the matrix A are also known, as listed in Equations 13.9, it is relatively easy to obtain

$$\tan \psi = \frac{\sin \psi}{\cos \psi} = \frac{a_{12}}{a_{11}}$$

$$\text{and} \quad \sin \theta = -a_{33}$$

$$\text{and} \quad \tan \phi = \frac{\sin \phi}{\cos \phi} = \frac{a_{23}}{a_{33}}$$

Each of the elements of the orientation matrix, A, are defined by the Equations 13.9, but the corresponding 'normalized' measurement of these elements is defined as a function of Equation 13.16 by Equation 13.23. Again, it should be emphasized that solving for P and A using these matrix methods, often results in some hemispheric ambiguity. These ambiguities, however, are usually eliminated by application boundaries or the appropriate placement and orientation of the Source.

In summary to this point, the three column vectors of the signal matrix E_o represent signals detected by the 3-axis Sensor. These signal vectors are the result of corresponding orthonormal excitations applied at the remote 3-axis Source. The 3x3 signal set at the Sensor is "normalized" and the resulting measure of the attenuation factor k provides the basis for a regulated open-loop measure of the distance. The "normalized" signal set M is the grist for the system process. This process provides the means for determining the relative position and the relative orientation of the Sensor frame, respectively, with respect to the frame of the relatively remote Source.

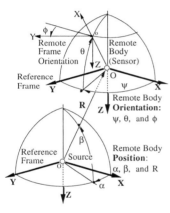

Figure 13.2
Six Degrees-of-Freedom

13.7   Quaternion Processes

As the reader of the earlier chapters knows very well by now, our primary purpose is to explore the use of quaternions in the analysis of rotations in R^3 as an alternative to the use of matrix rotation operators. So now we turn to considering the preceding problems from a quaternion point of view. We begin with a reminder of three important matters.

First, if a unit vector **u** defines the direction of the axis of a rotation in R^3, and α is the angle of that rotation, then the quaternion

q associated with that rotation is

$$q = \cos\frac{\alpha}{2} + \mathbf{u}\sin\frac{\alpha}{2}$$

Second, if q is a unit quaternion with complex conjugate q^*, then the quaternion operator $q^*\mathbf{v}q$ represents a coordinate frame rotation in which the vector $\mathbf{v}$ is now expressed as vector $\mathbf{w} = q^*\mathbf{v}q$ in the rotated frame.

Third, to this point we have used only vectors in our computations with quaternions — treating vectors as quaternions with real part zero. Now, however, it will be helpful to use the compact notation which matrices afford. That is to say, if A is a $3 \times n$ matrix, whose column vectors are $\mathbf{a}_1, \mathbf{a}_2, \cdots \mathbf{a}_n$, we will agree that the quaternion operation q^*Aq yields a $3 \times n$ matrix $B = q^*Aq$ whose column vectors $\mathbf{b}_i$ are given by $\mathbf{b}_i = q^*\mathbf{a}_i q$ for $i = 1, 2, \cdots, n$. It follows that if Q is the rotation matrix associated with the quaternion q, we have

$$B = QA = q^*Aq$$

With these considerations in mind, we return once again to our signal matrix E_o given in Equation 13.10 as

$$
\begin{aligned}
E_o &= SE_i \\
\text{where} \qquad S &= kAP^tCP = kQCP
\end{aligned}
$$

as in Equation 13.11. Here both A and P and therefore Q represent rotations in R^3. In terms of quaternions, let us associate with each of these rotation matrices the quaternions a, p and q, respectively. Then the signal matrix E_o may be written as

$$
\begin{aligned}
E_o &= SE_i \\
&= kQCPE_i \\
&= kq^*[C(p^*E_i p)]q
\end{aligned}
\qquad (13.24)
$$

$$
\begin{aligned}
\text{where} \qquad E_i &= \text{matrix whose columns are time-multiplexed} \\
&\qquad \text{excitations of the Source triad} \\
\text{and} \qquad q &= p^*a \quad \text{in quaternion notation}
\end{aligned}
$$

Earlier, we have shown that matrix Equations 13.18, 13.19 and 13.20 can be solved to give a closed form measure of the desired relative position and orientation of the Source and Sensor frames. It does seem to be the case, however, that a corresponding closed-form solution using quaternion rotation operators is not so easily obtained. In the next section we give a partial solution, in terms of

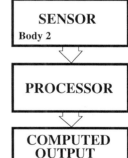

Figure 13.1
System Block Diagram

quaternions, for the tracking angles which determine the *direction* to the remote object. Closed-form quaternion-based solutions for the distance to the remote object, as well as its orientation angles, must await further investigations. It is likely, however, that the analysis must proceed along the lines indicated in the next section.

It should be noted at this point, however, that various rotation matrix and/or quaternion operators may still be used in a variety of configurations, even hybrid, depending upon the application. Therefore, using the results that we have obtained to this point, we will develop three alternative *iterative* quaternion processes which may be used to provide a measure of

1. Position only, or

2. Orientation only, or

3. Both Position and Orientation, as a

    (a) Simultaneous Process, or as a

    (b) Parallel Process

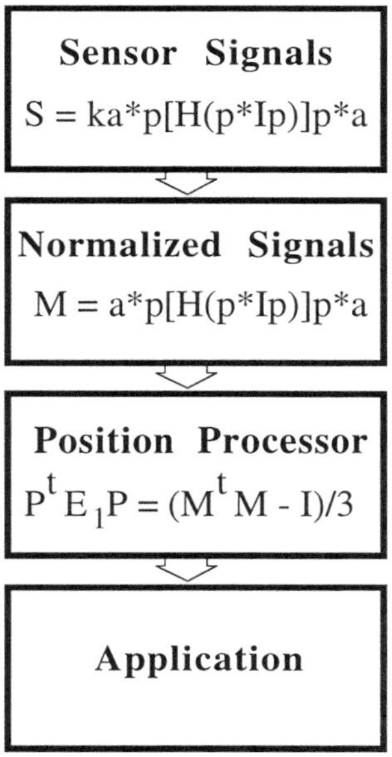

The parallel processor for determining Position and Orientation is time-efficient and is particularly appropriate in certain "real-time" computer graphics applications. Some of these applications will be discussed later, after we develop in some detail the processes listed above.

13.8 Partial Closed-form Quaternion Tracking Solution

We begin our quaternion solution for the tracking angles α and β by considering the matrix X of Equation 13.20. We have

$$X = P^t E_1 P$$

From Equation 13.21 we note that since the elements of the matrix M are known numerically from the sensed data, so are the elements of the matrix X. Further, for any vector $\mathbf{v}$, we know that if

$$p = p_0 + p_1 \mathbf{i} + p_2 \mathbf{j} + p_3 \mathbf{k}$$

is the quaternion which corresponds to the rotation matrix P we may write

$$P\mathbf{v} = p^* \mathbf{v} p$$

Sensor Signals

$S = ka*p[H(p*Ip)]p*a$

Normalized Signals

$M = a*p[H(p*Ip)]p*a$

Position Processor

$P^t E_1 P = (M^t M - I)/3$

Application

Figure 13.8
Processor for Position Only

Reference Equation 13.21

$$P^t E_1 P = (M^t M - I)/3$$

and

$$P^t \mathbf{v} = p\mathbf{v}p^*$$

It follows that we may write

$$
\begin{aligned}
X & = P^t E_1 P = \frac{1}{3}[M^t M - I] \quad \text{(computed)} \\
& = pVp^* & (13.25) \\
\text{where} \quad V & = E_1 P = E_1[p^* I p] & (13.26) \\
\text{and} \quad M & = 3x3 \text{ signal matrix (measured)} \\
I & = \text{a } 3x3 \text{ identity matrix} \\
\text{and} \quad E_1 & = \text{zero matrix, except } e_{11} = 1
\end{aligned}
$$

We now first expand the second equality on the right-hand side of Equation 13.21 for each of the column vectors of the Identity matrix

$$I = \begin{bmatrix} \mathbf{e}_1 & \mathbf{e}_2 & \mathbf{e}_3 \end{bmatrix}$$

We perform the quaternion operation $p^* \mathbf{e}_i p$ on each of the standard basis vectors $\mathbf{e}_i$. Each is then multiplied on the left by E_1 to produce vectors $\mathbf{v}_i$

$$\mathbf{v}_i = E_1[p^* \mathbf{e}_i p] \quad \text{for} \quad i = 1, 2, 3$$

We now compute these three new vectors as functions of the tracking quaternion components.

$$
\begin{aligned}
\mathbf{v}_1 & = E_1 p^* \mathbf{e}_1 p \\
& = E_1[(2p_0^2 - 1)\mathbf{e}_1 + 2(\mathbf{e}_1 \cdot \mathbf{p})\mathbf{p} + 2p_0(\mathbf{e}_1 \times \mathbf{p})]
\end{aligned}
$$

$$
= E_1 \begin{bmatrix} 2p_0^2 + 2p_1^2 - 1 \\ 2p_1 p_2 - 2p_0 p_3 \\ 2p_1 p_3 + 2p_0 p_2 \end{bmatrix} = \begin{bmatrix} a \\ 0 \\ 0 \end{bmatrix}
$$

Similarly,

$$
\begin{aligned}
\mathbf{v}_2 & = E_1 p^* \mathbf{e}_2 p \\
& = E_1[(2p_0^2 - 1)\mathbf{e}_2 + 2(\mathbf{e}_2 \cdot \mathbf{p})\mathbf{p} + 2p_0(\mathbf{e}_2 \times \mathbf{p})]
\end{aligned}
$$

$$
= E_1 \begin{bmatrix} 2p_2 p_1 + 2p_0 p_3 \\ 2p_0^2 + 2p_2^2 - 1 \\ 2p_2 p_3 - 2p_0 p_1 \end{bmatrix} = \begin{bmatrix} b \\ 0 \\ 0 \end{bmatrix}
$$

and finally,

$$
\begin{aligned}
\mathbf{v}_3 & = E_1 p^* \mathbf{e}_3 p \\
& = E_1[(2p_0^2 - 1)\mathbf{e}_3 + 2(\mathbf{e}_3 \cdot \mathbf{p})\mathbf{p} + 2p_0(\mathbf{e}_3 \times \mathbf{p})]
\end{aligned}
$$

Quaternion Operations on the Columns of a Matrix

When we show the quaternion rotation operator operating on a 3x3 matrix as in Equation 13.26, it is to be understood that the operation is to be performed on each of the column vectors which comprise the matrix, in turn.

$$= E_1 \begin{bmatrix} 2p_3p_1 - 2p_0p_2 \\ 2p_3p_2 + 2p_0p_1 \\ p_0^2 + 2p_3^2 - 1 \end{bmatrix} = \begin{bmatrix} c \\ 0 \\ 0 \end{bmatrix}$$

where, merely to simplify the notation, we let

$$\begin{aligned} a &= 2p_0^2 + 2p_1^2 - 1 \\ b &= 2p_2p_1 + 2p_0p_3 \\ c &= 2p_3p_1 - 2p_0p_2 \end{aligned}$$

Now, since $X = pVp^*$ as indicated in Equation 13.25, we must operate again on the matrix

$$V = \begin{bmatrix} a & b & c \\ 0 & 0 & 0 \\ 0 & 0 & 0 \end{bmatrix}$$

whose three columns are scalar multiples of each other, as indicated. It is not difficult to show that after performing this last quaternion operation, we get

$$X = \begin{bmatrix} a^2 & ab & ac \\ ba & b^2 & bc \\ ca & cb & c^2 \end{bmatrix} \tag{13.27}$$

where a, b, and c are functions of the tracking quaternion components indicated above.

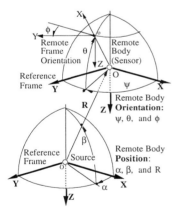

Figure 13.2
Six Degrees-of-Freedom

Again, the tracking quaternion we have designated as

$$p = p_0 + \mathbf{p} = p_0 + \mathbf{i}p_1 + \mathbf{j}p_2 + \mathbf{k}p_3$$

where each of the four quaternion components, namely p_0, p_1, p_2, and p_3, are functions of the two angles employed in the (as yet unspecified) angle/axis tracking sequence. We define the angle/axis sequence as before: first, a rotation through an angle α about the reference Z-axis, followed by a rotation through an angle β about the new y-axis. Then, as we have already shown at the end of Chapter 6, we have

$$\begin{aligned} p &= (\cos\frac{\alpha}{2} + \mathbf{k}\sin\frac{\alpha}{2})(\cos\frac{\beta}{2} + \mathbf{j}\sin\frac{\beta}{2}) \\ &= \cos\frac{\alpha}{2}\cos\frac{\beta}{2} \\ &\quad -\mathbf{i}\sin\frac{\alpha}{2}\sin\frac{\beta}{2} \end{aligned}$$

$$+\mathbf{j}\cos\frac{\alpha}{2}\sin\frac{\beta}{2} \tag{13.28}$$

$$+\mathbf{k}\sin\frac{\alpha}{2}\cos\frac{\beta}{2}$$

$$= p_0 + \mathbf{i}p_1 + \mathbf{j}p_2 + \mathbf{k}p_3$$

Now we return to Equation 13.27 and note that

$$x_{11} = a^2$$
$$\text{where} \quad a = p_0^2 + p_1^2 - p_2^2 - p_3^2$$

Upon substitution of the equivalent trigonometric expression from Equations 13.28 for the p_k's, and after invoking trigonometric identities (suggested in the margin), we get

$$a = \cos\alpha\cos\beta$$
$$\text{Therefore} \quad x_{11} = \cos^2\alpha\cos^2\beta$$

This result agrees with that obtained earlier in the matrix analysis of the tracking angles (see Equations 13.22).

In like fashion, Equation 13.27 tells us, for example, that

$$x_{23} = bc$$
$$\text{where} \quad b = 2p_2p_1 + 2p_0p_3$$
$$\text{and} \quad c = 2p_3p_1 - 2p_0p_2$$

Again, upon substitution of the equivalent trigonometric expression from Equations 13.28 for the p_k's and after invoking appropriate trigonometric identities, we get

$$x_{23} = bc = -\sin\alpha\sin\beta\cos\beta$$

as before. The other entries in the matrix X may be determined in like manner. From these expressions the tracking angles are determined just as before.

We will not at this time attempt to determine the distance to the remote body or its relative orientation using quaternion analysis. Rather, we turn instead to a quaternion-based iterative process for the system.

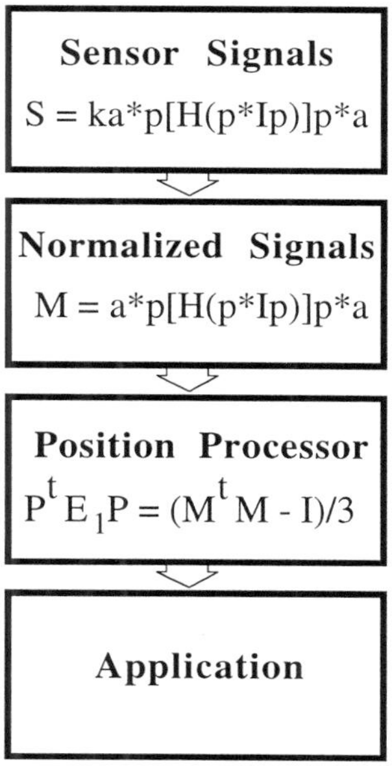

13.9 An Iterative Solution for Tracking

We now present an iterative process on Equation 13.21 which is rewritten in terms of the tracking quaternion p. The process will

Identities

Recall, the useful identities

$$\cos 2\theta = \cos^2\theta - \sin^2\theta$$
$$\sin 2\theta = 2\sin\theta\cos\theta$$

| **Sensor Signals** |
| S = ka*p[H(p*Ip)]p*a |

↓

| **Normalized Signals** |
| M = a*p[H(p*Ip)]p*a |

↓

| **Position Processor** |
| $P^t E_1 P = (M^t M - I)/3$ |

↓

| **Application** |

Figure 13.8
Processor for Position ONLY

converge to a measure of the tracking quaternion which defines the *direction to* the remote Sensor relative to the Source frame (see Figure 13.2).

We rewrite the right-hand side of Equation 13.20 partially in terms of the actual quaternion p, to give

$$X = P^t E_1 P = p(E_1 P)p^* \qquad (13.29)$$

The 2nd equality of the Equations 13.29 is now ready for the first step in the quaternion process.

**Reference
Equation 13.21**

$$P^t E_1 P = (M^t M - I)/3$$

We begin the iterative process by *estimating* the tracking quaternion, p, which we wish to determine. Using the estimated quaternion, denoted $\tilde{p}$, our strategy is to "undo" the indicated transformations which appear in Equation 13.29. If the estimated quaternion, $\tilde{p}$, is equal to the actual quaternion then the result, of course, will give the matrix E_1. If, however, the estimated quaternion is not equal to the actual quaternion, the resulting matrix E_1 will show errors. Our objective at this point is to determine how the errors which appear in E_1 are related to errors in the estimated tracking quaternion.

We define the *estimated* quaternion, $\tilde{p}$, in terms of the actual tracking quaternion, p. To do this we write

$$\tilde{p} = pd \qquad (13.30)$$

where the quaternion d is the transition or error-quaternion which relates the actual quaternion p and its estimate, $\tilde{p}$. The error-quaternion d is

Remember

The algorithm

$$X = P^t E_1 P$$
$$= \frac{1}{3}[M^t M - I]$$
$$= \text{f(measurements)}$$
$$= X^t$$
$$= \text{symmetric matrix}$$

$$d = d_o + \mathbf{d} \qquad (13.31)$$
$$\text{with} \qquad \mathbf{d} = \mathbf{i}d_1 + \mathbf{j}d_2 + \mathbf{k}d_3$$

We now operate on Equation 13.29 with this *best estimate* in an attempt to "undo" the effect of the first p operation.

$$
\begin{aligned}
X_1 &= \tilde{p}^* X \tilde{p} \\
&= \tilde{p}^* p(E_1 P)p^* \tilde{p} \\
&= d^* p^* p(E_1 P)p^* pd \\
&= d^*(E_1 P)d \qquad (13.32) \\
&\simeq (I - 2D)E_1 P \qquad (13.33)
\end{aligned}
$$

Equation 13.33 is Equation 13.32 expanded in matrix form with the assumption that $|\mathbf{d}|$ is *small*, that is, $|\mathbf{d}| \to 0$ and $d_0 \cong 1$. The vector components of the error-quaternion, namely **d**, are the elements of the error matrix

$$D = \begin{bmatrix} 0 & -d_3 & d_2 \\ d_3 & 0 & -d_1 \\ -d_2 & d_1 & 0 \end{bmatrix} \qquad (13.34)$$

We now take the transpose of both sides of Equation 13.33 and again express the tracking matrix P in terms of its equivalent tracking quaternion p, and we write

$$(X_1)^t = p[E_1(I + 2D)]p^* \qquad (13.35)$$

As before, we now operate on Equation 13.35 with $\tilde{p}$, the best estimate of p, which gives

$$\tilde{p}^*[(X_1)^t]\tilde{p} = (I - 2D)E_1(I + 2D)$$
$$= E_1 + 2E_1D - 2DE_1 - 4DE_1D \qquad (13.36)$$

Since the vector part of the tracking error quaternion d is assumed to be small compared to one, the last term on the right-hand side of Equation 13.36 is taken to be negligible. The remaining three terms on the right-hand side of 13.36 then form the matrix

$$E_1 + 2E_1D - 2DE_1 = \begin{bmatrix} 1 & -2d_3 & 2d_2 \\ -2d_3 & 0 & 0 \\ 2d_2 & 0 & 0 \end{bmatrix} \qquad (13.37)$$

Up to this point we have merely derived an error-matrix whose elements provide algorithms which define the errors in the most recent estimated quaternion, $\tilde{p}$, for the position-only Process. However, the same sequence of operations which resulted in Equations 13.22 define the actual (ongoing) process which must be performed on the right-hand side of Equation 13.21.

After performing these operations — also twice, as indicated above — on the right-hand side of Equation 13.21 each element in the matrix of Equations 13.22 may be assigned a numerical value. The numerical values of the elements of the resulting matrix provide values for the corresponding elements d_2 and d_3 in Error-matrix equation 13.22.

After successive iterations, the vector part, $|\mathbf{d}|$, hopefully, approaches 0, and the real part, d_o, approaches 1. As this happens, of course, the estimated quaternion, $\tilde{p}$, approaches the actual quaternion, p.

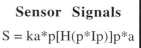

Sensor Signals
$S = ka*p[H(p*Ip)]p*a$

Normalized Signals
$M = a*p[H(p*Ip)]p*a$

Position Processor
$P^tE_1P = (M^tM - I)/3$

Application

Figure 13.8
Processor for Tracking ONLY

Reference Equation 13.21
$P^tE_1P = (M^tM - I)/3$

It might be argued that since the tracking transformation involves only two rotations, and if these two tracking angles are the conventional tracking heading and the tracking elevation, then the d_1 component of the tracking error-quaternion may be computed as, $d_1 = -d_2 d_3 / d_0$. However, recall that at the outset the vector part of the tracking error-quaternion was assumed small and $d_0 \simeq 1$. Therefore, the product of two small components should be exceedingly small which justifies that $d_1 \simeq 0$.

These computed values for d_2 and d_3 define the resulting tracking error-quaternion to be

$$\begin{aligned} d &= d_o + \mathbf{d} = d_o + \mathbf{j}d_2 + \mathbf{k}d_3 \\ &\simeq 1 + \mathbf{d} \end{aligned}$$

This computed value for the tracking error-quaternion, d, we now use in Equation 13.30 to compute the actual quaternion p. But the most we can claim is that this newly computed "actual" quaternion is merely a *more precise* estimate for the actual tracking quaternion, p. That is,

$$\begin{aligned} \text{Since} \quad \tilde{p} &= pd \\ \text{then} \quad p &= \tilde{p}d^* \end{aligned}$$

But, the right-hand side of this last equation is still an estimate, so the best we can say is that

$$p \simeq \tilde{p}_{new} = \tilde{p}_{old}d^*$$

and then repeat the process until the error-quaternion, $d \to 1$. In other words, repeating this process indefinitely or until the vector part of the tracking error-quaternion, d, becomes sufficiently small produces a precise measure of the tracking quaternion. We emphasize that only the direction portion of the Sensor position triplet is determined from this computed tracking quaternion. The value for R would, of course, come from the normalization process discussed above.

It should be noted that the foregoing process, which converges to a measure of the tracking quaternion q, is independent of the relative orientation between the Source and the Sensor.

13.10   Orientation Quaternion

We are now ready to develop an iterative process, using quaternion rotation operators, to determine the angles ψ, θ, and ϕ, which define the *orientation* of the remote Sensor frame relative to the Source frame. These three rotation angles are exactly those which comprise the Aerospace Sequence which we have discussed earlier. In the preceding pages we have used the matrix A to represent this sequence of rotations; the reader may wish to review the matrix form for the Aerospace Sequence as we discussed it in Section 4.4 of Chapter 4. From this form for the matrix A it is relatively easy to see that

$$\tan \psi = \frac{a_{11}}{a_{22}}$$
$$\sin \theta = -a_{13}$$
$$\tan \phi = \frac{a_{23}}{a_{33}}$$

Thus if the matrix A is known we easily can calculate the orientation angles ψ, θ, and ϕ. We also know that the components of the matrix A can be written in terms of the components of the associated quaternion (see Section 5.14). It follows that in order to find the orientation angles we need only to find the quaternion a which corresponds to the matrix A in Equation 13.23. As we have noted before, the elements of the matrix M are known from the sensed data, so that the elements of the matrix A are also known numerically. Thus we may calculate the orientation angles directly from this information.

Our interest here, however, is to see how quaternions may be used to solve this same problem. We remark further that in certain applications it is not necessary to solve for the orientation angles; it is sufficient simply to know the quaternion, a. So, here we present an iterative process on Equation 13.23 which will yield the desired orientation quaternion, a. The strategy we adopt here is similar to that employed in the determination of the tracking quaternion in the previous section, namely,

Estimate the orientation quaternion, a. Then, using this estimate in an inverse quaternion rotation operator, attempt to invert the expression for the orientation matrix A in Equation 13.23. If the estimate is error-free then the resulting matrix will be the identity matrix. If the estimate is in error, as will at first usually be the

Reference Equation 13.23

$$A = a^* I a$$
$$= \frac{1}{2} M (M^t M - 3I)$$

case, then these errors will be displayed as non-zero off-diagonal elements of the "identity" matrix. These error elements are then used to make the next estimate more precise, and the process is repeated.

The left side of Equation 13.23, namely A, is the symbol which represents the 3×3 orientation rotation matrix. The equivalent quaternion rotation operator, denoted a^*Ia in the first equality, is also a 3×3 matrix whose elements are functions of the components of the quaternion a. The second equality merely emphasizes that the nine elements which comprise A and a^*Ia have numerical values which are determined from the sensed data collected in the normalized measurement matrix M.

The unknown orientation quaternion, a, is estimated, the estimate being denoted $\tilde{a}$. We let $a = u\tilde{a}$, where u is the transition quaternion or, in this instance, the error-quaternion which relates the actual quaternion, a (which we wish to determine) and its estimate, $\tilde{a}$. With this definition of u, note that we then have

$$
\begin{aligned}
\tilde{a} &= u^*a \\
u &= a\tilde{a}^* \\
u^* &= \tilde{a}a^*
\end{aligned}
$$

**Reference
Equation 13.23**

$$
\begin{aligned}
A &= a^*Ia \\
&= \frac{1}{2}M(M^tM - 3I)
\end{aligned}
$$

As before we shall assume that the error-quaternion u is nearly the identity quaternion, so that $u_0 \approx 1$ and $u_1 \approx 0$, $u_2 \approx 0$, and $u_3 \approx 0$. Thus second order terms in u_1, u_2, u_3 may safely be neglected. We begin by operating on both sides of Equation 13.23, namely

$$
A = a^*Ia
$$

with the quaternion estimate $\tilde{a}$ to obtain

$$
\begin{aligned}
\tilde{a}A\tilde{a}^* &= \tilde{a}a^*Ia\tilde{a}^* \\
&= u^*Iu
\end{aligned}
$$

Now if $\tilde{a} = a$ we have $u = 1$ and we would obtain the identity matrix. But usually some errors will appear. If we have

$$
I = \begin{bmatrix} | & | & | \\ \mathbf{e}_1 & \mathbf{e}_2 & \mathbf{e}_3 \\ | & | & | \end{bmatrix}
$$

where $\mathbf{e}_1$, $\mathbf{e}_2$, and $\mathbf{e}_3$ are the standard basis vectors, we may calculate the columns of the matrix u^*Iu by noting that

$$
u^*\mathbf{e}_ku = (2u_0^2 - 1)\mathbf{e}_k + 2(\mathbf{e}_k \cdot \mathbf{u})\mathbf{u} + 2u_0(\mathbf{e}_k \times \mathbf{u})
$$

In view of our earlier comments on the components of the error-quaternion u we may write

$$u^* \mathbf{e}_k u \approx \mathbf{e}_k + 2(\mathbf{e}_k \times \mathbf{u}) \qquad \text{for} \qquad k = 1, 2, 3$$

The reader is urged to check the algebraic details which allow us to write

$$u^* I u = I + 2U$$
$$\text{where} \quad U = \begin{bmatrix} 0 & u_3 & -u_2 \\ -u_3 & 0 & u_1 \\ u_2 & -u_1 & 0 \end{bmatrix}$$

As we have said before, if $U = 0$ we have $\tilde{a} = a$ and our problem is solved. If not all of the u_k's are equal to zero then

$$\tilde{a} \neq a$$

and we must repeat the process. And since $\tilde{a} = u^* a$ we can use the last estimated $\tilde{a}$ and the newly computed value for the error quaternion u to compute a better estimate, $a = u\tilde{a}$ or

$$\tilde{a}_{(new)} = u\tilde{a}_{(old)}$$

Then we repeat the process until

$$u_0 \to 1 \quad \text{and} \quad |\mathbf{u}| \to \mathbf{0} \quad \text{then} \quad \tilde{a} \to a$$

It may be that *initially* the best estimate for the quaternion is the identity quaternion — ordinarily, the best estimate will be the most recent computed estimate. At any rate, the sequence of values which represent the estimated orientation quaternion, $\tilde{a}$, converges rapidly to the 'actual' orientation quaternion in this iterative process. During normal operations, the computed best estimated orientation quaternion, $\tilde{a}$, may (at each step in the iterative process) be used within some larger control process.

13.11   Position & Orientation

The process for getting both position and orientation can take two forms. Clearly, one could use the two separate processes, described above, which yield position only and orientation only, either as concatenated or as parallel processes.

Alternatively, we now present a process on Equation 13.16 which will simultaneously determine both the tracking and the orientation quaternions.   Recall the quaternion representation for the signal matrix in Equation 13.11 which after 'normalization' was written as

$$M = q^*[C(p^*Ip)]q$$

Reference Equation 13.11

$$S = ka^*p[C(p^*Ip)]p^*a$$

We now adopt the same processing strategy as used in the previous section, except that we estimate both the quaternion q and the tracking quaternion p, denoting these estimates as $\tilde{q}$ and $\tilde{p}$, respectively. Again, we relate the unknown actual quaternions with their estimates by means of appropriate transition or error quaternions, e and d, so that

$$\begin{aligned} q &= e\tilde{q} \qquad \text{or} \qquad \tilde{q} = e^*q \\ \text{and} \quad \tilde{p} &= pd \qquad \text{or} \qquad p = \tilde{p}d^* \end{aligned}$$

There are good reasons for choosing the *order* employed in the definition of these transition quaternions, as will become clear.

Reference Equation 13.16

$$\begin{aligned} M &= AP^tCP = QCP \\ &= a^*p[C(p^*Ip)]p^*a \\ &= q^*[C(p^*Ip)]q \end{aligned}$$

These estimates are now used to process the normalized signal matrix M. The strategy in this process is to invert the transformations indicated in Equation 13.16, using the improved quaternion estimates obtained after each iteration. Residual errors in the quaternion estimates appear as non-zero off-diagonal elements in the 'computed' coupling matrix. This iterative process, however, does converge to the expected diagonal coupling matrix C, and the estimated quaternions do converge to the actual quaternions we seek to determine.

As the first step in this process, we 'undo' the effect of the quaternion q in Equation 13.16, using its best estimate $\tilde{q}$. We do this by computing

An Exercise

The reader is urged to check the algebraic details which yield these results

$$\begin{aligned} M_1 &= \tilde{q}M\tilde{q}^* \\ &= \tilde{q}q^*[C(p^*Ip)]q\tilde{q}^* \\ &= e^*CPe \\ &= (I + 2E)CP \end{aligned} \qquad (13.38)$$

$$\text{where} \quad E = \begin{bmatrix} 0 & e_3 & -e_2 \\ -e_3 & 0 & e_1 \\ e_2 & -e_1 & 0 \end{bmatrix}$$

The next step is to "undo" (or invert) the effect of the tracking transformation, P, in Equation 13.38. To do this we first transpose both sides of Equation 13.38 which gives

$$
\begin{aligned}
M_2 &= M_1^t \\
&= P^t C (I - 2E) \\
&= p C (I - 2E) p^*
\end{aligned}
\tag{13.39}
$$

We now operate on both sides of Equation 13.39 using the best estimate of the quaternion p, namely, $\tilde{p} = pd$

Note

$$
\begin{aligned}
C^t &= C \\
(I + 2E)^t &= I - 2E
\end{aligned}
$$

$$
\begin{aligned}
M_3 &= \tilde{p}^* M_2 \tilde{p} \\
&= \tilde{p}^* p C (I - 2E) p^* \tilde{p} \\
&= d^* C (I - 2E) d \\
&= (I + 2D) C (I - 2E) \\
&= C + 2DC - 2CE - \underbrace{4DCE}_{negligible}
\end{aligned}
\tag{13.40}
$$

$$
= \begin{bmatrix}
2 & -2d_3 - 4e_3 & 2d_2 + 4e_2 \\
-4d_3 - 2e_3 & -1 & -2d_1 + 2e_1 \\
4d_2 + 2e_2 & 2d_1 - 2e_1 & -1
\end{bmatrix}
$$

$$
\text{where} \quad D = \begin{bmatrix}
0 & d_3 & -d_2 \\
-d_3 & 0 & d_1 \\
d_2 & -d_1 & 0
\end{bmatrix}
$$

We note that the elements of the matrix M_3 are the elements of the coupling matrix C, except for the indicated error terms in the off-diagonal positions.

The foregoing analysis gives us algebraic expressions for the elements of M_3 in terms of the components of the error quaternions, $d \simeq 1 + \mathbf{d}$ and $e \simeq 1 + \mathbf{e}$. However, since the elements of the measured matrix M are known numerically, so are the elements of the matrix M_3. If we denote these *numerically known* elements of M_3 as $m(ij)$, the reader should verify that we may write

Reference Equation 13.16

$$
\begin{aligned}
M &= AP^t CP = QCP \\
&= a^* p [C (p^* I p)] p^* a \\
&= q^* [C (p^* I p)] q
\end{aligned}
$$

note: $q = p^* a$

$$
\begin{array}{llll}
d_o &= 1 & e_o &= 1 \\
d_1 &= 0 & e_1 &= (m_{23} - m_{32})/4 \\
d_2 &= (2m_{31} - m_{13})/6 & e_2 &= (2m_{13} - m_{31})/6 \\
d_3 &= (m_{12} - 2m_{21})/6 & e_3 &= (m_{21} - 2m_{12})/6
\end{array}
$$

The new estimates for the quaternions p and q then are

$$
p = p_{new} = (p_{old})d \qquad q = q_{new} = (q_{old})e
$$

and the process is repeated either continuously or until the error becomes sufficiently small. Then, using the definitions stated with

Equation 13.16, the orientation quaternion, a, is computed as the quaternion product

$$a \;=\; pq \qquad\qquad (13.41)$$

Once that element of the orientation quaternion, a, have been determined, the orientation angles, ψ, θ, and ϕ, may be easily computed using the results presented in Section 7.8.

In conclusion, it should be clear that the foregoing six degree-of-freedom position and orientation transducer should have an important application in the burgeoning new technology called virtual reality. An important aspect of virtual reality is that the construction of a virtual situation requires the use of tools provided by computer graphics.

In the next and final chapter we develop the mathematical background upon which virtual reality applications are based.

Chapter 14

Computer Graphics

14.1 Introduction

The invasion of the personal computer into virtually every aspect of our lives has brought along with it a fascination with computer-generated imagery. A wide variety of software is on the market which enables novices to create images, even animated images, on the computer screen. And although the available software is often suitable for the most challenging requirements of the user, there are those instances when new software must be developed to achieve the unique objectives envisioned by the creative entreprenuer. Very likely that is what happened to the makers of the software the reader may currently be using.

So, in this chapter we present some of the mathematics of computer graphics, mostly using matrices, but with special attention given to quaternions and the quaternion rotation operator alternative, whenever rotations are required. We first introduce canonical transformations which are employed in the graphics-generating milieu, all in the context of homogeneous coordinates. We do this first in two-dimensions, for pedagogical reasons, since it is much easier to visualize what happens in R^2. In R^2 we will also consider some of the special problems associated with concatenation and the importance of order in sequences of transformations on point-sets. We then extend all of these basic ideas to R^3. Finally, we outline a strategy, based upon this background, for the development of the *Virtual Reality* required in the design of a Flight Simulator.

14.2   Canonical Transformations

There are three *canonical transformations* which are fundamental to the creation of two-dimensional images (graphics) which represent three-dimensional objects. These are

- Scale

- Translation

- Rotation

Transformations other than these, such as

- Skews

- Reflections

- Distortions

may be produced by merely inserting special entries in the three primary transformations, as may be easily verified.

14.3   Transformations in R^2

We first define the three canonical transformations in R^2. From this two-dimensional vantage point, we introduce homogeneous coordinates. We may then extend these results to the generation of images in R^3, along with the mathematics of perspective. In general we shall use P to designate the initial point-set and P' to denote the transformed point-set.

Matrix Operators

The indicated $2x2$ matrix transformations on point-sets in R^2 are simple but the reader should be alert to note that they manifest a fundamental difficulty. Same is true for point-sets in R^2. They are not all multiplicative
— a trouble which will be fixed when we adopt homogeneous coordinates.

14.3.1   Scale in R^2

By the *Scale Transformation* we mean that, given two scale factors, S_1 and S_2, we have the mapping

$$P = (x, y)   \longrightarrow   P' = (S_1 x, S_2 y)$$

 This transformation simply modifies independently the horizontal and vertical dimensions of an object in R^2. For example, if we let $S_1 = 3$ and $S_2 = 2$, the unit square shown in Figure 14.1 is transformed into a 3x2 rectangle, as indicated. The scale transformation

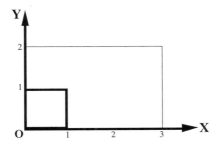

Figure 14.1: Scale

example may be conveniently represented in matrix form. In our example, we may write

$$
\begin{aligned}
P' &= SP \\
P' &= \begin{bmatrix} S_1 & 0 \\ 0 & S_2 \end{bmatrix} \begin{bmatrix} x \\ y \end{bmatrix}
\end{aligned}
$$

where

$$
S = \begin{bmatrix} 3 & 0 \\ 0 & 2 \end{bmatrix}
$$

$$
\begin{aligned}
P &= \text{original point-set} \\
P' &= \text{modified point-set}
\end{aligned}
$$

14.3.2   Translation in R^2

By *Translation* we mean that an entire set of points which comprise an object are shifted (without dimensional change) within the

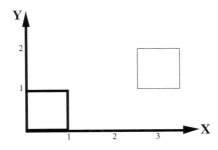

Figure 14.2: Translation

viewing frame, in accordance with the familiar rule

$$
P = (x, y) \quad \longrightarrow \quad P' = (x + h, y + k)
$$

Here h and k are the horizontal and vertical displacements.

Translation

In R^2 the *translation* transformation is an *additive* one, so that matrix addition is required.

Then, for the *translation* example shown in Figure 14.2 we may write in matrix form

$$P' \;=\; P + T$$

where

$$T \;=\; \begin{bmatrix} 2.5 \\ 1 \end{bmatrix}$$

$$\text{So} \quad P' \;=\; P + \begin{bmatrix} 2.5 \\ 1 \end{bmatrix}$$

where

$$P \;=\; \text{original point-set}$$
$$P' \;=\; \text{translated set}$$

14.3.3   Rotation in R^2

As was suggested in Section 2.5.2, by a *Rotation* we mean that an entire point-set is subjected to a *point* rotation about the origin

Rotation

For the *rotation* example shown in Figure 14.3 we write

$$P' \;=\; RP$$

where

$$R \;=\; \begin{bmatrix} \cos\theta & -\sin\theta \\ \sin\theta & \cos\theta \end{bmatrix}$$

$$P \;=\; \text{original point-set}$$
$$P' \;=\; \text{rotated set}$$

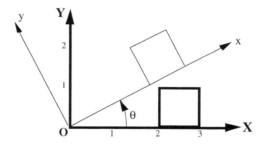

Figure 14.3: Rotation

through an angle θ, in accordance with the rule

$$P = (x, y) \quad \longrightarrow \quad P' = (x', y')$$

where
$$x' \;=\; x\cos\theta - y\sin\theta$$
$$y' \;=\; x\sin\theta + y\cos\theta$$

We write this transformation in matrix form, as shown in the margin. We ask at this juncture a question:

> *What if we intended that the square be rotated about its 'upper left' vertex? — or for that matter, how would one rotate the indicated square about any specified point?*

Using the transformations, R and T, just described, we simply translate to the origin, rotate, then translate back. We note, however, that S and R are both multiplicative operations, whereas T is an additive operation. To make these operations all multiplicative, we introduce *Homogeneous Coordinates*.

14.4 Homogeneous Coordinates

In the foregoing three canonical transformations, we note that the matrix products

$$RS \quad \text{and} \quad SR$$

make sense, whereas the products

$$RT \quad TR \quad TS \quad ST$$

are not defined, or do not produce sensible results. However, by using homogeneous coordinates, borrowed from Projective Geometry, we can concatenate the three canonical transformations multiplicatively in whatever order suits a particular application.

The *homogeneous coordinate* representation for the point (x, y) in R^2 is the triplet $(x, y, 1)$, that is

$$(x, y) \quad \longrightarrow \quad (x, y, 1)$$

For the Scale, Translation, and Rotation operators in terms of *homogeneous coordinates*, we will use the same symbols for their designation as before, namely, S, T, and R, respectively. We have

for Scale in R^2 Homogeneous Coordinates

$$
\begin{aligned}
S &= \text{Scale Operator} \\
&= \begin{bmatrix} S_1 & 0 & 0 \\ 0 & S_2 & 0 \\ 0 & 0 & 1 \end{bmatrix}
\end{aligned}
$$

so that

$$
\begin{bmatrix} x' \\ y' \\ 1 \end{bmatrix} = P' = SP = \begin{bmatrix} S_1 & 0 & 0 \\ 0 & S_2 & 0 \\ 0 & 0 & 1 \end{bmatrix} \begin{bmatrix} x \\ y \\ 1 \end{bmatrix} = \begin{bmatrix} xS_1 \\ yS_2 \\ 1 \end{bmatrix}
$$

for Translation in R^2 Homogeneous Coordinates

$$T \;=\; \text{Translation Operator}$$

$$=\; \begin{bmatrix} 1 & 0 & h \\ 0 & 1 & k \\ 0 & 0 & 1 \end{bmatrix}$$

so that

$$\begin{bmatrix} x' \\ y' \\ 1 \end{bmatrix} \;=\; P' \;=\; TP \;=\; \begin{bmatrix} 1 & 0 & h \\ 0 & 1 & k \\ 0 & 0 & 1 \end{bmatrix} \begin{bmatrix} x \\ y \\ 1 \end{bmatrix} \;=\; \begin{bmatrix} x+h \\ y+k \\ 1 \end{bmatrix}$$

for Rotation in R^2 Homogeneous Coordinates

$$R \;=\; \text{Rotation Operator}$$

$$=\; \begin{bmatrix} \cos\theta & -\sin\theta & 0 \\ \sin\theta & \cos\theta & 0 \\ 0 & 0 & 1 \end{bmatrix}$$

Inverses

The *inverses* of the homogeneous operators are defined as follows:

$$S^{-1} \;=\; \text{Inverse Scale}$$

$$=\; \begin{bmatrix} S_1^{-1} & 0 & 0 \\ 0 & S_2^{-1} & 0 \\ 0 & 0 & 1 \end{bmatrix}$$

$$T^{-1} \;=\; \text{Inverse Translation}$$

$$=\; \begin{bmatrix} 1 & 0 & -h \\ 0 & 1 & -k \\ 0 & 0 & 1 \end{bmatrix}$$

$$R^{-1} \;=\; \text{Inverse Rotation} \;=\; R^t$$

$$=\; \begin{bmatrix} \cos\theta & \sin\theta & 0 \\ -\sin\theta & \cos\theta & 0 \\ 0 & 0 & 1 \end{bmatrix}$$

so that

$$\begin{bmatrix} x' \\ y' \\ 1 \end{bmatrix} \;=\; P' \;=\; RP \;=\; \begin{bmatrix} \cos\theta & -\sin\theta & 0 \\ \sin\theta & \cos\theta & 0 \\ 0 & 0 & 1 \end{bmatrix} \begin{bmatrix} x \\ y \\ 1 \end{bmatrix}$$

$$=\; \begin{bmatrix} x\cos\theta - y\sin\theta \\ y\cos\theta + x\sin\theta \\ 1 \end{bmatrix}$$

14.5   An Object in R^2 Transformed

We next consider in some detail the question which was raised earlier, that is,

> *How do we concatenate these operators or transformations, and in what order, to achieve a specified objective?*

In this section we answer this question by applying alternative transformation sequences to a simple concrete example. Consider the unit square P and the rectangle P' as illustrated in Figure 14.4.

What we want to do is to devise a sequence of simple operations for the transformation which will take the point-set defined by P

into the point-set defined by P'. Here the transformed point-set P' is a 2×1 rectangle rotated clockwise approximately 30 degrees and translated such that its lower-right corner is at the point $(2,1)$, as shown. The facts stated in this fashion immediately suggests a *rotation* operator which gives a 30 degree clockwise rotation and also a *scaling* operator which takes a unit square into a (2×1) rectangle. The *translations* required in the complete sequence, however, require

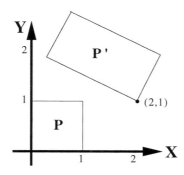

Figure 14.4: Operation on Unit Square

a bit more thought, because it is important that the concatenation of these various operations be done in the proper *order*. The overall strategy must be clear at the outset. We emphasize the importance of *concatenation order* in the next section.

14.6 Concatenation Order in R^2

At the outset, it is not entirely clear what operators should be in the desired transformation sequence, or in what order they should occur. It might seem at first glance that the first operation should be an appropriate scaling transformation, producing a 2×1 rectangle, followed by the 30 degree clockwise rotation.

The third operation required to achieve our goal should be an appropriate translation. However, this last translation from the previous rotated state might not be desirable because it could require some additional coordinate computations.

Alternatively, we might think a better plan would be to follow the scale transformation by a unit vertical translation. Our goal would then be achieved by a 30 degree clockwise rotation about the point $(2, 1)$. But, as we have observed, rotations about points other than the origin also require more translations. In Figure 14.5

Tools

The simplest set of distinct operations which might be used in a transformation to accomplish what we see in Figure 14.4 may be defined as follows:

$$S = \text{Scale operator}$$
$$= \begin{bmatrix} 2 & 0 & 0 \\ 0 & 1 & 0 \\ 0 & 0 & 1 \end{bmatrix}$$
$$R = \text{Rotation}$$
$$= \begin{bmatrix} \cos\theta & \sin\theta & 0 \\ -\sin\theta & \cos\theta & 0 \\ 0 & 0 & 1 \end{bmatrix}$$
$$\theta \approx -30 \text{ degrees}$$
$$T_1 = \text{Translation\#1}$$
$$= \begin{bmatrix} 1 & 0 & -1 \\ 0 & 1 & 0 \\ 0 & 0 & 1 \end{bmatrix}$$
$$T_2 = \text{Translation\#2}$$
$$= \begin{bmatrix} 1 & 0 & 2 \\ 0 & 1 & 1 \\ 0 & 0 & 1 \end{bmatrix}$$

Strategy

Develop a strategy which requires the simplest operations, remembering that rotations are about the origin. So, carefully choose the operations and their *ordering!*

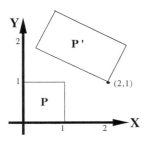

Transformation:
take P → P'

Some *transformation*, which consists of some *ordered* set of simple operations, will transform the Set of Points defined by P into the Set of points defined by P'.

we demonstrate that *order* in the sequence is extremely important. The unit square input point-set **P** in Figure 14.6 is transformed by the operator sequence $\mathbf{T_2SRT_1}$ to give an erroneous point-set $\mathbf{P'_1}$ as in Figure 14.5(a). And, the operator sequence in Figure 14.5(b) illustrates a transformation which also fails to meet our objective.

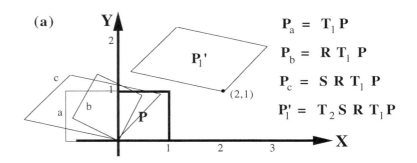

$$\mathbf{P_a = T_1\,P}$$
$$\mathbf{P_b = R\,T_1\,P}$$
$$\mathbf{P_c = S\,R\,T_1\,P}$$
$$\mathbf{P'_1 = T_2\,S\,R\,T_1\,P}$$

Tools

A set of simple operations sequence of transformation which are required to accomplish the stated objective in the figure above, are defined as follows:

S = Scale operator

$$= \begin{bmatrix} 2 & 0 & 0 \\ 0 & 1 & 0 \\ 0 & 0 & 1 \end{bmatrix}$$

T_1 = Translation#1

$$= \begin{bmatrix} 1 & 0 & -1 \\ 0 & 1 & 0 \\ 0 & 0 & 1 \end{bmatrix}$$

T_2 = Translation#2

$$= \begin{bmatrix} 1 & 0 & 2 \\ 0 & 1 & 1 \\ 0 & 0 & 1 \end{bmatrix}$$

R = Rotation

$$= \begin{bmatrix} \cos\theta & \sin\theta & 0 \\ -\sin\theta & \cos\theta & 0 \\ 0 & 0 & 1 \end{bmatrix}$$

$\theta \approx -30$ degrees

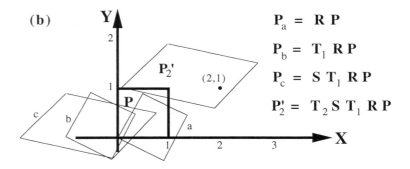

$$\mathbf{P_a = R\,P}$$
$$\mathbf{P_b = T_1\,R\,P}$$
$$\mathbf{P_c = S\,T_1\,R\,P}$$
$$\mathbf{P'_2 = T_2\,S\,T_1\,R\,P}$$

Figure 14.5: Concatenation Order

The *order* of operations in this sequence is very important, as we have demonstrated in this section. In Figures 14.5(a) and 14.5(b) the order of the operations were intentionally permuted, to demonstrate results that clearly do not meet our stated objective. We avoid these difficulties by now constructing a well-thought-out ordered sequence which we now describe.

The first operator is a translation T_1, which takes the set **P** one unit to the left, as in Figure 14.6(a). Next, in Figure 14.6(b), the transformation **S** scales this translated set **P** into the desired 2×1 rectangle. Third, in Figure 14.6(c), the transformation **R** rotates this rectangle 30 degrees clockwise. Finally, this rotated point-set is translated by $\mathbf{T_2}$ to the desired point $(2,1)$, as in Figure 14.6d. The composite transformation is, $\mathbf{P' = MP}$ where $\mathbf{M = T_2RST_1}$.

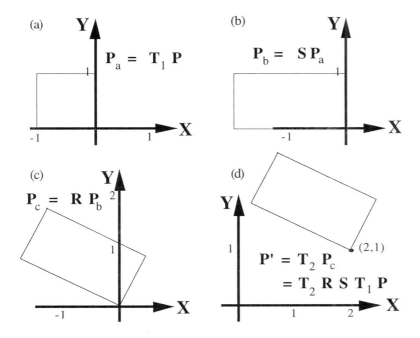

Figure 14.6: Transformation Sequence

Verify

As an exercise the reader should verify, using the tools defined on the preceeding page, that the sequences illustrated in Figure 14.5 indeed do produce their respective images.

14.7 Transformations in R^3

In a manner analogous to the 2-dimensional case, we introduce homogeneous coordinates for points in R^3. The *homogeneous coordinate* representation for a point (x, y, z) in R^3 is the 4-tuple $(x, y, z, 1)$, so that all points of objects in R^3 undergo the mapping

$$(x, y, z) \quad \longrightarrow \quad (x, y, z, 1)$$

The *homogeneous coordinate* representations for the operators Scale, Translation, and Rotation in R^3 again are designated using the same symbols, namely, S, T, and R, respectively. We have

for Scale in R^3 Homogeneous Coordinates

$$
\begin{aligned}
S &= \text{Scale Operator} \\
&= \begin{bmatrix} S_1 & 0 & 0 & 0 \\ 0 & S_2 & 0 & 0 \\ 0 & 0 & S_3 & 0 \\ 0 & 0 & 0 & 1 \end{bmatrix}
\end{aligned}
$$

so that

$$
\begin{bmatrix} x' \\ y' \\ z' \\ 1 \end{bmatrix} = S \begin{bmatrix} x \\ y \\ z \\ 1 \end{bmatrix} = \begin{bmatrix} S_1 & 0 & 0 & 0 \\ 0 & S_2 & 0 & 0 \\ 0 & 0 & S_3 & 0 \\ 0 & 0 & 0 & 1 \end{bmatrix} \begin{bmatrix} x \\ y \\ z \\ 1 \end{bmatrix}
$$

Thus, under a scaling transformation with scale factors, S_1, S_2, and S_3, we have the mapping

$$
P = (x, y, z, 1) \quad \longrightarrow \quad (S_1 x, S_2 y, S_3 z, 1) = P'
$$

for Translation in R^3 Homogeneous Coordinates

$$
\begin{aligned}
T &= \text{Translation Operator} \\
&= \begin{bmatrix} 1 & 0 & 0 & x_0 \\ 0 & 1 & 0 & y_0 \\ 0 & 0 & 1 & z_0 \\ 0 & 0 & 0 & 1 \end{bmatrix}
\end{aligned}
$$

so that

$$
\begin{bmatrix} x' \\ y' \\ z' \\ 1 \end{bmatrix} = T \begin{bmatrix} x \\ y \\ z \\ 1 \end{bmatrix} = \begin{bmatrix} 1 & 0 & 0 & x_0 \\ 0 & 1 & 0 & y_0 \\ 0 & 0 & 1 & z_0 \\ 0 & 0 & 0 & 1 \end{bmatrix} \begin{bmatrix} x \\ y \\ z \\ 1 \end{bmatrix}
$$

Thus, under a translation transformation with displacements, x_0, y_0, and z_0, we have the familiar mapping

$$
P = (x, y, z, 1) \quad \longrightarrow \quad (x + x_0, y + y_0, z + z_0, 1) = P'
$$

for Rotation in R^3 Homogeneous Coordinates

$$
\begin{aligned}
R &= \text{Rotation Operator} \\
&= \begin{bmatrix} r_{11} & r_{12} & r_{13} & 0 \\ r_{21} & r_{22} & r_{23} & 0 \\ r_{31} & r_{32} & r_{33} & 0 \\ 0 & 0 & 0 & 1 \end{bmatrix}
\end{aligned}
$$

so that

$$
\begin{bmatrix} x' \\ y' \\ z' \\ 1 \end{bmatrix} = R \begin{bmatrix} x \\ y \\ z \\ 1 \end{bmatrix} = \begin{bmatrix} r_{11} & r_{12} & r_{13} & 0 \\ r_{21} & r_{22} & r_{23} & 0 \\ r_{31} & r_{32} & r_{33} & 0 \\ 0 & 0 & 0 & 1 \end{bmatrix} \begin{bmatrix} x \\ y \\ z \\ 1 \end{bmatrix}
$$

Here the 3×3 submatrix with the r_{ij} entries must represent an appropriate rotation matrix in R^3. That matrix, of course, is peculiar to the particular application.

Most applications in R^3 will involve some sequence of the transformations we have just described, as was the case in R^2. At the risk of being overly pedantic, we now give a simple detailed example which uses each of these three transformations. We are given a cube, each of whose sides has length 2, which is centered at the point $(4, 2, 3)$ with all edges parallel to the axes of the reference coordinate frame. Our objective is to

> *Transform the cube into a vertical $2 \times 2 \times 6$ parallelepiped, still centered at $(4, 2, 3)$; then rotate the object 60 degrees clockwise about an axis which contains the center, and which is parallel to the vector* $\mathbf{u} = (1, 1, 1)$.

To do this, we first translate the cube to the origin, then scale it to give the desired parallelepiped. We then rotate the object appropriately, and finally translate it back to the original point.

In this simple application we need only to apply these transformations to the vertices of the cube. In order to facilitate the required matrix algebra, we characterize the cube by a 4×8 matrix,

$$P = [\mathbf{v}_k | \ k = 1, 2, \cdots, 8]$$

Here, the elements of the kth column, $\mathbf{v}_k$, are the homogeneous coordinates of the kth vertex of the given cube, each of which are known. Moreover, use of the index k suggests possible ordering of these vertices for imaging purposes.

We first translate the cube to the origin, by means of the transformation, T, where

$$TP = \begin{bmatrix} 1 & 0 & 0 & -4 \\ 0 & 1 & 0 & -2 \\ 0 & 0 & 1 & -3 \\ 0 & 0 & 0 & 1 \end{bmatrix} P = P_1$$

$$= \text{ cube translated to the origin}$$

Next, the translated cube, as characterized in matrix P_1 is subjected to the Scaling Transformation, S, to make a vertical parallelepiped centered at the origin. The eight new vertices are given by the

Inverse Operators in R^3

The homogeneous *inverse* operators are defined as follows:

S^{-1} = Inverse Scale

$$= \begin{bmatrix} S_1^{-1} & 0 & 0 & 0 \\ 0 & S_2^{-1} & 0 & 0 \\ 0 & 0 & S_3^{-1} & 0 \\ 0 & 0 & 0 & 1 \end{bmatrix}$$

T^{-1} = Inverse Translation

$$= \begin{bmatrix} 1 & 0 & 0 & -x_0 \\ 0 & 1 & 0 & -y_0 \\ 0 & 0 & 1 & -z_0 \\ 0 & 0 & 0 & 1 \end{bmatrix}$$

R_4^{-1} = Inverse Rotation = R_4^t

$$= \begin{bmatrix} & & & 0 \\ & R_3^t & & 0 \\ & & & 0 \\ 0 & 0 & 0 & 1 \end{bmatrix}$$

expression

$$SP_1 = \begin{bmatrix} 1 & 0 & 0 & 0 \\ 0 & 1 & 0 & 0 \\ 0 & 0 & 3 & 0 \\ 0 & 0 & 0 & 1 \end{bmatrix} P_1 = P_2$$

$$= \text{a vertical parallelepiped}$$
$$\text{centered at the origin}$$

Next we need to rotate this parallelepiped about an axis (through the center of the object), which is parallel to the vector $\mathbf{u} = (1, 1, 1)$. This may be accomplished by a sequence of two *frame rotations* such that the x-axis defines the required axis of rotation. We follow this by the required 60 degree rotation about that axis.

The first two frame rotations, $R_2 R_1$, which remind us of the familiar Tracking sequence, take the frame x-axis into the required axis of rotation. This is followed by a rotation R_3 about this axis, so that

$$P_3 = R_3 R_2 R_1 P_2$$

gives the final orientation of the parallelepiped.
Here

$$R_2 R_1 = \begin{bmatrix} \sqrt{2}/\sqrt{3} & 0 & -1/\sqrt{3} & 0 \\ 0 & 1 & 0 & 0 \\ 1/\sqrt{3} & 0 & \sqrt{2}/\sqrt{3} & 0 \\ 0 & 0 & 0 & 1 \end{bmatrix} \begin{bmatrix} \sqrt{2}/2 & 1/2 & 0 & 0 \\ -1/2 & \sqrt{2}/2 & 0 & 0 \\ 0 & 0 & 1 & 0 \\ 0 & 0 & 0 & 1 \end{bmatrix}$$

makes the x-axis the required axis of rotation, and

$$R_3 = \begin{bmatrix} 1 & 0 & 0 & 0 \\ 0 & 1/2 & -\sqrt{3}/2 & 0 \\ 0 & \sqrt{3}/2 & 1/2 & 0 \\ 0 & 0 & 0 & 1 \end{bmatrix}$$

rotates the object 60 degrees, clockwise.

Finally, we translate this rotated parallelepiped back to the original point, by the operation

$$P_4 = T^{-1} P_3 \quad \text{where}$$

$$T^{-1} = \begin{bmatrix} 1 & 0 & 0 & 4 \\ 0 & 1 & 0 & 2 \\ 0 & 0 & 1 & 3 \\ 0 & 0 & 0 & 1 \end{bmatrix}$$

returns the transformed cube, with its new scaling and orientation, to its original location $(4, 2, 3)$.

In summary, the properly ordered sequence which accomplishes this stated objective is:

$$P_4 = T^{-1}R_3R_2R_1STP = MP$$

where M is the indicated product. In practice we would ordinarily compute this matrix product before operating on the point-set.

As in R^2, we introduced homogeneous coordinates in order to make the translation, scaling, and rotation operators multiplicative — but again, the price we pay for this is that the algebraic dimensionality has gone from three to four, and all points of objects in R^3, are defined by

$$(x, y, z) \quad \longrightarrow \quad (x, y, z, 1)$$

14.8   What about Quaternions?

To this point, all transformations were represented strictly in terms of matrices and matrix algebra. In this section we ask

Are quaternions as easy to apply in these applications as the conventional rotation matrix?

It is true that in order to make translation (which is intrinsically an additive process) multiplicative, along with scaling and rotation, we had to introduce homogeneous coordinates. Homogeneous coordinates, as we have noted, are 4-tuples, as are quaternions. This suggests that there might be a way of doing scaling and translation in terms of some sort of quaternion operator. However, at this point there seems to be no such way; quaternions and their rotation operators are algebraically incompatible with homogeneous coordinates. However, there is at least one situation in which quaternions are helpful.

In the foregoing example we required the rotation of an object about a *specified axis*, with direction $\mathbf{v} = (1, 1, 1)$, through a *specified angle*, 60 degrees. The quaternion required, in this instance, is immediately known to be

$$q = \cos\frac{\pi}{6} + \mathbf{u}\sin\frac{\pi}{6} \quad \text{where} \quad \mathbf{u} = \frac{\mathbf{v}}{|\mathbf{v}|}$$

The components of q can then be used, quite efficiently, to determine the required rotation matrix (see Equation 7.17). The resulting matrix may then be used in a homogeneous coordinate format. An important question which remains, however, is

> *How can we produce appropriate two-dimensional screen images of these three-dimensional objects?*

In the following sections, we define some of the projections which are commonly used to produce these images.

14.9 Projections $R^3 \rightarrow R^2$

Now that we have defined the operators which enable us to establish and/or change the position and/or orientation of three-dimensional objects, we next consider methods and options for projecting an image of this object on a two-dimensional screen. There are two general classes of planar projections, namely,

Parallel and *Perspective* (not parallel). In such projections, we emphasize that Lines *always map to* Lines.

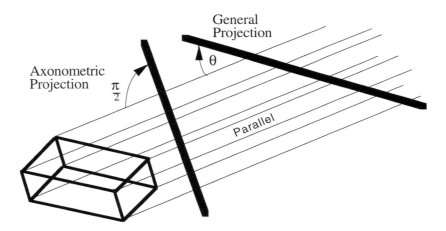

Figure 14.7: General and Axonometric Projections

By a *Parallel* **projection** we mean any projection in which the lines of projection (taken to be the viewing direction) are parallel, as in Figure 14.7.

By a *Perspective* **projection** we mean a projection in which the
lines of projection intersect in at least one point, called a point
of perspective (see Figure 14.8).

In the following two sections we present three sub-classes of each of
these two kinds of projections.

14.9.1   Parallel Projections

By a *General projection* we mean a parallel projection onto a plane
where the orientation of the plane, relative to the viewing
direction, is unrestricted.

By an *Axonometric projection* we mean a parallel projection onto
a plane in which the orientation of the plane is normal to the
viewing direction. See Figure 14.7.

By an *Isometric projection* we mean an axonometric projection such
that the viewing direction is parallel to the vector $\mathbf{v} = [1, 1, 1]$
defined in the object frame. Thus the three direction cosines
of the viewing direction are equal.

14.9.2   Perspective Projections

A perspective projection is classified as a *one-point*, *two-point*, or
three-point projection, depending upon the number of focal points
(points of perspectives) it has.

A *one-point perspective* has a *Surface* of the object parallel to the
projection plane. The entrance to the tunnel, illustrated in
the margin, is parallel to the projection plane.

Isometric Projection

The Isometric Projection is in
a direction which shortens each
axis in the viewing frame equally.

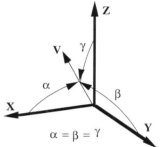

One-Point Perspective

The One-Point Perspective has a
point at infinity, straight ahead.
To illustrate this notion, a tunnel
is shown with a person near the
end of the tunnel. The entrance to
the tunnel is parallel to the projection plane.

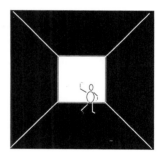

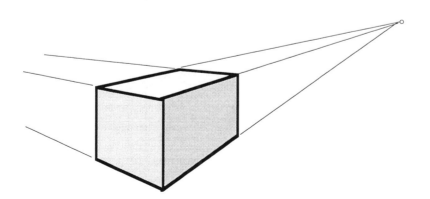

Figure 14.8: Two-Point Perspective

A *two-point perspective* has an *Edge* (but no surface) of the object parallel to the projection plane. In Figure 14.8 the vertical edges are parallel to the projection plane.

A *three-point perspective* has *no Edge* of the object parallel to the projection plane. This is the perspective one gets, say, when flying low over a city with tall buildings.

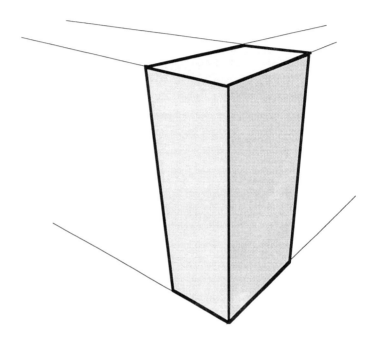

Figure 14.9: Three-Point Perspective

For our purposes the computer screen will always be the local projection screen. The images of three-dimensional objects on the computer screen will manifest, in general, a three-point perspective. At any one instant, of course, the perspective depends upon the relative disposition of the view-frame and the object in the world-frame. An object in motion, of course, may at times encounter positions where an edge or even a primary surface is parallel to the projection screen.

14.10 Coordinate Frames

The mathematics for a perspective transformation and for the creation of images for *Virtual Reality* applications requires that we define the following independent orthonormal frames.

- *World Frame* — The *world frame* is a primary reference frame, with respect to which all other coordinate frames are defined. The relative positions and orientations of these frames may thus be directly or indirectly established.

Virtual Reality

By Virtual Reality we mean the creation of an artificial environment which simulates, for the viewer, some actual environment.

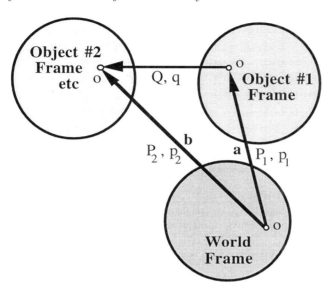

Figure 14.10:   Objects in World Frame

- *Object Frame* — Each object is defined as a point-set in *an object frame*. Object frames are independent of the world frame and of each other, and are not necessarily fixed in time. The parameters P and p, illustrated in Figure 14.10, define the position and orientation of an object with respect to the world frame. Given these parameters for two objects, their relative position and orientation, Q and q, may be determined.

- *Image Frame* — The *image frame* (or *projection frame*), defined in greater detail in the next section, is simply a coordinate frame whose xy plane contains the planar image we seek, and whose z-axis contains the observers Eye.

A significant *Virtual Reality* application, which involves these frames, is that of a Flight Simulator, whose mathematical model we will develop next. These frames, along with the position of the observer's eye, are directly or indirectly defined in the world frame by means of vectors $\mathbf{a}$, $\mathbf{b}$, $\mathbf{e}$, and $\mathbf{u}$, as illustrated in Figure 14.11. First, however, we must introduce the *perspective transformation* needed to take the observed points into the *Image Plane* — into some proper and realistic perspective.

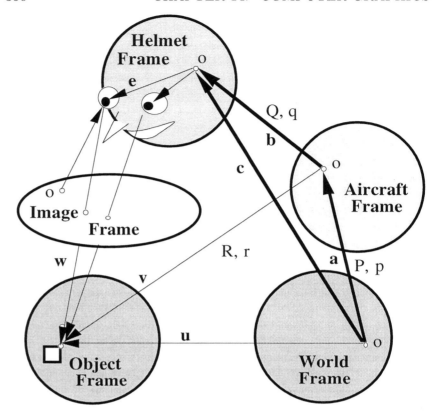

Figure 14.11: Flight Simulator Coordinate Frames

14.10.1   Perspective — Simple Case

We begin with the simple case illustrated in Figure 14.12. We place
the eye at the origin of the world frame, with the viewing direction
taken as the positive $\mathbf{Z}$-axis. We take the image plane to be D units
above and parallel to the $\mathbf{XY}$-plane. Suppose that in homogeneous
coordinates we have $\mathbf{p} = [x, y, z, 1]$. Then, from similar triangles, it
is clear that $\mathbf{p}$ in the world frame maps into

$$\mathbf{q} \;=\; [\frac{D}{z}x, \frac{D}{z}y, D, 1] \;=\; \frac{D}{z}[x, y, z, \frac{z}{D}]$$

in the image plane. In matrix notation this Perspective Transformation may be written

$$\mathbf{q} \;=\; Q\mathbf{p} \;=\; \frac{D}{z} \begin{bmatrix} 1 & 0 & 0 & 0 \\ 0 & 1 & 0 & 0 \\ 0 & 0 & 1 & 0 \\ 0 & 0 & 1/D & 0 \end{bmatrix} \begin{bmatrix} x \\ y \\ z \\ 1 \end{bmatrix}$$

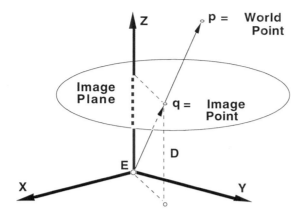

Figure 14.12:   A Point in Perspective

14.10.2   Parallel Lines in Perspective

Another simple case which provides further insight into the perspective transformation is the view one gets of parallel lines. The horizon in the world frame (where all parallel lines meet) is represented by a finite set of points in the image plane — a very useful notion borrowed from projective geometry. See Figure 14.13. Note in particular that the images of all parallel lines meet at points on the horizon in the image plane.

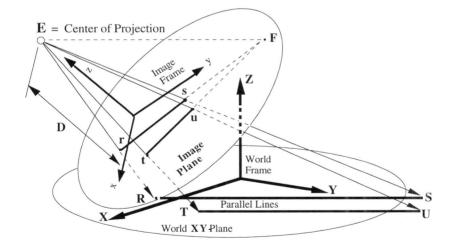

Figure 14.13:   Parallel Lines in Image Plane

The line which represents the horizon in the image plane is readily determined geometrically by finding the intersection of the image plane with a plane through point E parallel to the world **XY** plane.

We now consider the sequence of operators required for the more general case, that of projecting arbitrary images which are defined in their respective frames, first into the world frame and then into the image frame — in perspective.

14.10.3   Perspective in General

Free Vectors

We emphasize that it is helpful to view the vectors, *in the mind's eye*, as *free* vectors, especially after the various vector operations.

For the general case, we define the *Image Frame* or *Projection Frame* as a right-handed xyz orthonormal coordinate frame, where the xy plane is the *Image Plane*. See Figure 14.14. The unit vector **N** is normal to the *Image Plane*. The point E is the *Center of Projection*, defined in the *World Frame* with the "eye" at point E "looking"

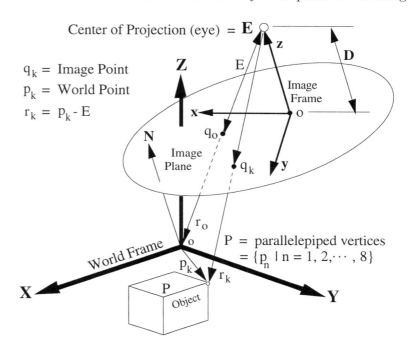

Center of Projection (eye) = **E**

q_k = Image Point
p_k = World Point
$r_k = p_k - E$

P = parallelepiped vertices
= $\{p_n \mid n = 1, 2, \cdots, 8\}$

Figure 14.14:   Object P in the World Frame

in the negative **N** direction. Point E is on the z-axis of the image frame and is D units above the Image Plane. The *orientation* of the image frame with respect to the world frame is specified by a rotation operator R, whereas the *position* of the image frame is governed by the parameters E and D. See Figure **??**, in the margin.

The object of our concern is again a parallelepiped defined by the point-set $P = \{p_k \mid k = 1, 2, \cdots 8\}$ in the world frame. To create the image of P we find the image q_k of each p_k, then join these image points with appropriate line segments.

Our strategy is first, translate the origin of the world frame to the point E, using a translation T; then, using the rotation operator R, rotate the the translated world frame so that its orientation coincides with that of the image frame. This reduces the required perspective transformation to the simple case of the previous section. The composite transformation $M = QRT$ is then our general perspective transformation.

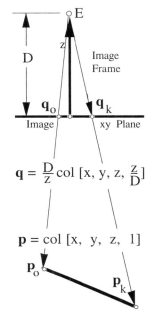

$$\mathbf{q} = \frac{D}{z}\,\text{col}\,[x,\ y,\ z,\ \tfrac{z}{D}]$$

$$\mathbf{p} = \text{col}\,[x,\ y,\ z,\ 1]$$

World Point Image

An image point, $\mathbf{q}_k$, of a world point, $\mathbf{p}_k$, each of which is defined in the World frame, is shown projected in *perspective* into the Image Plane.

In matrix notation, we have

$$\mathbf{p}_k\ =\ \text{a point defined in the World Frame.}$$

For T = Translation Matrix

$$T\mathbf{p}_k\ =\ \begin{bmatrix} 1 & 0 & 0 & -E_x \\ 0 & 1 & 0 & -E_y \\ 0 & 0 & 1 & -E_z \\ 0 & 0 & 0 & 1 \end{bmatrix}\mathbf{p}_k$$

which translates the World Frame origin to the point E.

For R = Rotation Matrix or Quaternion Operator

$$RT\mathbf{p}_k\ =\ \begin{bmatrix} & & & 0 \\ & R & & 0 \\ & & & 0 \\ 0 & 0 & 0 & 1 \end{bmatrix} T\mathbf{p}_k \tag{14.1}$$

$$\text{or}\quad =\ \begin{bmatrix} & & & 0 \\ & q^*(T\mathbf{p}_k)q & & 0 \\ & & & 0 \\ 0 & & 0 & 0 & 1 \end{bmatrix} \tag{14.2}$$

which rotates the World frame to the Image Frame. We are now ready for the Perspective transformation.

We let Q = Perspective Operator

$$=\ \frac{D}{z_k}\begin{bmatrix} 1 & 0 & 0 & 0 \\ 0 & 1 & 0 & 0 \\ 0 & 0 & 1 & 0 \\ 0 & 0 & 1/D & 0 \end{bmatrix}$$

which then takes points $\mathbf{p}_k$ into the points $\mathbf{q}_k$ in the image plane.

That is, $\mathbf{q}_k\ =\ \dfrac{D}{z_k}\begin{bmatrix} 1 & 0 & 0 & 0 \\ 0 & 1 & 0 & 0 \\ 0 & 0 & 1 & 0 \\ 0 & 0 & 1/D & 0 \end{bmatrix} RT\mathbf{p}_k \tag{14.3}$

$$= QRT\mathbf{p}_k \tag{14.4}$$

$$= \text{the corresponding point in the image plane}$$

Taking the entire set $P = \{\mathbf{p}_k | k = 1, 2, \cdots 8\}$ through this sequence gives an image of the remote object, *in perspective*.

We note that $\mathbf{q}_k$ is of the form

$$\mathbf{q}_k = [x_k \frac{D}{z_k}, y_k \frac{D}{z_k}, D, 1]$$

and that only the ordered pairs formed from the first two components of each 4-tuple are required to make the image in the image plane.

14.11   Objects in Motion

We now simulate the motion of an object P, such as the aircraft shown in Figure 14.15. Here the aircraft is defined by the point-set $P_0 = \{\mathbf{p}_k | \ k = 0, 1, \cdots n\}$, whose geometric center is at the origin. We will simulate the flight of the aircraft *kinematically*, that is, by subjecting the point-set P_0 to an appropriate sequence of incremented translations and/or rotations. At any step in the sequence, we let P be the point-set which represents the current position and orientation of the reference object, P_0, say, at time t. We let P' be the incremented point-set which describes this object at time, $t + \Delta t$.

Further, at any step in the sequence, the *position* of the object is specified by a translation operator T which takes the world frame origin to the current geometric center, say, the point (x_0, y_0, z_0). The matrix form for T is

$$T = \begin{bmatrix} 1 & 0 & 0 & -x_0 \\ 0 & 1 & 0 & -y_0 \\ 0 & 0 & 1 & -z_0 \\ 0 & 0 & 0 & 1 \end{bmatrix} \tag{14.5}$$

In like manner, the *orientation* of the object is specified by a rotation operator, R, defined by

$$R = \begin{bmatrix} & & & 0 \\ & R_0 & & 0 \\ & & & 0 \\ 0 & 0 & 0 & 1 \end{bmatrix} \tag{14.6}$$

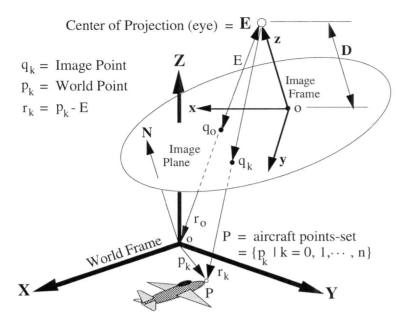

q_k = Image Point
p_k = World Point
$r_k = p_k$ - E

Figure 14.15:   Object P in Motion

Thus the current state of the object, in *position* and *orientation* in the world frame, is described by the point-set, P, where

$$P = TRP_0$$

In what follows, we consider how these operators, T and R, may be modified in order to show the object in motion — in translation and/or rotation, incrementally.

14.11.1   Incremental Translation Only

We simulate an object moving incrementally in a direction specified by a *velocity* vector $\mathbf{v} = [v_x, v_y, v_z]$ using an incremental translation operator defined by the matrix

$$\Delta_T \;=\; \begin{bmatrix} 1 & 0 & 0 & -v_x \Delta t \\ 0 & 1 & 0 & -v_y \Delta t \\ 0 & 0 & 1 & -v_z \Delta t \\ 0 & 0 & 0 & 1 \end{bmatrix} \;=\; \begin{bmatrix} 1 & 0 & 0 & -\Delta_x \\ 0 & 1 & 0 & -\Delta_y \\ 0 & 0 & 1 & -\Delta_z \\ 0 & 0 & 0 & 1 \end{bmatrix} \quad (14.7)$$

for some very small time increment, Δt. The velocity vector $\mathbf{v}$ may vary with successive increments. A *new* translation is then defined in terms of the current or old translation by

$$T_{new} = \Delta_T T_{old}$$

Incrementals

Motion of objects in computer graphics is always accomplished by a sequence of small incremental values for the parameter Δt. Adopting the notation used in computer science, we write successive translation operators as

$$T_{new} = T_{\Delta t} T_{old} = \Delta_T T_{old}$$

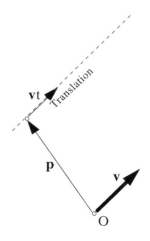

$$\mathbf{v} = [v_x, v_y, v_z]$$
Object Translation

Note

In the equation, on the right, we
assume no rotation is required.

$\Longrightarrow$

for each successive Δ_T. That is, for a given velocity vector $\mathbf{v} = [v_x, v_y, v_z]$, and for each successive Δ_T, the new incrementally translated point-set is given by

$$P' = \Delta_T P = \Delta_T T_{old} P_0 = T_{new} P_0$$

The new incrementally translated *image* is then obtained from P' using the perspective transformation of the last section.

14.11.2   Incremental Rotation Only

In free space an object rotates, in general, about its center of mass. So, for our purposes we assume that the orientation of the object P, or more specifically the orientation of its coordinate frame, is defined about this center by either a rotation matrix R or a quaternion r. These two alternatives are each considered separately in what follows.

Motion Simulation

The precise simulation of the forced motion of an object, particularly one with unequal principal moments, requires input from an appropriate mathematical model which takes into account the laws of dynamics which govern these matters. For example, rotation of the parallelpiped about a principle diagonal introduces a *couple* which is not taken into account in a *kinematic* rotation.

Because the current position of the object to be rotated is not, in general, at the origin of the world frame, the center of the object must first be translated to the origin. The object is then rotated about this center, and translated back to its former location. If Δ_R is the incremental rotation matrix, and P is the current point-set (*not* P_0), then the new incrementally rotated point-set P' is given by

Note

If the *linear* velocity of the object is not zero then incremental translation will also occur during this computational interval.

$$P' = TRT^{-1}P \tag{14.8}$$

$$\text{where} \quad R = R_{new} = \Delta_R R_{old} \tag{14.9}$$

$$\text{Here} \quad T^{-1} = \begin{bmatrix} 1 & 0 & 0 & -x_o \\ 0 & 1 & 0 & -y_o \\ 0 & 0 & 1 & -z_o \\ 0 & 0 & 0 & 1 \end{bmatrix} \tag{14.10}$$

$$T = \begin{bmatrix} 1 & 0 & 0 & x_o \\ 0 & 1 & 0 & y_o \\ 0 & 0 & 1 & z_o \\ 0 & 0 & 0 & 1 \end{bmatrix} \tag{14.11}$$

$$\text{and} \quad R = \begin{bmatrix} & & & 0 \\ & R_{new} & & 0 \\ & & & 0 \\ 0 & 0 & 0 & 1 \end{bmatrix} \tag{14.12}$$

The details of the rotation transformation R have not yet been specified. However, if the specified angular rotation rate of the object is ω about an axis, $\mathbf{u}$, say, in the body frame, then the use of a quaternion rotation operator immediately comes to mind.

14.11.3 Incremental Rotation Quaternion

The orientation of the object P_0 we define with a quaternion, r. As before, the position of the object is dictated by the translation, T. The current *rotated* state P is then

$$P = T(r^* P_0 r) \tag{14.13}$$

We increment the orientation, using an incremental quaternion, call it s, where $s = \cos \omega \Delta t + \mathbf{u} \sin \omega \Delta t$. Then the incremented quaternion is given by r_{new} where $r_{new} = r_{old} s$.

Finally, the incremented point-set, P', is computed as

$$\begin{aligned} P' &= T(s^* r_{old}^* P_0 r_{old} s) \\ &= T(r_{new}^* P_0 r_{new}) \end{aligned} \tag{14.14}$$

In practice, the incremental rotation angle, $\omega \Delta t$, is very small, so we may write

$$\begin{aligned} s &\approx 1 + \mathbf{u} \omega \Delta t = 1 + \boldsymbol{\Delta} \\ \text{where} \quad \boldsymbol{\Delta} &= \Delta \mathbf{u} = \Delta_x \mathbf{i} + \Delta_y \mathbf{j} + \Delta_z \mathbf{k} \\ \text{and} \quad |\boldsymbol{\Delta}| &\ll 1 \end{aligned} \tag{14.15}$$

Rotation Simulation

It is important to note that if the rotation of an object is about some specified axis **and is not a small angle rotation** then to use a matrix rotation operator may not be the optimal choice. Given the axis and the angle of rotation, finding the rotation matrix is possible (see Equation 7.16) but tedious.

Alternatively, the components of this new incremented quaternion, $r = r_0 + \mathbf{i} r_1 + \mathbf{j} r_2 + \mathbf{k} r_3 = r_{new}$, may be used to define the elements of an equivalent new composite rotation matrix, $R_{new} = R_{new}(r_0, r_1, r_2, r_3)$, as in Equation 7.7. This equivalent rotation matrix, whose elements then are known functions of the components of the corresponding quaternion, r_{new}, will incrementally modify the last rotation state of the point-set P. That is to say, for the incrementally rotated point-set, P', using this matrix, we may write

$$P' = T R_{new} P_0$$

14.11.4 Incremental Rotation Matrix

Suppose, on the other hand, the incremental rotation is specified as some linear combination of incremental *body axis* rates, ω_x, ω_y, and ω_z, about the principal axes of the body, P. Using Equation 4.4 and small-angle approximations, we get an incremental rotation matrix, Δ_R, of the form

$$\Delta_R = \begin{bmatrix} 1 & \Delta_3 & -\Delta_2 & 0 \\ -\Delta_3 & 1 & \Delta_1 & 0 \\ \Delta_2 & -\Delta_1 & 1 & 0 \\ 0 & 0 & 0 & 1 \end{bmatrix} \tag{14.16}$$

Rotation Increment

In writing the matrix, Δ_R, we have used the small angle approximations

$$\begin{aligned} \sin \theta &= \theta \\ \cos \theta &= 1 \\ \text{and} \quad \theta^n &= 0 \quad n \geq 2 \end{aligned}$$

for small angle, θ.

where
$$\Delta_1 = \omega_x \Delta t$$
$$\Delta_2 = \omega_y \Delta t$$
$$\Delta_3 = \omega_z \Delta t$$

The off-diagonal entries in this matrix are small-angle rotations about the indicated body axes. Again we write

$$R = R_{new} = \Delta_R R_{old}$$

and so
$$P' = TRT^{-1}P = TRP_0$$

The *simultaneous* motion in both translation and rotation will involve the appropriate incremental combination of the foregoing. The need for this generalized motion in both translation and ro-

Figure 14.16:   Model of Aircraft

tation becomes more apparent as one further develops the graphics software necessary to make the *virtual reality* required by a Flight Simulator more real.

In the next section we introduce a simplified control-law for the implementation of an aircraft flight control system. And finally we suggest an enhanced *virtual reality* perspective which offers the possibility of depth perception of objects within the cockpit by simulating binocular vision.

14.12   Aircraft Kinematics

The motion of the aircraft (see Figure 14.16) is generally modelled or simulated in accordance with a set of differential equations which represent the aircraft. These equations usually are written in terms of the appropriate aerodynamic coefficients which characterize the aircraft of interest. If we followed this approach, we could study such matters as the stability and control of a specific aircraft.

These concerns, however, are beyond the scope of this overview. Instead we consider the relative position and orientation of an Aircraft in the World Frame as represented in Figure 14.17. Here, *P* is the position matrix, and the orientation operator is defined by the quaternion p. The vector **a** locates the remote Aircraft in the World Frame. The vector **u** defines an arbitrary point of interest in

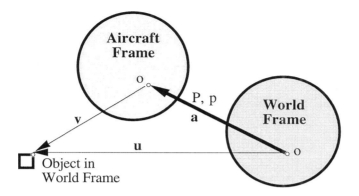

Figure 14.17: Aircraft and World Frames

the world frame, and the vector

$$\mathbf{v} = \mathbf{u} - \mathbf{a}$$

defines this same world point in the aircraft frame.

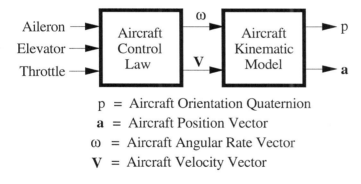

p = Aircraft Orientation Quaternion
a = Aircraft Position Vector
ω = Aircraft Angular Rate Vector
V = Aircraft Velocity Vector

Figure 14.18: Kinematic Flight Control System

We adopt the kinematic model shown in Figure 14.18 where with a simple *Control Law* (a throttle and a "joy-stick") we can control the aircraft *linear* velocity **V**, and the *angular* velocity,

$$\omega = \mathbf{i}\omega_x + \mathbf{j}\omega_y + \mathbf{k}\omega_z$$

of the aircraft, both defined in the aircraft body frame. In our simplified kinematic model we also make the aircraft "turn rate",

$\dot{\psi}$, somehow *linearly* consistent with

$$\begin{aligned}
\dot{\psi} &= f(\mathbf{V})\tan\phi & \text{that is,}\\
&= f(\mathbf{V})\phi & \text{or}\\
&= c\mathbf{V}\phi & \text{for } |\phi| \le \pi/4
\end{aligned}$$

and for some proportionality constant, c. This simple expression gives turn rate, $\dot{\psi}$, as a *linear* function of aircraft bank angle, ϕ, and will suffice for modest aircraft maneuvers. We now sketch a strategy for the analysis of important concerns encountered in the design of the remainder of the Flight Simulator System. This system simulates a view of the external world environment as seen by the pilot of the aircraft.

So, consider the pilot's viewing environment as illustrated in Figure 14.19. It is important to note that the vector magnitudes $|\mathbf{b}|$ and $|\mathbf{e}|$ are in general negligibly small compared to the magnitudes $|\mathbf{a}|$, $|\mathbf{u}|$, $|\mathbf{v}|$, or $|\mathbf{w}|$. This means that if $\mathbf{b}$ is negligible, the vec-

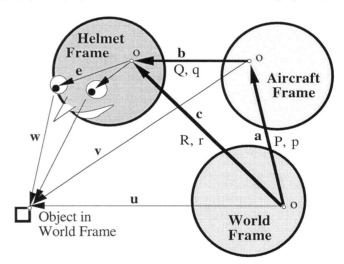

Figure 14.19: The Pilot Helmet-Sight System

tor $\mathbf{c}$ approaches $\mathbf{a}$, and the view-point vector, $\mathbf{v}$, from each eye approaches $\mathbf{v}$. Therefore, for world points, the vector $\mathbf{w}$ may be regarded as being equal to the vector $\mathbf{v}$, or that the remote *position* of the helmet frame and that of the aircraft frame may be regarded as one and the same.

But it is not quite so simple for relative *orientations*. Although the *motion* of the aircraft is influenced in important ways by the orientation of the aircraft, the appropriate *view* of the world points from the aircraft depends not only upon the orientation of the aircraft, but also upon the orientation of the helmet frame.

The relationship between these two frames has already been discussed in the *Six degree-of-freedom Transducer* described in Chapter 13. This transducer represents an important component in the development and implementation of new Virtual Reality concepts and applications. Such transducers are currently used by the various branches of The United States Armed Services in a variety of Visually-Coupled Control Systems. In one such application the position and orientation of the Pilot's head (Helmet), and therefore her Line-of-Sight (LOS), is measured with respect to the Aircraft Frame (Cockpit). This Transducer and Sight, together called the Helmet-Mounted Sight (HMS), completely relates the Pilot's Head Frame (LOS) and the Aircraft Frame.

In the next section we consider the relative position and/or orientation of several objects, as we continue to sketch the development and implementation of an Aircraft Flight Simulator. This simulator entails a computer generated virtual reality for a pilot who is wearing a helmet equipped with the Helmet-Mounted Sight (HMS). Along with a kinematic model which will govern the aircraft flight patterns, we now consider possible strategies for creating Computer Generated Images of virtual reality, for use in such a Flight Simulator.

Transducer

A *Transducer* is a device which converts an excitation in one component into useful information in another component. In this case, a sequence of polarized excitations applied to a *Source* produces *position* and *orientation* information at the *Sensor* as discussed in Chapter 13.

14.13   n-Body Simulation

The process which takes the point-sets which are defined in the cockpit frame into an appropriate *binocular* view merely requires that we use the ideas presented thus far to the two different values of the vector **e**, one for each eye. These two values for the vector **e** define the location of each eye in the Q,q frame, with respect to the HMS Sensor. These, in turn, result in two distinct view-point vector sets which are used to construct the image for the device assigned to its respective eye.

The image generator for each eye may be a distinct, dedicated CRT-type imaging device which is fixed to the Helmet. As the person wearing the helmet changes her viewing direction, the sensor of the HMS detects this change and the computer-generated view of the World landscape changes accordingly. See Figure 14.20.

As stated above, because the magnitudes of the vectors $|\mathbf{e}|$, and $|\mathbf{b}|$, are usually negligible compared to the world vectors $|\mathbf{w}|$, etc., a separate and distinct image need not be computed for those images

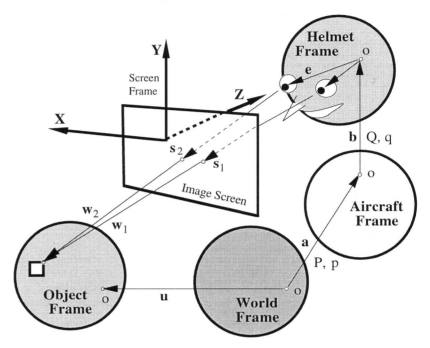

Figure 14.20: Frames for Virtual Reality

defined by the relatively remote world vectors $|\mathbf{w}|$. On the other hand, those vector sets which represent instrumentation, and their time-variation, in the cockpit are computed separately for each eye. This affords a depth perception which contributes to the realism of the view.

This chapter merely outlines an approach to the potential concerns one might encounter in the design and development of instrumentation which will yield a virtual reality environment. Hopefully, however, this overview presents enough to enable the reader to understand how such a development might proceed.

Exercises for Chapter 14

1. Verify the images in Figure 14.5 and although the two look quite similar, note and explain their differences.

2. With reference to the unit square input point-set $\mathbf{P}$ which is illustrated in Figure 14.6, find the resulting output point-set $\mathbf{P}' = \mathbf{MP}$ where the operator sequence $\mathbf{M} = \mathbf{T}_1\mathbf{SRT}_2$.

3. After interchanging the operators $\mathbf{R}$ and $\mathbf{S}$ in the transformation $\mathbf{M}$ in the 2. above, find the output point-set $\mathbf{P}$'.

Further Reading and Some Personal References

- Altman, Simon L., Rotations, Quaternions, and Double Groups, Oxford Science Publications, Oxford University Press, 1986

- Barfield, W., and Furness III, Thomas A., Virtual Environments and Advanced Interface Design, Oxford University Press, 1995

- Eves, Howard, An Introduction to the History of Mathematics, Fifth Edition, The Saunders Series, 1982

- Foley, J.D., and Van Dam, A., Fundamentals of Interactive Computer Graphics, Addison-Wesley Publishing Company, Inc., 1982

- Housner, G.W., and Hudson, D.E., Applied Mechanics Dynamics, D. Van Nostrand Company, Inc., 1959

- Kuipers, J.B., The SPASYN, a new transducing technique for visually-coupled control systems, Proceedings of a symposium on Visually-Coupled Systems, AMD TR-73-1, September 1973

- Kuipers, J.B., Object tracking and orientation determination means, system and process, U.S. Patent 3,868,565, February 25, 1975

- Kuipers, J.B., Tracking and determining orientation of object using coordinate transformation means, system and process, U.S. Patent 3,983,474, September 26, 1976

- Kuipers, J.B., Apparatus for generating a nutating electromagnetic field, U.S. Patent 4,107,858, April 12, 1977

- Kuipers, J.B., SPASYN, an electromagnetic relative position and orientation tracking system, IEEE Transactions on Instrumentation and Measurement, vol. IM-29, no. 4, December 1980

- Kuipers, J.B., Method and apparatus for determining remote object orientation and position, U.S. Patent 4,742,356, May 3, 1988

- Kuipers, J.B., Characterization and Application of Quaternions for Enhanced Computer Processing Algorithms, SBIR Final Report, AL/CF-SR-1994-0014, US Air Force Materiel Command, Wright-Patterson Air Force Base, Ohio 45433-6573, July 1994

- Pence, Dennis, Spacecraft Attitude, Rotations and Quaternions, UMAP Unit 652, The UMAP Journal, vol. 5, no. 2, 1984

- Pio, R.L., Euler angle transformations, IEEE Transactions on Automatic Control, vol. AC-11, no. 4, October 1966

- Raab, F.H., Blood, E.B., Steiner, T.O., and Jones, H.R., Magnetic Position and Orientation Tracking System, IEEE Transactions on Aerospace and Electronic Systems, vol. AES-15, no. 5, September 1979

- Robinson, A.C., On the Use of Quaternions in Simulation of Rigid Body Motion, Technical Report 58-17, Aeronautical Research Laboratory, Wright-Patterson Air Force Base, Ohio, December 1958

- Weeks, Jeffrey R., The Shape of Space: How to Visualize Surfaces and Three-Dimensional Manifolds, Marcel Dekker, Inc., 1985

Index